# MATHEMATICAL PHILOSOPHY

*BY THE SAME AUTHOR*

SCIENCE AND RELIGION: The Rational and the Superrational.
*—The Yale University Press.*

THE NEW INFINITE AND THE OLD THEOLOGY.
*—The Yale University Press.*

THE HUMAN WORTH OF RIGOROUS THINKING.
*—The Columbia University Press.*

# MATHEMATICAL PHILOSOPHY

## A STUDY OF FATE AND FREEDOM

*Lectures for Educated Laymen*

BY

CASSIUS J. KEYSER, Ph.D., LL.D.

ADRAIN PROFESSOR OF MATHEMATICS IN COLUMBIA UNIVERSITY

NEW YORK

E. P. DUTTON & COMPANY

681 FIFTH AVENUE

*First Printing, . . Feb., 1922*
*Second Printing, . July, 1924*
*Third Printing, . June, 1925*

*Printed in the United States of America*

TO

My Wife

# PREFACE

For more than two score years I have meditated upon the nature of Mathematics, upon its significance in Thought, and upon its bearings on human Life. In the following course of lectures I have endeavored to present, in the language current among educated men and women, some of the maturer fruits of that study.

Though the course is designed primarily for students whose major interest is in Philosophy, I venture to hope that the lectures may not be ungrateful to a much wider circle of readers and scholars:

To the growing class of such professional mathematicians as are not without interest in the philosophical aspects of their science.

To the growing class of such teachers of mathematics as endeavor to make the spirit of their subject dominate its technique.

To the growing class of those natural-science students who are interested in the logical structure and the distinctive method of mathematics regarded not only as a powerful instrument for natural science but also and especially as the prototype which every branch of science approximates in proportion as its basal assumptions and concepts become clearly defined.

To the innumerous but precious tribe of those literary critics who know that the art of Criticism owes its first allegiance to the eternal laws of thought.

To such psychologists as are interested either in the

psychology of mathematics or in the mathematics of psychology.

To such sociologists as desire to conceive the nature of our humankind justly—in accord with the mathematical principle of " logical types " or dimensionality.

To the rapidly increasing class of engineers who are learning to conceive engineering worthily, as the science and art of directing the civilizing energies of the world to the advancement of the welfare of all mankind including posterity.

Finally, to all readers who desire to acquire a fair understanding of such genuinely great mathematical ideas as are accessible to all educated laymen and to come thus into touch with the universal spirit of the science which Plato called divine.

In closing this preface I desire to record my gratitude to Mr. John Macrae, vice-president of E. P. Dutton & Company, for his generous encouragement in this enterprise.

CASSIUS JACKSON KEYSER.

*Columbia University,*
*New York*, January 11, 1922.

## CONTENTS

### LECTURE I

INTRODUCTION



### LECTURE II

POSTULATES



### LECTURE III

BASIC CONCEPTS

## LECTURE IV

### Doctrinal Interpretations



## LECTURE V

### Another Geometric Interpretation



## LECTURE VI

### Non-geometric Interpretation



## LECTURE VII

### Essential Discriminations

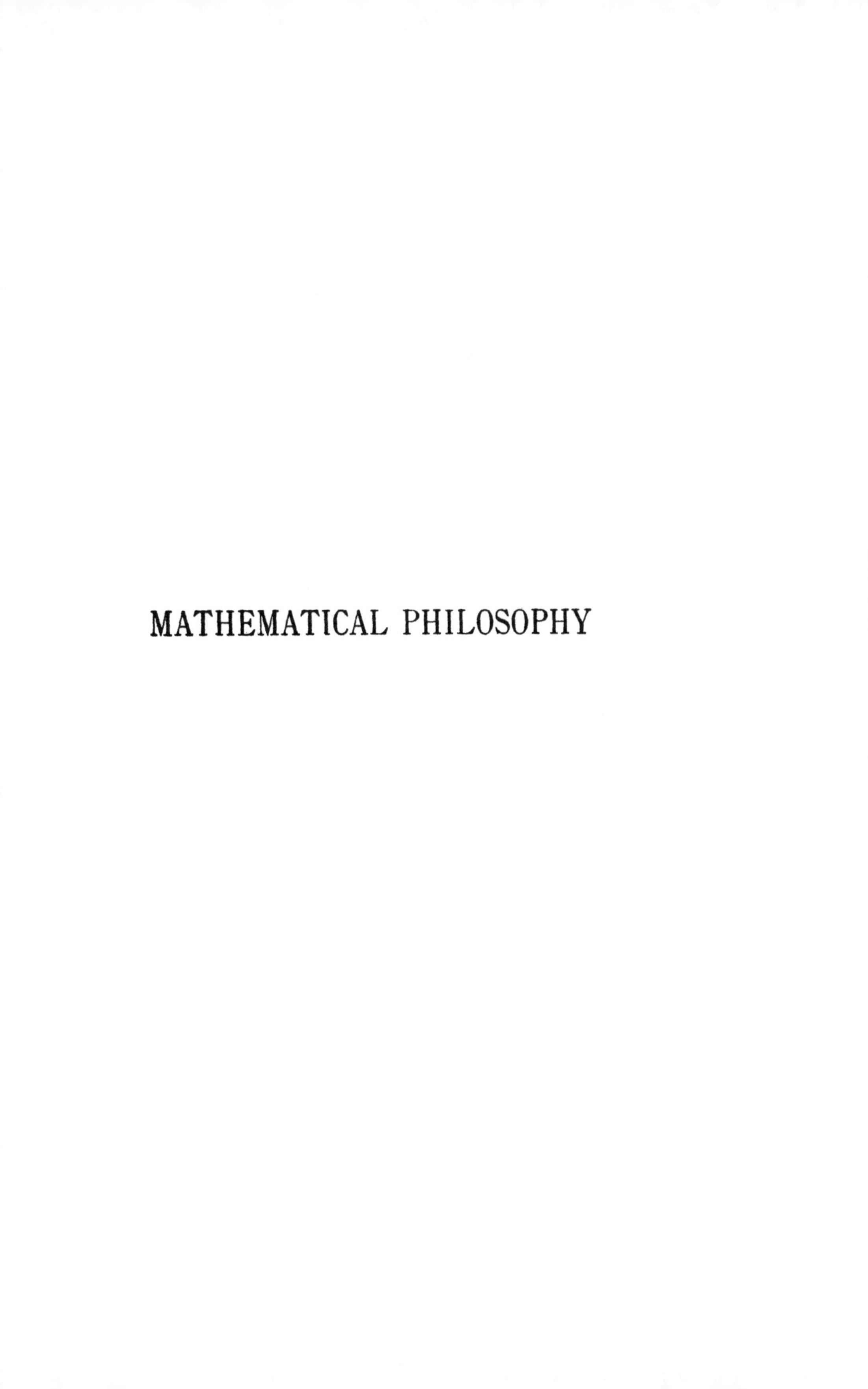

MATHEMATICAL PHILOSOPHY

# MATHEMATICAL PHILOSOPHY

## LECTURE I

### Introduction

INTELLECTUAL FREEDOM AND LOGICAL FATE—MATHEMATICAL OBLIGATIONS OF PHILOSOPHY AND EDUCATION—COMMON HUMANITY AND INDIVIDUALITY—HUMANISTIC AND INDUSTRIAL EDUCATION—MAN NOT AN ANIMAL—ETHICS NOT A BRANCH OF ZOOLOGY—EXCELLENCE AND THE MUSES—LOGIC THE MUSE OF THOUGHT—THE HEROIC TRADITION IN PHILOSOPHY—RADIANT ASPECTS OF AN OVER-WORLD

It is the aim of the following lectures to point out, in a manner suitable for you as students of Thought, and to submit for your consideration, some of the more essential and more significant relations between Mathematics and Philosophy. Each of these great terms is to be understood in its most embracing sense. Mathematicians sometimes speak contemptuously of philosophy; and philosophers sometimes speak contemptuously of mathematics. The contempt thus manifested does not spring from mathematics in the former case, nor from philosophy in the latter; in both cases it springs out of ignorance—philosophical ignorance of mathematicians and mathematical ignorance of philosophers. No doubt philosophically unenlightened mathematicians and mathematically unenlightened philosophers will quarrel in the future as in the past; but in the future as in the past, the quarreling and the sneering will be the quarreling and sneering of men and not of the great subjects they represent and misrepresent; for between the spirit of mathematics and the spirit of philosophy there is no discord, no antagonism, no strife; they are by their natures friendly rivals in the pursuit of truth and light; they are companions in excellence; they are comrades in the service of wisdom.

I have said that the "aim" of these lectures is to disclose fundamental connections between mathematics and philosophy. What I have described as their "aim" is not so much the aim, or end, as a means. For it will become increasingly evident as we advance that the work we are to be engaged in is fundamentally the study of Fate and Freedom—logical fate and intellectual freedom. I mention the matter here because you ought to have it consciously in mind from the beginning. You should bear it in mind at every stage of the discussion, even in connections where so warm an interest may seem remote. A preliminary word of explanation is therefore desirable.

We are going to deal with ideas—with their characters, with their meanings, with their relations. Now, an idea is in itself an eternal thing and the relations of an idea with other ideas are eternal. An idea is just what it is and it is unalterable; a relation among ideas is just what it is and it is unalterable. We do, indeed, often speak as if such were not the case; we habitually speak as if ideas and their relations were temporal affairs, impermanent, mutable, malleable, capable of growth, of modification, of decay, of destruction, as when we say, for example, that we have "changed" our ideas or that such-and-such an idea has "grown" in importance or has "become" sterile or is "dead." It is, I fancy, hardly necessary to say that all such ways of speaking are figurative,—convenient no doubt, often pleasing, sometimes very effective, yet thoroughly figurative,—and that, if taken literally, they quickly and inevitably lead to scientific and philosophic disaster. You or I may abandon an idea that we have held and we may adopt an idea that is new to us; the "old" one and the "new" one may closely resemble each other; they may indeed be identical in some respect and may even be called by the same name; but neither of them has been transmuted into the other; each of them remains and will remain just what it was. Let me illustrate the eternality of ideas and of their relations by means of a simple example. You know that in discourse ideas are represented by symbols—by words or other signs. Consider the symbols 2, 7, 9, +, and =; each of them stands for an idea familiar to all of us. The symbols are man-made; but the things they stand for, though they were discovered by man, are not man-made; they are increate, as Milton would say, and indestructible, and the like is true of their relations; one of these is expressed by the statement (*a*) $2+7=9$; the statement expressing the relation is a creature of man, but the relation itself is not—man discovered it, but he did not make it—it is a thing increate and indestructible, the same yesterday, today and forever. The truth of what I have just now said is very evident, but the illustration is arithmetical. Is the eternality equally evident in the case of all other ideas and their relations? No, it is not equally evident, but it is none the less true. Shall we take another example? Let us take one that is very far from being specifically arithmetical. Consider the statement:

(*b*) If something $S$ has the property $p$ and whatever has the property $p$ has the property $p'$, then $S$ has the property $p'$.

You observe that the statement expresses a certain *relation* among certain ideas—the idea, for example, denoted by "something," that denoted by "property" (or quality or mark), that denoted by "whatever," and so on. The denoting terms are indeed man-made, but the ideas denoted are not, they are merely man-discovered and man-known; and the statement expressing the relation is a creature of man, but the relation itself, though man discovered it, was not created by him: it is an unoriginated thing, immutable, universal, timeless. The illustration is very general, very abstract and very cold. Perhaps you prefer something warmer, more specific, more concrete. Well, it is easy to find such, for the foregoing general statement is infinitely rich in concrete applications. Let me instance one of them, one that is sufficiently warm; it is indeed one that goes to the very heart of our human ethics—not to our ethics as it is, but as it ought to be and as no doubt it will be. The application is this, namely:

> If human beings are by nature civilization-builders, or "time-binders," and if all time-binders, or civilization-builders, are both inheritors from the toil of bygone generations and trustees for the generations to come, then we humans stand in the double relationship —debtors of the dead, trustees of the unborn—thus uniting past, present and future in one living, growing reality.

The infinite and eternal significance of that fact may, I trust, be left for your meditation.

Without more talk and without danger of misunderstanding, we may, I believe, now speak of ideas as constituting a world—the world of ideas. With that world all human beings as human have to deal—there is no escape; it is there and only there that foundations are found—foundations for science, foundations for philosophy, foundations for art, foundations for religion, for ethics, for government and education; it is in the world of ideas and only there that human beings as human may find principles or bases for rational theories and rational conduct of life, whether individual life or community life; choices differ but some choice of principles we must make if we are to be really human—if, that is, we are to be rational—and when we have made it, we are at once bound by a destiny of consequences beyond the power of passion or will to control or modify; another choice of principles is but the election of another destiny. The world of ideas is, you see, the empire of Fate.

Is the human Intellect, then, a slave? No: it is free; but its freedom is not absolute; it is limited by fact and by law—by the laws of thought, by the immutable characters of ideas and by their unchanging eternal relationships. Intellectual freedom is freedom to think in accord with the laws of thought, in accord with the natures of ideas, in accord with their interrelations, which are unalterable. And no variety of human freedom—no institution erected in its sacred name—if it does not conform to the eternal conditions of intellectual freedom—can stand.

What I have now said is, I hope, a sufficient preliminary intimation of what I mean by saying that our work in these lectures is to be fundamentally a study of Freedom and Fate.

Your major interest is in philosophy; mine is in mathematics. You have besides, I trust, a lively, if only a minor, interest in mathematics, as I have had from the days of my youth a genuine interest, albeit a subordinate one, in the concerns of philosophy and especially in the philosophy of mathematics. It is, I believe, a happy circumstance that your interest and mine in these great subjects are thus complementary instead of coincident or antagonistic; for in this relation of interest there is implied a corresponding relation of attainment, limitation, outlook and temper; and this relation, if we bear it in mind, will be favorable in important ways to the prosperity of our enterprise; for example, it should, on the one hand, have the effect of restraining me from adducing too lightly or too freely, with too little explanation, mathematical considerations with which you may justly feel I have no right to suppose you familiar; and, on the other hand, when you discover, as you will doubtless frequently discover, that I have fallen into error because of my philosophical limitations, it will, I hope, make you feel it your duty to "impose upon me the just retribution," in accordance with the saying of Plato that "The just retribution of him who errs is that he be set right."

It need hardly be said that no one should follow this course in the hope of thereby acquiring mathematical knowledge or skill in the usual sense of these terms. I assume that what is mainly responsible for your presence here is a desire and a hope of a different kind: you desire to gain insight into the essential nature of mathematics regarded as a distinctive type of thought; you desire to acquire knowledge of what is characteristic and fundamental in mathematical method; you hope to gain acquaintance with some of the great mathematical concepts, with such of the dominant concepts as are accessible to laymen; you desire to win a just sense of the spiritual significance of the science; in a word, your quest is for such an understanding of it as will help you to view mathematics in a vast perspective—in relation, that is, with the other sciences and arts and the other modes and forms of human activity. Such, I take it, are the ends that define our task. I should indeed be unhappy if I did not hope that the lectures, though they have been fashioned with controlling reference to the task indicated, will at the same time serve in some measure to extend your acquaintance with the existing body of mathematical doctrine. But it is to be understood that this result, if the lectures produce it, will be incidental and subsidiary to their main purpose; for they are not designed to teach a recognized *branch* of mathematics whether elementary or more advanced. Mathematical students having little or no interest in the philosophy of their science must be frankly counseled to repair to other courses for the kind of instruction they desire. And students of philosophy should not indulge themselves in the vain hope of acquiring mathematical knowledge by merely "philosophizing" about the subject or by pensively gazing upon its general aspects from an external point of view. From time immemorial, there has been but one way to become a mathematician and there will never be another: it is a way interior to the subject and involves years of assiduous toil. Short cuts to mathematical scholarship there is none, whether the seeker be a philosopher or a king.

How much mathematical training is essential to the qualification of one who may hope to follow the lectures profitably? It is natural that you should wish to ask that question at this point. The question is important and the answer easy and short: so much mathematical training—so much knowledge of algebra, geometry and trigonometry—as a capable student can acquire in one collegiate year. Compared with the existing science of mathematics such knowledge is very meagre, a bare beginning; but, taken absolutely, it is much; in respect of content or mere information as distinguished from insight and power, it is far more than Thales had, or Pythagoras or even Plato or even Galileo. It would be very convenient if I might assume more; projective geometry, for example, and some acquaintance with analytical geometry —which should remind you of Descartes, and with the calculus—which should remind you of Leibniz; for I shall be obliged occasionally to employ ideas drawn from these and other branches of mathematics, and shall have to interrupt and delay the discussion a little in order to explain the ideas as the necessity arises for using them. Perhaps I should add that, for understanding the lectures, a certain intellectual maturity, logical acumen, open-mindedness and philosophical insight are not less essential than the stated minimum of mathematical knowledge.

I desire to invite you now to a somewhat comprehensive consideration of a much larger question, one of greater difficulty and far greater importance—a question of both general and permanent interest. The question is: How much mathematical training—how much mathematical knowledge, discipline, and habit—may be reasonably regarded as indispensable to the proper equipment of a philosopher? It may well be that you will be qualified to give a better answer at the end of the course than that which I am about to submit here at the beginning of it. Nevertheless, I am disposed to think that a preliminary discussion of the matter will be of some service. A complete discussion would involve many considerations differing greatly in weight. I shall ask your attention to such of them as seem to me cardinal and decisive.

The first consideration grows out of the fact that a philosopher is a human being. It is immediately evident that the proper equipment of a philosopher must include as much mathematical training as is essential to the appropriate education of men and women as human beings. How much is that? Be good enough to note what the question precisely is. I am not asking how much mathematical discipline is essential to a "liberal education" for this fine term, though clearly defined long ago by Aristotle in terms of spiritual interest and attitude, has in our day lost its significance even for the majority of academic folk, who ought to be ashamed of the fact. That great man, the late Lord Kelvin, used to tell his students that among the "essentials of a liberal education is mastery of Newton's *Principia* and Herschel's *Astronomy*." On the other hand, such educators as Matthew Arnold, John Henry Newman, Thomas Huxley, though differing infinitely in their outlooks upon the world and in their estimates of worth, yet unite in denying Kelvin's contention impetuously or even with scorn. Let us so frame our question as to avoid that debate. The question is: How much mathematical discipline is essential to the appropriate education of men and women as human beings? This exceedingly important question admits of a definite answer and it admits of it in terms of a supremely important and incontestable general principle. A clue to the principle is found in the phrase I have just now employed: education of men and women *as human beings*. Before stating the principle, it will be convenient to give it a name. I shall call it the Principle of Humanistic Education as distinguished from what has come to be designated in our day as Industrial Education. I say "as distinguished from" because the two varieties of education, whether they be compared with respect to the conceptions which lie at the heart of them or with respect to the motives which actuate and sustain them, are widely different. In order to set the principle in a clear light, let me indicate briefly the obvious facts lying at its base and leading naturally to its formulation.

What the individuals composing our race have in common falls into two parts: a part consisting of those numerous instincts, impulses, traits, propensities and powers which we humans have in common, not only with one another, but with many of the creatures constituting the world of animals—a subhuman [1] world; and a second part consisting of such instincts, impulses, traits, propensities and powers as are distinctively human. These latter, we may say, constitute our Common Humanity. They present, indeed, an endless variety of detail, but in the long course of man's experience with man he has learned to group them, in accordance with their principal aspects, into a small number of familiar classes. And accordingly, the nature of our common humanity is fairly well characterized by saying that human beings as such possess in some recognizable measure such marks as the following: a sense for language, for expression in speech —the literary faculty; a sense for the past, for the value

[1] See Lecture XX for a discussion of Korzybski's concept of Man in terms of Time—a conception according to which humans are not animals.

of experience—the historical faculty; a sense for the future, for prediction, for natural law—the scientific faculty; a sense for fellowship, coöperation, and justice—the political faculty; a sense for the beautiful—the artistic faculty; a sense for logic, for rigorous thinking—the mathematical faculty; a sense for wisdom, for world harmony, for cosmic understanding—the philosophical faculty; and a sense for the mystery of divinity—the religious faculty.

Such are the evident tokens and the cardinal constituents of that which in human beings is human. It is essential to note that to each of the senses or faculties in virtue of which humans are, not animals, but a higher class of beings, there corresponds a certain type of distinctively human activity—a kind of activity in which all human beings, whatever their stations or occupations, are as humans obliged to participate. Like the faculties to which they correspond, these types of activity, though they are interrelated, are yet distinct. Each of them has a character of its own. Above each of the types there hovers a guardian angel—an ideal of excellence—wooing our loyalty with a benignant influence superior to every compulsive force and every authority that may command. Nothing more precious can enter a human life than a vision of these angels, and it is the revealing of them that humanistic education has for its function and its aim. Stated in abstract terms the principle is this: Each of the great types of distinctively human activity owns an appropriate standard of excellence; it is the aim of humanistic education to lead the student into a clear knowledge of these standards and to give him a vivid and abiding sense of their *authority* in the conduct of life. Ethics is not a branch of Zoology.

It is plain that this conception stands in sharp contrast with the central idea of industrial education. For humanistic education has for its aim, as I have said, the attainment of excellence in the things which constitute our *common* humanity. On the other hand, industrial education is directly and primarily concerned with our *individualities*. It might, therefore, be more appropriately called individualistic education. It regards the world as an immense camp of industries where endlessly diversified occupations call for special propensities, gifts, and training. Accordingly its aim, its ideal, is to detect in each youth as early as may be the presence of such gifts and propensities as tend to indicate and to qualify him for some specific form of calling or bread-winning craft; then to counsel and guide him in the direction thereof; and finally, by way of education, to teach him those things which, in the honorable sense of the phrase, constitute the tricks of the trade.

What are we to say of it? The answer is obvious. Industrial education, rightly conceived, is essentially compatible with the humanistic type; it may breathe the humanistic spirit; the two varieties of education are essential to constitute an ideal whole, for human beings possess both individuality and the common humanity of man. Industrial education, when thus regarded as supplementary to humanistic education, is highly commendable; but when it is viewed as an equivalent for the latter or as an ideal substitute for it, it is ridiculous, contemptible and vicious. For the fact must not be concealed that a species of education which, in producing the craftsman, neglects the man, is, in point of kind and principle, precisely on a level with that sort of training which teaches the monkey and the bear to ride a bicycle, or the seal to balance a staff upon its nose or to twirl a disc.

These considerations are no doubt obvious. I should not dwell upon them at so great length but for the fact that in the excitement and confusion of our industrial age the most obvious of important facts and the most evident of important principles are so commonly lost sight of that they require to be cited again and again and again. Nowhere is the confusion of the time more evident than in the somewhat noisy and sometimes acrimonious discussion that has been recently and still is going on throughout our country regarding the value of mathematics as a subject in secondary and collegiate education. The instigators of the discussion, those, that is, who advocate so reducing mathematical requirements as practically to abolish the subject from curricula of general education, are not malicious nor insincere; many of them, I do not doubt, are well-meaning citizens. And if their rather voluminous discourses are often singularly lacking in coherence, in clarity and in depth, the defects are not due to evil intentions but rather, I suspect, to confusion and a lack of just that sort of discipline which the subject the authors are engaged in depreciating is peculiarly qualified to give. Perhaps we should not be astonished. If the saying of Sir Oliver Lodge be true that "the mathematical ignorance of the average educated person has always been complete and shameless," one ought not, I suppose, to be too much astonished if in a vast, crude, formless, sprawling democracy like ours, a way to educational "leadership" is sometimes found by men whose innocence, not only of mathematics but of the other great subjects, including the principles of education, is well-nigh complete and shameless. And yet, despite familiarity with the phenomenon, it is sometimes a bit hard to avoid astonishment and even a loss of patience. Not long ago a high-placed counselor of a well-known college of liberal arts challenged me, with defiant confidence and unfeigned solemnity, to give any good reason why college students should be required to pursue a course in algebra rather than one in some practical art, say the art of cooking mutton chops. On receiving such a challenge from a grown man, what should a grown man do? Confess his astonishment? Betray an exhaustion of patience? Fly to the easy refuge of ridicule? Any such reaction would probably have been misunderstood. In dealing with a solemn question, no matter how stupid, it is usually the wiser course to treat it with respect if possible. I might have responded, in the fine words of Professor Whitehead,[1] that

> "Algebra is the intellectual instrument which has been created for rendering clear the quantitative aspects of the world. . . . Through and through the world is infected with quantity. To talk sense, is to talk in quantities. It is no use saying that the nation is large,—How large? It is no use saying that radium is scarce,—How scarce? You can not evade quantity. You may fly to poetry and to music, and quantity and number will face you in your rhythms and your octaves. Elegant intellects which despise the theory of quantity are but half developed. They are more to be pitied than blamed."

It did not seem to me, however, that one capable of issuing such a challenge as that to which I have alluded

[1] A. N. Whitehead: *The Organization of Thought.* Cambridge University Press.

could feel the weight of such a response, and I did not make it. It is, you observe, a response in terms of *quantity*. Quantity is indeed omnipresent in our world; but so, too, is *quality,* and of the two things, the latter is perhaps the more universal in its appeal. Algebra is indeed essential to the theory of quantity and the theory of quantity is essential to the subjugation of natural resources to the use of man; of quality, on the other hand, algebra is not a science but, though it is not a science of quality, it *has* a quality, a human quality, to which it owes its high rank in the spiritual hierarchy of human disciplines. And so I endeavored, with poor success I fear, to answer the challenge in terms of quality. I invoked the principle which in this lecture I have been calling the principle of humanistic education. I sought, that is, to make it clear that, in contrast with the practical arts, the science of algebra as a discipline possesses a certain quality by virtue of which, if the subject be rightly administered, the student is gradually brought into the presence of one of those great standards of excellence by which, as we have seen, distinctively human activity in all its principal types is to be guided and judged. The standard to which I refer is, as you have doubtless surmised, the standard of excellence in the quality of thinking as thinking—the standard which mathematicians are accustomed to call Logical Rigor—clarity, that is, precision and coherence.

And now the mention of that great term may serve to reassure you, should you have begun to suspect that in the course of this rather long excursion I may have forgotten the question initiating it. The question is: How much mathematical training is essential to the appropriate education of men and women as human beings? I have said that the question admits of a definite answer in terms of a supreme and incontestable principle. I have stated the principle as well as I can and have tried to signalize its importance for a general theory of education. It remains to apply it to the specific question before us. The task is not difficult. It is plain that one of the great types of distinctively human activity—perhaps the greatest and most distinctively human type—is what is known as Thinking—the handling of ideas as ideas—the formation of concepts, the combination of concepts into higher and higher ones, discernment of the relations subsisting among them, embodiment of these relations in the forms of judgments or propositions, the ordering and use of these in the construction of doctrine regarding life and the world—in a word, the whole complex of activity involved in the discourse of Thought. It is essential to the argument I am making to keep steadily in mind that this kind of activity, our sense for it, our faculty for it, the need to which it ministers, the joy it gives, and the obligation it imposes are part and parcel of what we have been calling our common humanity as distinguished, on the one hand, from that which is animal in man, and, on the other, from such special propensities or other marks as give the differing specimens of humankind their respective individualities. Thinking is not indeed essential to life, but it is essential to human life. All men and women as human beings are inhabitants of the *Gedankenwelt*—citizens, so to speak, of the world of ideas, native citizens of the world of thought. And now what shall we say is the prototype of excellence in thinking? What is the hovering angel wooing our loyalty to what is best in thinking? What is the muse of life in the world of ideas? An austere goddess, high, pure, serene, cold towards human frailty, demanding perfect precision of ideas, perfect clarity of expression, and perfect allegiance to the eternal laws of thought. In mathematics the name of the muse is familiar: it is Rigor—Logical Rigor, which signifies a kind of silent music, the still harmony of ideas, the intellect's dream of logical perfection.

Can the dream be realized? I am well aware that most of the things which constitute the subject-matter of our human thinking—that most of the things to which our thought is drawn by interest or driven by the exigencies of life—are naturally so nebulous, so vague, so indeterminate that they cannot be handled in strict accordance with the rigorous demands of logic. I am aware that these demands can not be *fully* satisfied even in mathematics, the logical science par excellence. Nevertheless I contend that, as the *ideal* of excellence in thinking, Logical Rigor is supremely important, not only in mathematical thinking, but in all thinking and especially in just those subjects where precision is least attainable. For without this ideal, thinking is without a just standard for self-criticism, and without light upon its course; it is a wanderer, like a vessel at sea without compass or star. Were it necessary, how easy it would unfortunately be to cite endless examples of such thinking from the multitudinous writings of our time. Indeed, if the pretentious books produced in these troubled years by men without logical insight or a sense of logical obligation were gathered into a heap and burned, they would thus produce, in the form of a bright bonfire the only light they are qualified to give. "Logic," it has been said, "is the child of a good heart and a clear head." We know, however, that an evil heart is not essential to a fool and that, on the other hand, few heads are naturally so clear as not to require discipline.

Now, it so happens that the term mathematics is the name of that discipline which, because it attains more nearly than any other to the level of logical rigor, is better qualified than any other to reveal the prototype of what is best in the quality of thinking as thinking. And so, in accordance with the principle of humanistic education, we have to say that the amount of mathematical training essential to the appropriate education of men and women as human beings and essential, therefore, to philosophers as human beings, is the amount necessary to give them a fair understanding of Rigor as the standard of logical rectitude and therewith, if it may be, the spirit of loyalty to the ideal of excellence in the quality of thought as thought.

Such is my answer to the question that has detained us so long. It is, you observe, a qualitative answer in terms of a great ideal and a sovereign principle of education. If I must add a word touching the strictly quantitative aspect of the question, if I must, that is, attempt to indicate the extent of courses and the length of time necessary and sufficient to yield the required quality and degree of training, I do so with less confidence and far less interest. For so much, so very much, depends on the pupil's talent and the quality of instruction. A considerable degree of native mathematical ability is much more common than is commonly supposed. Born mathematical imbeciles are rare. Youth of fair mathematical talent constitute an immense majority. I venture to say, regarding the question of time and the extent of courses, that, for pupils of fair mathematical endowment, a collegiate freshman year or even a high school senior year of geometry and algebra, if the subjects be administered in the true mathematical spirit, with due regard to precision of ideas and to the exquisite beauty of perfect demonstration, is sufficient to give a fair vision of the ideal and standard of sound thinking.

Herewith, I have come to the end of what I desired to say respecting the mathematical equipment essential to a philosopher in so far as its measure depends upon the fact that philosophers are human beings. It remains to enquire what further mathematical attainments are to be regarded indispensable to the proper equipment of a philosopher as a philosopher. It is evident that the answer must be sought in the nature of the philosopher's vocation. It would be presumptuous in me, a student of mathematics, to offer to teach you, who are students of philosophy, the nature of your vocation, but I may remind you of it for it is necessary to have it clearly in mind if we are to see its bearings upon the question in hand. No one, I suppose, has conceived the philosopher's vocation more justly and nobly or characterized it more clearly and truly than Plato, as no other has drawn, with such clarity and charm, with so perfect a union of finesse and amplitude, so beautifully and so truly, the spiritual portrait of the genuine philosopher. You are, of course, familiar with the characterization and the portrait, which together give for all time a vision of the great ideal: what genuine philosophy is, and the philosopher ought to be. I wish to remind you of such elements of it as our present task requires.

The genuine philosopher, says Plato, "has magnificence of mind"; there is in him "no secret corner of illiberality"; he is "noble, gracious, the friend of truth, justice, courage, temperance"; he aims at being "a spectator of all time and all existence," and so he is a lover and seeker of "wisdom," which does not consist of sense-impressions nor of "the tempers and tastes of the motley multitude" nor of fickle "blinking opinion" begotten of time-born appearances and events destined to the doom of things that perish in "the sea of change," but consists in knowledge of things that abide—of true being—of whatsoever in the world is eternal: pursuit of such wisdom is the philosopher's vocation, sustained by the twofold hope of coming at length into the full-shining presence of the Beautiful, the True, and the Good and of bringing light from them into the lives of the children of men.

From that conception of the genuine philosopher's vocation and character, what conclusion follows regarding his obligation to mathematics? An important conclusion, as I hope to show if you agree with me in thinking that we ought to ascertain what it is.

Let me say at the outset that there are two pretty obvious considerations which I do not intend to insist upon, although they are not without relevance and weight. One of them is that which conceives mathematics as being itself a branch of philosophy; the other relates to the familiar contention of Plato, that mathematical discipline is indispensable as a preparation for what he conceived to be the philosopher's distinctive task—that of Dialectic.

As to the former consideration, one might argue, pertinently and confidently, that both historically and in accordance with the foregoing conception of philosophy, Logic is one of its branches; that mathematics not only employs logic as an instrument but is, in fact, identical with it, mathematics (as traditionally viewed) being related to logic (as traditionally viewed) as the trunk and branches of a tree are related to its roots; that, consequently, mathematics, being identical with logic, is not external to philosophy but is, strictly speaking, one of its principal divisions; and that, accordingly, philosophers, if they are not to be ignorant of one of the chief departments of their own subject, are obliged to be, not merely mathematical *dilettanti,* but mathematical students, serious explorers of the science. Theoretically, the argument is sound, which is the highest quality of argument as such. I do not, however, as I have said, intend to press it, because it imposes on the student of philosophy an obligation that he cannot fully meet; his obligations are many, too many and too great; he may not reasonably hope to win the proper competence of a mathematician in a subject where the developments, still rapidly progressing in numerous directions, have already reached proportions so great that no man, though he have the wide-reaching arms of a Henri Poincaré, can contrive to embrace them all.

Turning now to the second one of the two considerations mentioned a moment ago, let me guard against the danger of being misunderstood. You are aware that, in the view of Plato, what is peculiar to philosophy is dialectic—"the coping stone of the sciences"; you are aware that dialectic is the sole means by which the philosopher may gain a knowledge of "what each thing" in the hierarchy of being "essentially is," and by which he may gain, at length, as he ascends the scale, a vision of things supreme—absolute justice, absolute beauty, absolute truth, absolute good; you are aware that the successful employment of dialectic requires not only native "magnificence of mind," but also a long course of preparation in the subsidiary sciences; you are aware that, according to Plato, the most indispensable of these sciences are arithmetic and geometry: the former because arithmetic, not as the mere practical art of calculation but as a discipline in the logical nature of pure number, "lays hold of true being"; and the latter because "the knowledge at which geometry aims is knowledge of the eternal." Such is in brief, as you know, the famous contention of Plato respecting the importance of mathematical discipline as a preparation for philosophy. There can be no doubt that the contention is perfectly just. Why, then, do I not stress it in this connection? The reason is that the mathematical discipline insisted on by Plato is more than covered by the mathematical training I have already urged as essential to the appropriate education of the philosopher as a human being, and that we are here considering such further mathematical attainments as are essential to him as a philosopher. Before leaving this theme, however, I desire to point out a different aspect of it and in connection therewith to speak very briefly, in passing, of a matter which I have discussed elsewhere,[1] to which I hope to return at a later stage of these lectures and which, I believe, has a very important bearing upon the question before us.

After some years of reflection, I am convinced that the great Platonic Absolutes, whose "perception by pure intelligence" brings us, says Plato, to "the *end* of the intellectual world"—have indeed their proper locus *beyond* it. I am convinced that, instead of being genuine concepts amenable as such to the logical processes valid in the intellect's world, the Platonic Absolutes are radiant *ideals* of concepts, shining from above them like downward-looking aspects of an over-world; transcending

[1] *Science and Religion,* also *The New Infinite and the Old Theology.* Yale University Press.

every type of excellence in which intellectual progress is possible, they appear as ideals supernal—as stars beyond the sky. I need not say that the Absolutes, thus regarded, retain their glory unimpaired and their previous value as sources of light and inspiration. We should not, however, fail to see clearly that, if they be thus regarded, the philosopher is thereby confronted by a new challenge, a new problem, a new field of study or, perhaps I should say, by an old one seen as new. For, if the Absolutes are not *in* the intellectual world but are beyond it; if they be, in fact, not concepts, but ideals of concepts, shining downward from above them, then obviously their origin, the manner and genesis of their appearance, and their significance for life, must be sought in the nature and function of that strange and familiar spiritual process omnipresent among the activities of the intellectual world and known as Idealization. And now the point I am aiming at and to which I invite your special attention is this: In the study of this great subject—the nature and function of Idealization—the philosopher and especially the theologian as philosopher—for rational theology, rightly conceived, is the science of Idealization—will have need of mathematical discipline surpassing the Platonic requirement and surpassing what I have deemed essential to the education of the philosopher as a human being. For the term "idealization" is the generic literary term for what in science and especially in mathematics is known as generalization by means of the method or process of limits. In mathematics, particularly in the modern theory of the Real Variable, in connection with the generalization of the number concept, the essential nature of Idealization, the pattern of it as the process and method of directing the attention from within a given domain of operation to the existence and the character of outlying domains, comes into perfect light. It is in mathematics and not elsewhere that Idealization is beheld in its purity; and unless the philosopher becomes familiar with it there in its purity, his endeavor to study the great process elsewhere, amid the many disguises half concealing its subtle ramifications throughout the shadowy world of general thought, will encounter serious difficulties, if not defeat.

The considerations I have now advanced, though they are subordinate, are weighty, and I commend them as worthy of your further reflection. Let us proceed, without further delay, to the heart of the matter.

We have seen that the genuine philosopher "has magnificence of mind"; that there is in him "no secret corner of illiberality"; that his vocation requires him to be "a spectator of all time and all existence"; and that the wisdom he seeks is the wisdom which consists in knowledge of whatsoever is eternal. It is these great things—the highest distinctive marks of the genuine philosopher—that determine the character of his mathematical obligations and enable us to measure them. For what is mathematics? What is that science which Plato[1] called "divine," which Goethe called "an organ of the inner higher sense," which Novalis called "the life of the gods," and which Sylvester called "the Music of Reason"? The question is not intended to call for a complete description of the science, much less for a definition of it. What it seeks is a partial description. I wish merely to draw your attention to one feature of mathematics—to that feature of it which all competent judges agree in signalizing as the chief aspect of the science viewed as an enterprise. The aspect in question I endeavored to point out some years

[1] See *Memorabilia Mathematica* by Professor Moritz.

ago in the following words: "As an enterprise, mathematics is characterized by its aim, and its aim is to think rigorously whatever is rigorously thinkable or whatever may become rigorously thinkable in course of the upward striving and refining evolution of ideas."[1] The same feature has been recently indicated, even more clearly perhaps and somewhat poignantly, in a striking utterance by Mr. Bertrand Russell. "Pure logic, and pure mathematics (which is the same thing), aims at being true, in Leibnizian phraseology, in all possible worlds and not merely in this higgledy-piggledy job-lot of a world in which chance has imprisoned us."[2]

You know, at least in a general way, that in pursuit of that enterprise and aim through the centuries, the mathematical spirit has achieved immense results and that today the science of mathematics, as a body of permanent knowledge regarding things eternal, is a veritable continent of expanding doctrine. And so it is pertinent to ask: How can one aspiring to be a philosopher, unless he explores that growing continent of knowledge respecting what is "true of all possible worlds," be in any proper sense "a spectator of all time and all existence"? You may wish to reply that, owing to his other obligations, the philosopher cannot make the exploration fully; that indeed, owing to the nature of the continent, he cannot, without exploring it step by step, gain even so much as a clear knowledge of its contour and relief; that, however, notwithstanding the endless diversity of the things that are there, they have a certain essential character in common; that for the philosopher's vocation, knowledge of

[1] *Human Worth of Rigorous Thinking*, p. 3. Columbia University Press.

[2] *Introduction to Mathematical Philosophy*. The Macmillan Company, New York.

that common character is sufficient; and that such knowledge does not demand exploration of the continent in all its length and breadth and height and depth, but may be gained by examination of representative parts and especially of the elements which fundamentally compose the whole.

That reply, if we rightly interpret the meaning of the terms, is just. But their meaning is momentous. The mathematical knowledge which they tell us is sufficient for the purposes of the philosopher is neither slight nor simple nor easy to gain. The questions it must answer determine its nature and its scope. What are the idiosyncrasies of mathematics as a body of content? As a system of methods? As a type of activity? As a distinctive enterprise among the great kindred enterprises of the human spirit? If the science be logical, what are its relations to Logic? If it be beautiful, what are its relations to Art? If it employ hypothesis, observation and experiment, what are its relations to Natural Science? If it be purely abstract and conceptual, what are its relations to the concrete world of Sense? If it be theoretic, what are its relations to Practical Life? If it be self-critical, what are its relations to the science and art of Criticism? If it be a wisdom respecting infinite and eternal things, what are its relations to Philosophy and to Religion? If it have limitations, what are its relations to the dream of Universal Knowledge? To the challenge of these great questions and their kind, no one having "magnificence of mind," no one called to be "a spectator of all time and all existence," can fail to respond. And so we see that the mathematical obligations of the philosopher confront him with two difficult close-related Problems: the problem of *definition* and the problem of *evaluation:*

he must endeavor to ascertain what mathematics essentially is and endeavor to estimate, in the terms of spiritual Worth, the rank and the dignity of the science in the hierarchy of knowledges and arts.

It is a radical error to regard these kindred tasks of definition and evaluation as belonging to the proper function of mathematicians as such. The term "mathematics" is the name of an immense class of logically related terms and most of these the mathematician must indeed define, but the term "mathematics," which names the class, is not among them; the class is not a member of itself, for no class can be; the name "mathematics" is not a mathematical term; the mathematician would be none the less a mathematician, had he never heard of it; it is a philosophical term, used by mathematicians as a convenience but never as a necessity. The proper activity, the distinctive function, of the mathematician is to mathematicize, as that of a swimmer is to swim; or that of a farmer, to farm; or that of a poet, to make poetry; or that of a trader, to trade. And it is as little the business of the mathematician to define and evaluate the peculiar type of his proper activity as it is that of the swimmer or the farmer or the poet or the trader to do the like for his. The philosopher, therefore, may not rightly look to mathematicians as such for a definition of mathematics nor for any appraisement of its significance or its worth.

Is it not true, nevertheless,—you may wish to ask—that nearly all real advancement made in the course of the centuries in these tasks of definition and appraisement has been made by mathematicians? The answer is yes, even if we do not forget or underrate the relevant contributions of Plato and Aristotle, for knowing, as they did, what was known then of mathematics, they must be counted among the mathematical scholars of their day. It must be noted, however, that, though the advancement in question was made by mathematicians, it was made by them, not in their character as mathematicians, but in their capacity as philosophers. There is nothing in the fact to astonish. For a man is greater than any occupation, and a mathematician, like a physician or lawyer or poet or statesman or farmer, may be—indeed he must be, in some measure—a philosopher as well. It is not, then, strange or a matter for wonder that there have been mathematicians who, in relation to their proper subject taken as a distinctive whole, have sometimes taken the attitude and played the rôle of philosopher. Nay, even *within* the subject, in relation to its parts, the rôle is very common; for whenever a mathematician, having acquired competence in two or more branches—say algebra and geometry—pauses to compare them, seeking to ascertain the essential nature of each, what they have in common, their respective worths and their joint significance as forms of activity, his interest and his attitude have then become for the time, whether long or short, those of the philosopher. The fact is that such minor alternations of the scientific and the philosophic interests may be constantly witnessed even in the activity of such mathematicians as ignorantly affect to spurn philosophy and to scorn its achievements; but they are not aware of it.

Of the two tasks with which, as we have seen, the mathematical obligations of the philosopher confront him, the task of definition is far more advanced than that of evaluation; and, though the work of the former is not yet complete, we know much better today what mathematics is than what it is worth. That it should be so is natural, for a just appraisement of worth depends, of course, upon the nature of the thing appraised. We are, therefore, not surprised to find that researches concerning the essential nature of mathematics have been prosecuted, especially in recent times, far more resolutely and systematically than such as aim at a critical estimate of its significance and value. In Plato and in Aristotle, as you know, research of both kinds produced results of great importance. I shall not speak of the great Greek mathematicians for their interest centered, not in the philosophy of their subject, but in the science of it. They were swimmers mainly—not non-aquatic students of swimming. It seems incredible that, after Plato and Aristotle, no important contribution to the philosophy of mathematics was made in the course of twenty hundred years. Yet that is the fact. Even the brilliant and exquisite *De L'Esprit Géometrique* of Pascal is thoroughly Aristotelian. The great revival had to await the appearance of Leibniz —of him who said, *"Ma métaphysique est toute mathematique."* As students of philosophy, you know that throughout his life this marvelous man was haunted by a magnificent dream—the dream of "a universal mathematics." In his manifold endeavors to make the dream come true is found the origin of that great critico-constructive movement which has done more than all previous centuries to disclose the essential nature of rigorous thought and which, after notable vicissitudes of fortune, is known today, in all scientific countries of the world, under the characteristic name of Symbolic Logic.

The leading names of its pioneers and contributors—Leibniz, Lambert, De Morgan, Boole, Jevons, Schröder, Peirce (C. S.), MacCall, Frege, Peano, Russell, Whitehead, Hilbert, Huntington, Couturat, and others—sufficiently indicate its international interest and the variety of genius to which it appeals. The growing literature of the subject is large. Fortunately, it is not necessary, except for the historian, to examine it all, for it has been refined, assimilated, and, all but the later developments, superseded in the monumental work of Whitehead and Russell—*Principia Mathematica*—the present culmination of the movement. This work, however, which has not yet been completed, the philosopher must examine *minutely* if he would understand, as a philosopher ought to understand, the fundamental nature of mathematics as disclosed in the best light that has been thrown upon it and especially if he would realize the hope of being able to improve the light, which is not yet perfect. The symbols are at first repellent; they tend to frighten but are not in fact difficult to master.

> They are things of so frightful mien
> That to be hated need only be seen.
> But often seen, familiar with their face,
> We endure them first and then embrace.

Theoretically, the symbols are not essential, a sufficiently powerful god could get along without them; but practically they are indispensable as instruments for economizing our intellectual energy.[1]

No kind of work, whether philosophic or scientific, can be severer in its demands. None surpasses it in respect of the toil involved, nor in patience, nor in depth of penetration, nor in subtlety, nor imagination, nor analytic finesse, nor in the demand it makes upon the *constructive* faculty, and none can give to the competent student a serener vision of eternal things. If on this

[1] In relation to the early history and importance of symbolism do not fail to read Professor David Eugene Smith's beautiful essay, "Ten Great Epochs in the History of Mathematics," in *Scientia,* June, 1921.

account it seems to you, as it may seem, a little strange that the majority of mathematicians have little interest in such work and are not familiar with it, it is sufficient to reflect that, though its results as results are strictly scientific, strictly a part of mathematics, they are deeply tinged with philosophic interest and owe their discovery primarily to the spirit of philosophic enquiry. In mathematics, as in other subjects, fashions change; it is, moreover, so large a subject that a student is obliged by his limitations to specialize in a branch of it or in a group of branches; and it so happens that a large majority of mathematicians are disqualified,—some of them by breeding, more of them by temperament,—for study or research in that branch which deals with the foundations of their science as a whole. Such disqualification is not to be imputed to them as a fault; often no doubt,—oftener than not, perhaps,—it is only a defect of a quality; at all events, a mathematician may not be rightly blamed for the temperamental bent of his scientific interests. The same may not be said of those who are inclined to depreciate other interests than their own. I refer to the type of mathematician,—such as you may sometimes meet,—who, as if to mitigate his sense of guilt for being consciously innocent of symbolic logic and so to protect his self-respect, will occasionally ask you, in a somewhat disparaging tone, to tell him, if possible, of any important service rendered by symbolic logic or of any important proposition established by it or of any important method devised by it for the use of mathematicians. If you disregard the spirit in which such questions are sometimes asked, it is easy to answer them in a way satisfactory to any candid and competent enquirer. The answer, as I conceive it, is, in brief, as follows:

(1) Symbolic logic has established the thesis that all existing mathematics (and presumably all potential mathematics) is literally a logical outgrowth of a few primitive ideas, and a few primitive propositions, of logic; and, that, accordingly, logic and mathematics are spiritually one in the sense in which the roots, the trunk and the branches of a tree are physically one: a proposition which, though philosophical and not mathematical, is, in respect of human significance, unsurpassed.

(2) In course of the work establishing the foregoing proposition, symbolic logic has discovered and rigorously demonstrated a long sequence of theorems respecting propositions, classes, and relations, which theorems constitute an immense new body of genuinely mathematical doctrine underlying mathematics as commonly understood and they are open to inspection by all critics, whether friendly or unsympathetic.

(3) Symbolic logic has not promised nor pretended to devise methods to facilitate mathematical research except research in mathematical foundations; in such research the effectiveness of the methods employed is patent in the results.

(4) Finally, symbolic logic is simply the latest fruit of the critical spirit in mathematics—fruit of the refinement,—the inevitable refinement,—of that spirit which has led to so many mathematical developments familiar to all mathematicians,—the postulational method, for example, the birth of non-Euclidean geometries, the theory of manifolds including the hyperspaces, the so-called arithmetization of mathematics, and similar phenomena throughout the history of the science. To depreciate symbolic logic is to oppose the

progress of the *spirit* of constructive criticism and that means opposition to the progress of science; for Cousin's famous *mot* is just: *La critique est la vie de la science.*

In saying that the philosopher's mathematical obligations require him to familiarize himself with the methods and results of symbolic logic, I have not quite finished the tale. One point remains to be stressed. Before presenting it, let me remind you of a certain fairly obvious distinction which Bergson [1] has emphasized and has elevated, rightly I believe, to the level of an important principle of knowledge. I may best make it clear by an example. You *know,* as we say, how to move your arms. This knowledge is not a part of, and is not derived from, your "scientific" knowledge of physiology, anatomy and physics, though this knowledge, too, may tell you much respecting the motion in question. The latter knowledge is indirect and external—a knowledge from without; the former is immediate and internal—a knowledge from within; it is a living instinct—of the essence of your life; the other is only a superadded understanding. Complete or perfect knowledge of any thing involves both of these kinds of knowledge. In the illustration I have used, the thing to be known is a part of the knower—the mobile arm is yours and its life is yours. But most objects of knowledge are not thus parts of the knower. Of such objects complete knowledge, even if we suppose the element of "understanding" to be perfectible, is unattainable; for to attain it, to gain the other element,—the instinctive element, the inner kind of knowlege,—would

[1] "Introduction a la Métaphysique." *Revue de Métaphysique et de Morale,* Vol. 11, (1903).

require the knower to make the object's life an intimate part of his own; and this, it is plain, cannot be done perfectly. But—and here is a fact of the utmost importance—it can be done approximately. Do you ask, how? The answer is: By the noetic agency of sympathy or love; by the means which Bergson has so finely described as "intellectual sympathy" with the object's life. Your thought, I fancy, runs ahead of my speech and already sees the bearing of the point upon the philosopher's obligations to mathematics. In a sense more than figurative, this science has a life of its own. Else how could it grow? To acquire such knowledge of the science as the philosopher's vocation demands, to know it from within as it instinctively knows itself, he must acquire such intellectual sympathy with it as will enable him to feel its proper life as part of his own. Sympathy so living and intimate,—embracing the instincts, and feeling the impulses and moods, of an alien life,—is not easily acquired. In the case of mathematics, collegiate courses in algebra, geometry and trigonometry cannot give it, except to the born mathematician, who has it already; neither can it be given adequately by symbolic logic for this study is too meditative for the purpose, too introspective, being more concerned to "understand," than to "live," the life of mathematics. No, if the student of philosophy would acquire that kind of knowledge of mathematics which can come to him only through intellectual sympathy with its life, he must share its life; he must penetrate it deeply enough to feel the touch and thrill, the push and sweep, of its conquering tide; he must at least plunge into Analytical Geometry and the Infinitesimal Calculus where the science first won, and its votaries first win, a worthy sense of its power and its destiny.

In the light of the foregoing considerations, the mathematical obligations of the philosopher appear to be heavy. They are heavy; but they are not too heavy for those whose native talents qualify them for a vocation demanding "magnificence of mind." It is consoling to know that a student who faithfully keeps the obligations will have two great rewards: the joy of an insight and a power not to be otherwise gained; and the joy of representing and perpetuating a noble tradition of his kind,—the tradition, I mean, of mathematical competence as illustrated by the heroes of philosophy in every important age. In relation to that tradition, it is indeed true, as you know, that there have been many philosophers of great learning, some of them important thinkers, whose ignorance of mathematics has been virtually complete, and these have differed widely in kind; of their mathematical ignorance some of them have not been aware; some have deeply regretted it and humbly confessed it—our own beloved William James, for example; in some it has been not only complete but shameless as well, even haughty and defiant, as in Sir William Hamilton and Schopenhauer, whose false and malicious diatribes against mathematics I have dealt with elsewhere,[1] and in case also, I am sorry to say, of Benedetto Croce,[2] whose fine literary and artistic culture and true elevation of spirit have not availed to restrain him from speaking with strange confidence and very disparagingly of a science which his fellow countrymen, by brilliant research, have done so much to honor and which he has not qualified himself to understand even slightly.

It is edifying to compare such representatives of phi-

[1] *Human Worth of Rigorous Thinking*, p. 290.
[2] *Logic as the Science of the Pure Concept.*

losophy with its towering heroes, its men of "summit-minds": with Plato, for example, who knew perfectly the mathematics of his time, whose sense and revelation of its spiritual significance has never been surpassed, and whose influence in his own and all succeeding ages has given his name a permanent place in mathematical history; and with Aristotle, whose discussions of such fundamental questions as the nature of mathematical definition, hypothesis, axiom, postulate, and subject matter, are of high value even today and whose great contributions to logic must now be regarded, in the light of modern symbolic logic, as being, though he did not know it, genuine contributions to mathematics; and with Descartes, discoverer of important mathematical propositions, and chief inventor of analytical geometry,—second in scientific power to only one among mathematical methods; and with Leibniz, father of modern symbolic logic and co-inventor with Newton of the infinitesimal calculus, "the most powerful instrument of thought yet devised by the wit of man";[1] and with Spinoza to whose lot it fell to try the great experiment,—inevitable in the history of thought,—of clothing ethical theory,—highest of human interests,—with the strength and beauty of mathematical rigor and form, and, in trying it, to exemplify in a singularly noble way, the fact that illustrious failures fall to the lot of none but illustrious men; and with other great philosophic personalities, if I did not fear to weary you in naming them, who by their mathematical competence worthily represent the heroic tradition.

In closing this initial lecture, I desire to indicate in a general way the sort of topics with which the following lectures will deal. The endless number of the ideas, or

[1] See the preface of Professor W. B. Smith's *Infinitesimal Calculus.*

notions, or concepts,—as they are variously called,—which enter as components into the stately edifice of mathematics, though they are all of them, in a sense, indispensable to it, yet differ very widely in respect of their place and rank, their dignity and structural service. Examination of the great edifice makes it evident that some of them,—a relatively small number of them,—have the distinction of being related to it as central supporting pillars. Among the chief of these are the concepts denoted by the terms: Function—Propositional Function—Implication—Proposition—Class—Relation—Postulate System—Doctrinal Function—Doctrine—Variable—Limit—Number—Finitude—Infinity—Transformation—Group—Invariance. It is with such pillar-concepts,—which are obviously not coordinate in rank,—that I purpose to deal, and I shall deal with them primarily as concepts, explaining them with constant regard to clarity, with a minimum of technical symbols, and with a view, not alone to their mathematical meanings, but to their significance and use in outlying fields of thought. But I shall not endeavor to expound, in the proper sense of the term, the great technical doctrines that have grown up about them as subject matter, for such exposition would demand, as you know, not merely one course, but many courses, of lectures. You will rightly infer that, though proof or demonstration may not be entirely absent, it will not be permitted to detain us too long, much less to dominate the discussions.

Let me say, finally, that the course is not designed to be, in the stricter and narrower sense of the term, a course in the philosophy of mathematics. It aims at being at once something less and something more: *less,* in that it does not endeavor to begin with the most ultimate of logical principles and to build upon them, little step by step, with infinite patience, the solid masonry of the mathematical edifice; *more,* in that it is a good deal concerned with the mentioned task of evaluation—with disclosing the relations of mathematics to other great forms of intellectual activity and especially its bearings upon the universal interests of the human spirit.

# LECTURE II

## Postulates

CONCRETE DEFINITION OF POSTULATE SYSTEM—THE PROTOTYPE OF PRINCIPLES OR PLATFORMS—THE ANCIENT "CRAFT OF GEMETRY"—THE SWORD OF THE GADFLY—CLARITY OR SILENCE—MUNICIPAL LAWS AND THE LAWS OF THOUGHT.

THE introductory lecture has served, I hope, to indicate in a general way the aim, the spirit, and the scope of our undertaking. In deciding to begin the work proper with a study of the great concept denoted by the familiar term —Postulate System—I have been guided by three considerations: (1) every question arising in what is strictly called the philosophy of mathematics—in the study, that is, of its logical foundations—is connected more or less closely, directly or indirectly, with that concept, which is thus the central ganglion of mathematical philosophy; (2) by means of the concept in question and without unnecessary delay, I desire to set in clear light another concept, intimately related to it, to which I have given the name—Doctrinal Function—and which, if I am not mistaken, has great philosophic importance; (3) postulate systems as employed in mathematics, appear there in perfect light as systems of principles underlying and supporting definite bodies of thought, and so they serve as a model, as an ideal prototype, for the inspiration, the guidance and the criticism of *every* rational enterprise, whether of philosophy, of science, or of life in general.

A subject so fundamental, many-sided, and far-reaching will naturally detain us for some time. The wisdom we seek is golden, but it cannot be gained by any of the get-rich-quick methods characteristic of our industrial and neurasthenic age; the way to it is a little long and I may as well warn you that in these lectures I intend to pursue it in a leisurely fashion. The study is not so "entertaining" as a "movie" nor so easy as the life of "maggots in a cheese" or that of summer birds in a valley of fruits. It demands some patience, hard work and endurance. It will quickly weary such as are content with a little phraseological facility in matters they do not understand, but not those whose curiosity is deep and genuine, for they will be sustained by the dignity of the task and the joy of the game.

Let us now enter upon it. What are we to understand by the term postulate? You are aware that a branch of mathematics (or, for that matter, of mechanics or of physics or of any other science), if the branch be ideally constructed, is autonomous: it consists, that is, of a body of propositions of which a few are assumed—not proved in the branch but taken for granted there—and the rest are deduced from them as logical consequences. To students of philosophy, I need not say that to suppose *all* the propositions of an autonomous theory to be proved in it, plainly involves circularity and a contradiction in terms. In accordance with current usage, which I intend to follow in this matter, any proposition thus taken for granted in a given branch is called a postulate, or assumption, or axiom, or primitive proposition, or fundamental hypothesis, of the branch; these terms being used interchangeably according to the taste of the author. It has not always been so; the term axiom, for example, was long used to denote "self-evident proposition," which is a kind of proposition that modern mathematicians have not been able to discover. But I shall not detain you with an historical account of the terms, interesting and instructive as their history is. It gives me pleasure to say, however, that, if you feel drawn thereto, as I hope you do, you will find much more than an ample clue to it in the introduction to Dr. T. L. Heath's superb edition of Euclid's *Elements* where these terms and kindred matters are set in the bright light of critical commentary from the days of Plato down to the present time. In passing, let me add, by way of indicating an opportunity, that this work of Dr. Heath, like other works of his, attains a high degree of excellence in a type of activity in which our American mathematical scholarship has been singularly lacking; not because American mathematicians have lacked facilities or ability, for these they have not lacked, but because the universities in which they have received their training and have done their work have not yet acquired the requisite atmosphere and spirit.

A postulate is one thing; a system of postulates is another. In defining the former, I have by no means defined the latter. It is not easy to do so with logical precision: it is, I mean, not easy to give an *abstract* definition of the generic concept denoted by the term, postulate system; and I shall not attempt it at this point, for it presupposes study of the concept as actually revealed in mathematics and so has its proper place at the end of the study. Here, at the beginning, we must be content with definition by example, with what Professor Enriques, in his *Problems of Science,* has called *concrete* definition, which is nothing more mysterious than the practice, familiar alike in science and in ordinary life, of telling the meaning of a general term by pointing out one or more of the many objects imperfectly representing it, and saying, "there, there, that is what it means, or that and that." I wish it were practicable in this course to deal adequately with Definition as a separate topic, with its varieties, its functions and its history. It is, I think, an admirable subject for a scholarly dissertation. In such an undertaking the student would find many helpful suggestions in the treatment of definition by Enriques in the work just now mentioned; in certain passages of *Science and Hypothesis* by Poincaré; in some remarkably keen observations found in Pascal's immortal essays, *De L'Esprit Géometrique,* which I cited in the preceding lecture; in the above-mentioned work of Dr. Heath; in the literature of symbolic logic; and, as I need not say to you, who are students of philosophy, in the Metaphysics and Posterior Analytics of Aristotle, not to mention the Platonic Dialogues where philosophy in our western-world first becomes fully conscious that the way to wisdom—to knowledge of things eternal—is not the way of song, however glorious, nor that of sophistry, however pretentious, but is the way of logic, and where accordingly, despite the presence there of many mystical elements, the spirit of Definition, which is the spirit of clear thinking and determinate speech, becomes in Socrates a conquering sword. And this leads me to say, in passing, that in these our democratic times of free speech when everyone, no matter how ignorant or foolish, is a licensed prophet, and blatant sophists abound on every hand, there is no way in which you as teachers of philosophy can render greater service than by carrying on the work of the great Gadfly

—constraining men by relentless logical criticism to a choice of one or the other of two alternatives: coherency and clarity of speech or—silence. Today, the mind of the world is a weltering sea of wild passions and wilder opinions. It can not be calmed by municipal law, but it can be by disciplining men to a decent respect for the eternal laws of thought. And that is the supreme obligation of philosophy as the guardian of Reason.

A few moments ago I said that, in the beginning of the study of postulate systems, we must be content to define the notion concretely—by means, that is, of examples. Accordingly, I am going to spread before you presently a definite system of postulates and invite you to examine it as a geologist might examine a specific rock formation; or as a student of poetry might examine a specific poem; or a student of law, the constitution of the Soviet republic or that of the United States. From the large variety of postulate systems recently invented for various mathematical branches, I have selected, as a specimen for our initial study, the system devised by the late Professor Hilbert and found in his famous *Foundations of Geometry*. It is one of several systems invented in our time to serve as logical bases of Euclidean Geometry. Though it is not intrinsically superior to its rivals, whether in geometry or in other branches, I have selected it in preference to them for two reasons. One of them is that, *practical* arithmetic not being a science, Euclidean Geometry is the oldest and most familiar branch of mathematics, as well as being historically the most interesting and even romantic.

"The clerk Euclide on this wyse hit fonde
Thys craft of gemetry yn Egypte londe

In Egypte he tawghte hyt ful wyde,
In dyvers londe on every syde.
Mony erys afterwarde y understonde
Yer that the craft com ynto thys londe.
Thys craft com into England, as y you say,
Yn tyme of good Kyng Adelstone's day."

From which we see that even in the old island home of our beautiful English tongue the Greek "Craft of Gemetry" has been known for a thousand years. The second reason for my selecting Hilbert's system is that it is the most famous of all existing postulate systems, save one only—that of Euclid. Hilbert's acquired its great fame immediately, not entirely by its merits, for these, as already said, are not superior to the merits of some other systems, but largely through the fame of its author, which was world-wide. If you ask why I have chosen it instead of Euclid's system, which surpasses all others in fame, the answer is that, though Euclid's system was good enough to withstand more than two thousand years of criticism, it is now known, as we shall see later, to have some grave imperfections—most of them sins of omission. The postulates of Hilbert's system are called axioms by him—"axioms of geometry." As, however, the term axiom as employed by him is exactly equivalent to the term postulate as I have defined it, I shall be doing him no injustice in uniformly referring to his system as a system of *postulates,* thus avoiding the term axiom as likely to suggest the unavailable notion (so-called) of "self-evident truth." The postulates of Hilbert fall into six sets: postulates of connection; of order; of parallels; of congruence; of continuity; of completeness. I give them as found in the authorized English translation of Hilbert's book by Professor Townsend.

Physically, the book, as you observe, is small and light; but spiritually it is big and weighty. Except for some harmless abbreviations of statement, the postulates together with the definition of certain terms occurring in them are as follows:

*Postulates of Connection*

(1) Two distinct points determine a straight line.

(2) Any two points of a straight line determine it.

(3) Three non-collinear points determine a plane.

(4) Any three non-collinear points of a plane determine it.

(5) If two points of a line are in a plane, every point of the line is in the plane.

(6) If two planes have one common point, they have another.

(7) Every straight line contains at least two points; every plane at least three non-collinear points; and space at least four points not lying in a plane.

*Postulates of Order*

(8) If $A$, $B$, $C$ are points of a straight line and $B$ is between $A$ and $C$, then $B$ is between $C$ and $A$.

(9) If $A$ and $C$ are two points of a straight line, there is a point $B$ between $A$ and $C$, and a point $D$ such that $C$ is between $A$ and $D$.

(10) Of any three collinear points, one, and but one, is between the other two.

(11) Any four collinear points, $A$, $B$, $C$, $D$, can be so arranged that $B$ shall be between $A$ and $C$ and between $A$ and $D$, and that $C$ shall be between $A$ and $D$ and between $B$ and $D$.

DEFINITIONS.—A pair of points, $A$ and $B$, on a line, is a *segment* $AB$ or $BA$; $A$ and $B$ are the *segment's ends*; the points between $A$ and $B$ are the *segment's points*.

(12) Let $A$, $B$, $C$ be three non-collinear points and let $a$ be a line of their plane but not containing any of them. If $a$ contains a point of segment $AB$, it contains a point of segment $BC$ or of segment $AC$.

*Postulate of Parallels*

(13) If a straight line $a$ and a point $A$, not in $a$, be in a plane $\alpha$, there is in $\alpha$ one and only one straight line containing $A$ but no point of $a$.

*Postulates of Congruence*

(14) If $A$ and $B$ are two points on a straight line $a$, and if a point $A'$ be on a straight line $a'$, then on either side of $A'$ there is one and but one point $B'$ such that the segment $AB$ is congruent to the segment $A'B'$. Every segment is congruent to itself.

(15) If a segment $AB$ is congruent to a segment $A'B'$ and to a segment $A''B''$, then $A'B'$ is congruent to $A''B''$.

(16) If segments $AB$ and $BC$ of a straight line $a$ have no common point but $B$, and if segments $A'B'$ and $B'C'$ of a straight line $a'$ have no common point but $B'$, then, if $AB$ and $BC$ are respectively congruent to $A'B'$ and $B'C'$, $AC$ is congruent to $A''C''$.

DEFINITIONS.—If $O$ be a point of a straight line $a$, the points of $a$ on a same side of $O$ constitute a *half-ray emanating from* $O$; a pair of half-rays, $h$ and $k$, emanating from a point $O$ and not being parts of a same straight line is an *angle* $(h, k)$; $O$ is the angle's *vertex*, and $h$ and $k$

its *sides;* its *interior* is the class of points such that, if $A$ and $B$ be any two of them, segment $AB$ contains no point of $h$ or $k$; its *exterior* is composed of all other points of the plane except $O$ and the points of $h$ and $k$.

(17) Given an angle $(h, k)$; a line $a'$ in a plane $\alpha$; a point $O$ of $a'$; and in $\alpha$ a half-ray $h'$ emanating from $O$; then in $\alpha$ and emanating from $O$ there is one and but one half-ray $k'$ such that the angle $(h', k')$ is congruent to $(h, k)$ and that the interior of $(h', k')$ is on a given side of $a'$.

(18) If the angle $(h, k)$ is congruent to $(h', k')$ and to $(h'', k'')$, then $(h', k')$ is congruent to $(h'', k'')$.

(19) If, in the triangles $ABC$ and $A'B'C'$, $AB$, $AC$ and angle $BAC$ are respectively congruent to $A'B'$, $A'C'$ and angle $B'A'C'$, then the angles $ABC$ and $ACB$ are respectively congruent to the angle $A'B'C'$ and $A'C'B'$.

*The Postulate of Continuity (or of Archimides)*

(20) Let the point $A_1$ be between any two given points $A$ and $B$ of a straight line $a$. Let the points $A_2$, $A_3$, $A_4$, . . . of $a$ be such that $A_1$ is between $A$ and $A_2$, $A_2$ is between $A_1$ and $A_3$, and so on, and that the segments $AA_1$, $A_1A_2$, $A_2A_3$, . . . are mutually congruent. Then in the point series there is a point $A_n$ such that $B$ is between $A$ and $A_n$.

*Postulate of Completeness*

(21) To a system of points, lines and planes it is not possible to add other elements such that the system thus generalized shall form a new geometry in which all the postulates of the foregoing five sets are valid.

Such is a list of the postulates devised by Hilbert to serve as a foundation of Euclidean geometry. I regret having had to detain you so long in the rather arid business of presenting so long a list in detail. My apology is the importance of having the list definitely before us. In closing this lecture, let me recommend that, as a preparation for the next one, you familiarize yourselves with the postulates and in doing so, that you read enough of Hilbert's book to see how carefully the theorems are deduced from the postulates and how inevitably they follow therefrom.

# LECTURE III

## Basic Concepts

PROPOSITIONAL FUNCTION AND DOCTRINAL FUNCTION—MARRIAGE OF MATTER AND FORM—ITS INFINITE FERTILITY—PROPOSITIONS AND DOCTRINES THE OFFSPRING—VERIFIERS AND FALSIFIERS—SIGNIFICANCE AND NON-SENSE—A QUESTION ASKED BY MANY AND ANSWERED BY NONE.

ALL postulate systems have certain properties or features in common. In connection with the Hilbert system, I desire to draw your attention to such of these features as will lead us to form a certain conception which I think highly important and to which I have given the name—Doctrinal Function.

As a preliminary, I must explain briefly a closely related term—Propositional Function—invented by Bertrand Russell; it is, perhaps, the weightiest term that has entered the nomenclature of logic, or mathematics, in the course of a hundred years. It has the rare distinction of being, as we shall see, a perfect name for a supreme concept. Every one is familiar with the *ordinary* notion of a function—with the notion, that is, of the lawful dependence of one or more variable things upon other variable things, as the area of a rectangle upon the lengths of its sides, as the distance traveled upon the rate of going, as the volume of a gas upon temperature and pressure, as the prosperity of a throat specialist upon the moisture of the climate, as the attraction of material particles upon their distance asunder, as prohibitionary zeal upon intellectual distinction and moral elevation, as rate of chemical change upon the amount or the mass of the substance involved, as the turbulence of labor upon the lust of capital, and so on and on without end. This familiar notion of mutual dependence and mutual variation thus exemplified in every turn and feature of life and the world, is indeed a powerful concept; it is, in a sense, the sole subject matter of science; its scientific name—function—was first pronounced, it is said, by Leibniz; in modern mathematical analysis, it has played a dominant rôle, giving both name and character to certain great branches, as the theory of functions of real variables and the theory of functions of complex variables. Yet, powerful as it is, this Leibnizian conception, as employed in traditional mathematics, is far inferior in scope to that denoted by propositional function, which indeed embraces the former as a special case. What, then, are we to understand by this great term?

The answer, describing rather than strictly defining, is that a propositional function is any statement containing one or more real variables, where, by a real variable, is meant a name or other symbol whose meaning, or value as we say, is undetermined in the statement but to which we can at will assign in any order we please one or more values, or meanings, now one and now another. I fear that what I have just said is too general to be quite intelligible. The idea can be made sufficiently clear, however, by some simple examples—by concrete definition—provided you will understand that the examples are to the general concept in question as a burning match to a world-conflagration or as a few water drops to a boundless ocean. If we denote the real variables by such symbols as $x$, $y$, $z$, $w$, etc., then for simple examples of what is meant by propositional function we may cite the following quite at random: $x$ is a man; $x$ is a lover of $y$; $x$ is the specific gravity of $y$; $x$ is a noble citizen intemperately desiring to impose abstinence on $y$; $x$ has been divinely appointed by $y$ to subjugate $z$; $2x-3y=10z+w$; $\sin x=\cos y$; $x$ denied that $y$ said that $z$ confessed to being the author of $w$; $x$ knows that $y$ voted against $z$ on account of jealousy of $W$; and so on *ad infinitum*. How many variables may enter a propositional function? As many as we please. How many such functions are there? Their name is legion—the host of them is literally infinite. Even so, you may wish to say, the examples are not impressive. Nevertheless, the concept they represent, each in its little way, is sovereign—"like Jupiter among the Roman gods, first without a second."[1] Its majesty, its power, its subtlety, the immeasurable depth and range of its significance can not be perceived and felt at once, but only more and more with days and months and years of reflection. You will reflect upon it a very great deal if ever you enter seriously upon the study of symbolic logic.

Let us reflect a little upon it now. There will be occasion to resume its consideration at a later stage. At present, I wish merely to direct your attention to the very significant fact that propositional functions, though they have the *forms* of propositions, are *not* propositions. It is of the utmost importance to bear that in mind. A proposition is a statement that is true or else false. That is why propositions are so important—they, and not

[1] Gladstone.

human hearts, are the residences—the dwelling places—of those curious things called Truth and Falsehood. A propositional function, owing to the presence in it of variables, is neither true nor false. The statements $2+7=9$, $3+7=9$, are propositions, one of them true, the other one false; but the statement, $x+y=9$, is neither true nor false; it is not a proposition but is a propositional function.

You see at once that to derive propositions from a propositional function it is *necessary* to replace the latter's variables with what we may call constants, or values—with terms of definite meaning; but such substitution, though necessary, is not sufficient, for it is always possible to substitute such constants as will give, not a proposition, but nonsense. Suppose, for example, that our given function is the statement, $x$ is an integer less than 5. Now, the *class* of all integers less than 5 is a constant—a definite somewhat. Substituting it for the variable $x$, we get the statement, the *class* of all integers less than 5 is an integer less than 5. This statement is neither a propositional function nor a proposition; it is nonsense—nonsense consisting in talking of a class of things as if a given class could conceivably be one of the things composing it; as if the class, for example, of locomotives were itself a locomotive; or as if the class of prohibitionary moralists were itself a holy constituent thereof; or as if the class of apples or of asses were itself an apple or an ass. Such "talking" is sheer chattering, as if there were no such things as laws of Thought. It is evident that a propositional function is a *matrix* of the propositions derivable from it by substitution and has the same *form* as the propositions it thus moulds. This latter fact should be noted carefully for in logic—that is to say, in mathematics—form is all-important—so important indeed that some critical thinkers have ventured to call mathematics the science of Form.

The constants that convert a given propositional function into nonsense may be called *inadmissible* constants for that function; all other constants may be called *admissible* constants for the function since they convert it into propositions. It is worthy of note, in passing, that the line of cleavage between the admissible and the inadmissible constants for a given function is not always sharply defined. You can readily construct or find functions of $x$ in respect of which it may be doubtful whether certain constants—the sweetness of sugar, for example, or the glory of renown—are admissible or not. You stand here before an open and inviting field for research, the problem being to determine criteria for deciding, in the case of any propositional function, what constants in the universe of constants are admissible and what ones are not. The situation may be likened to that of physical organisms, for there are plants and there are animals, but in the case of some living organisms there is at present no means of deciding to which division of the kingdom they belong.

The admissible constants for a given function fall into two classes: those converting it into true propositions and those converting it into false ones. It is convenient to call the constants of the former class *verifiers* of the function; and those of the latter class *falsifiers* of it. The verifiers of a function are said to *satisfy* it and are called the *values* of its *variables;* and the propositions derived from a function by substituting values of its variables for these are called *values* of the *function.* Thus, you see that a propositional function is itself a variable—

albeit of a different type from the variables it contains—having for its values the true propositions derivable from it by means of its verifiers.

With the foregoing ideas and distinctions in mind, let us return to the Hilbert postulates and ask: Are they propositions or propositional functions? To answer, it is necessary and sufficient to ascertain whether or not they contain variables. We observe at once the presence in them of certain *substantive* terms—"point," "straight line," "plane," and "space"—which seem to denote the things about which the postulates talk, their subject-matter—and certain *relational* terms—"between" and "congruent"—which have the air of denoting definite fundamental relations among the "points" or figures composed of them. We must now ask: Do these terms denote constants—things of unique and definite meaning—or do they play the rôle of variables? Euclid does indeed, as you know, give what he calls "definitions" of point, line and plane, but in his proofs and constructions he makes no use whatever of the so-called definitions, which he ought to have called *descriptions* designed merely to indicate what *he* meant by the terms; or, better, he ought to have omitted the definitions as *logically* useless. As to the term, space, it does not, as it should not, occur in Euclid's *Elements*. By examining Hilbert's book, you will find that he does not attempt either to define or to describe any of the above-mentioned six terms, except, of course, in so far as they are defined—restricted in their possible meanings—by having to satisfy, or verify, the postulates. The omission of all other definition of them is deliberate. And so our question is reduced to this: Does the requirement that the things denoted by the six terms—"point," "straight line," etc.—make the terms constants, assign to each of them a unique and definite meaning? The answer is No: each of the terms admits of many, infinitely many, different definite meanings satisfying the postulates. The answer will be justified at a later stage of our discussion. For the present, I ask you to assume its correctness. We may, therefore, now state, in answer to our main question, that the six terms are not constants, but variables, and that, accordingly, the postulates are not propositions, as they are wont to be called, but are propositional functions. As you reflect upon this fact, you will find that its importance is immeasurable, not only for philosophy in its narrower sense, but for Criticism [1] in the widest sense, in all its fields. In a future lecture, I shall return to the matter of estimating the fact's general importance. For the present, let us follow its strictly logical and philosophical leading.

We have to say at once that the postulates of the system we are examining as a representative specimen of postulate systems in general, are neither true nor false, being propositional functions. The same must, of course, be said of all the theorems deduced or deducible from them as their logical consequences or implicates, for all such theorems, being statements involving the same variables as are present in the postulates, are propositional functions and are, therefore, neither true nor false. At this point, I cannot refrain from pausing long enough to point out how the most vitally fundamental fact in logical theory appears here with startling vividness in new light. Suppose that in the postulates we replace the seven terms—" point," " straight line," " plane," etc.—respectively, by any meaningless vocables whatever, as

[1] In this connection the reader should consult Professor F. C. S. Schiller's very suggestive article "Doctrinal Functions" in *The Journal of Philosophy, Psychology and Scientific Methods, Vol. XVI., 1919.*

*loig, boig, ploig,* etc., so that postulates (1) and (3), for example, shall read: (1) Two distinct loigs determine a boig; (3) Any three loigs not in a same boig determine a ploig. Imagine the other postulates to be similarly restated. Then, of course, all the theorems and indeed the entire Hilbert book will discourse explicitly about loigs, boigs, ploigs, etc., and nothing else. Do not fail to note now, once for all, that as thus restated, the theorems and postulates are *related* precisely as before—the former being logical consequences of the latter and deducible therefrom without even the slightest change in the reasoning. The fact which thus leaps naked into view is that logical deduction,—mathematical demonstration,—*all* valid proof in no matter what subject-matter,—depends *entirely* upon the *form* of the premises, or postulates, and not at all upon any specific meanings we may assign to their undefined, or variable, terms or symbols. What is meant by propositional *form?* The question has been often asked but never answered. I ask it here merely to signalize its importance. It is exceedingly difficult. I hope we may return to it later. At present, let us go on with the central thread of this lecture.

We have seen that the Hilbert postulates and all the theorems logically deducible from them are propositional functions. So important a fact ought not to be concealed, not even from the physical eye. To lay it bare, it is sufficient to replace in the postulates the terms, there playing in disguise the rôle of variables, with proper symbols for variables; substituting, let us say, the symbols $v_1$, $v_2$, $v_3$, $v_4$, respectively, for the substantive, or element-naming, terms,—" point," " straight line," " plane," " space,"—and for the relational terms,—" between," "congruent,"—the symbols $R_1$ and $R_2$. Then postulate (1) will read: Two distinct $v_1$'s determine a $v_2$. For another example, postulate (8) will read: If $v_1'$, $v_1''$, $v_1'''$ are $v_1$'s of a $v_2$ and $v_1''$ has the relation $R_1$ to $v_1'$ and $v_1'''$ then $v_1''$ has $R_1$ to $v_1'''$ and $v_1'$. It is obvious that all of the postulates and theorems admit of such restatement. I strongly recommend that, as a very enlightening exercise, you thus restate all of the postulates, a few of the theorems, and rewrite the proof of at least one of the latter.

Having thus dragged into solar light the fact,—hitherto evident only in the psychic light of understanding,—that our postulates and theorems involve variables, let us now think of the postulates and theorems as constituting a Whole—a definite Body of logically related propositional functions. Not one of them is true; not one of them is false. What is true is that *the postulates imply the theorems.* But this statement of implication, though it is a proposition and is a true one—is not a part of the Whole; it is not contained in the Body of functions; were we to put it in, it would stand there alone as an intruder, being neither one of the postulates nor one of the theorems, neither a premise nor a conclusion, neither an implier nor an implied; it is a philosophical proposition *about* the Whole but is not a member of it; it is a critical commentary upon it but not upon itself; it is a judgment,—a just and important judgment,—regarding the Body of propositional functions, but is wholly external to it.

This definite Body of logically compendent propositional functions, if one will but meditate upon it, is a truly wonderful thing—a great indestructible shining Form of forms—"poised in eternal calm" above the changeful things of the world of sense. What shall we call it? It is evidently one of many, for every postulate system gives rise to such a Form and many of these systems, as we shall see, are essentially different. Shall we call it Euclidean Geometry as Hilbert called it with the world's consent? A part of our future task is to show that it has neither more nor less to do with geometry as this term has been understood from time immemorial than with a thousand other things. Shall we say it is a Doctrine of a certain kind? No; for a doctrine must have a specific subject-matter, which our Form has not; it must consist of propositions, which our Form does not; it must be true or else false, but our Form is neither.

What, then, shall we say it is? What, pray, ought our Form,—our definite autonomous Body of propositional functions,—to be called? Observe that if we replace the variables in its postulated functions by admissible constants, we thus obtain a body of propositions matching, in one-to-one fashion, *all* the functions of our Body of functions; we thus obtain, that is, a *doctrine*, for the body of propositions has a specific subject-matter and is true or false according as the substituted constants are all of them verifiers, or some of them falsifiers, of the postulated functions. Obviously, we may thus obtain various doctrines from our Body of functions by substituting various sets of admissible constants for the variables in the postulated functions. It is obviously natural to call the true doctrines thus derivable the values of the Body of functions.

It is now as plain as the noon-day sun what the answer to our question must be: our Body of logically related propositional functions, since it is a thing having doctrines for its values must be named a Doctrinal

Function. The same name must, of course, apply to the function body consisting of the postulates of any other postulate system together with the theorems logically deducible from them. It can hardly escape your attention that just as a propositional function has true propositions for its values, a doctrinal function has true doctrines for its values; that just as we viewed a propositional function as the matrix of all the propositions (true or false) derivable from it by substitution of admissible constants, so we may view a doctrinal function as the matrix of all the doctrines (true or false) derivable from it in like manner; and that just as a given propositional function and the propositions derivable from it are identical in form, so a given doctrinal function and the doctrines derivable from it are the same in respect of form; they are *isomorphic*, as we say. In marriage with subject-matter, a Doctrinal Function becomes the matrix of an infinite family of doctrines; the children inherit the form of the mother.

It will be convenient to say that we are *interpreting* a given doctrinal function whenever we derive from it, in the way now familiar, one of its values, or true doctrines; and these values, or true doctrines, may be conveniently called *interpretations* of the function.

## LECTURE IV

### Doctrinal Interpretations

A MOTHER OF DOCTRINES MISTAKEN FOR HER ELDEST CHILD—INFINITELY MANY INTERPRETATIONS OF ONE DOCTRINAL FUNCTION—ORDINARY GEOMETRY BUT ONE OF THEM—OTHER INTERPRETATIONS GEOMETRIC, ALGEBRAIC AND MIXED—IDENTITY OF FORM WITH DIVERSITY OF CONTENT—DISTINCTION OF LOGICAL AND PSYCHOLOGICAL—PROJECTIVE GEOMETRY THE CHILD OF ARCHITECTURE—A SCIENCE BORN OF AN ART—INFINITE POINTS AND THE MEETING OF PARALLELS—POLE-TO-POLAR TRANSFORMATIONS—LOGICAL USE OF PATHOLOGICAL CONFIGURATIONS.

In the following discussion, I shall assume that you have before you the Hilbert postulates as restated in terms of the variable-symbols, $v_1$, $v_2$, $v_3$, $v_4$, $R_1$ and $R_2$. It will be convenient to call the doctrinal function consisting of these postulates and their consequent theorems the " Hilbert doctrinal function " and to denote it by $H\Delta F'$. Now be good enough to note very carefully that, if we omit from the postulates all reference to points not in a given plane, the remaining postulates together with their theorematic consequences constitute another doctrinal function and that this is included in $H\Delta F'$. Let us denote the minor function by $H\Delta F$. The purpose of this lecture is to present or rather to indicate some of the infinitely many values, or interpretations, of these two functions; to indicate, that is, some of the true doctrines having the functions for their common mould.

One of the interpretations of $H\Delta F'$ is the familiar doctrine which results from letting the symbols, $v_1$, $v_2$, $v_3$, $v_4$, $R_1$, $R_2$, denote, respectively, *point*, *straight line*, *plane*, *space*, *between* and *congruent*, or *equal*, taken in the sense in which they have been taken from pre-Euclidean days,—in the sense in which they (or some of them) are "described" by Euclid in the *Elements*,—in the sense in which Hilbert takes them in his *Foundations* as shown by the drawings or figures he there employs and which is doubtless responsible for his calling his book *The Foundations of Geometry*. This special interpretation of $H\Delta F'$,—this special value of that function,—this special doctrine, which I shall denote by $D_1'$,—is, you observe, the ordinary Euclidean Solid Geometry, or geometry of three dimensions, with which we all of us gained some acquaintance in high school or college despite the somewhat rough or uncritical way in which it was there presented as for beginners. The corresponding interpretation of $H\Delta F$ is the yet more familiar Euclidean geometry of the plane, a two-dimensional geometry. Denote it by $D_1$. I shall take both $D_1'$ and $D_1$ for granted, assuming them, whenever it is convenient to do so, in future discussion.

Let me now direct your attention to another geometric interpretation of the two functions—to one which, though it is near-lying and fairly obvious, has not, so far as I am informed, been published. In order to present it intelligibly, I must, by way of preparation, make you acquainted with the concepts of projective straight line, projective plane and projective space, for, as you will recall, I have not assumed on your part a knowledge of Projective Geometry. It will be sufficient for our purpose to introduce them in the rough traditional way instead of the very refined way employed by Veblen and Young, for example, in their Projective Geometry, which is based upon a postulate system appropriate for projective geometry.

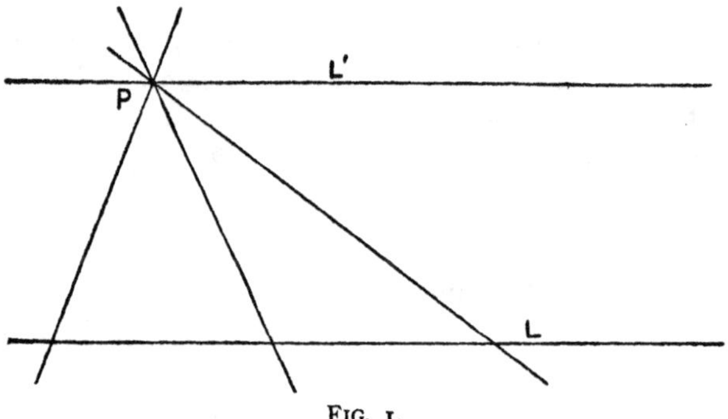


FIG. 1.

Let the figure be in a Euclidean plane—the kind of plane belonging to $D_1$. All lines of the plane that contain a given point $P$ constitute a *pencil* of lines; $P$ is the pencil's *vertex*. All the points of a line $L$ constitute a *range* of points; $L$ is the range's *base*. It is plain that each point of range $L$ is on one line of pencil $P$; and that, reciprocally, each line of $P$ has one point of $L$, with a single exception,—$L'$, parallel to $L$, contains no point of $L$. To remove this exception to the one-to-one correspondence, otherwise perfect, there is made in projective geometry an agreement or convention: namely, that each line has (at an infinite distance) a so-called " ideal " point, or point at infinity, and that the " ideal " points of any two parallel lines are coincident. We thus get, as you see, a new sort of straight line and of plane and of space, which we describe by calling them respectively *projective* straight line, projective plane and projective space. The adjective has fine propriety, but that need not here detain us. You can readily prove, or you may assume, that the "ideal" points of the projective plane constitute a straight line—called the "ideal" line, or line at infinity; and that the locus of the "ideal" points of projective space is a plane—called the "ideal" plane, or plane at infinity. I can not pause here to justify the convention. It is amply justified by its consequences, for which, if you be interested, you must repair to projective geometry,—invented by the engineer, Desargues, a contemporary of Descartes and Pascal,—quickly forgotten—reinvented, in France again, about one hundred years ago—perhaps the most beautiful branch of mathematics.

We may now proceed to the promised new interpretation of our doctrinal functions. As $H \Delta F$ is simpler than $H \Delta F'$, let us first deal with the former.

Let $\pi$ denote a projective plane. Let a chosen point $O$ be the vertex of a pencil of lines of $\pi$; call each line of the pencil an $O$-line. Note that every other pencil of $\pi$ contains one and but one $O$-line. Now let us in thought remove from $\pi$, once for all, the $O$-pencil. We thus remove one and but one line from every other pencil. We may conveniently call the pencils, thus bereft of a line, *pathopencils* as being defective or, so to speak, pathological. We have taken from $\pi$ one and but one pencil of lines. Our field of operation consists of all that is left. Denote the field by $\Phi$. We are going to give $H \Delta F$ an interpretation in $\Phi$; the interpretation, as you will see, will be a doctrine about certain things in $\Phi$—a geometry of the field. The interpretation results from assigning to the $v$'s and the $R$'s in the postulates of $H\Delta F$ the following meanings, or constant values: $v_1$ is to mean a line of $\Phi$; $v_2$, a pathopencil of $\Phi$; $R_1$ is to mean " between " in the sense that, if $A$, $B$, $C$ be three lines of a pathopencil, B will be considered to be between $A$ and $C$, if $A$ (or $C$) must rotate through the position of $B$ to coincide with $C$ (or $A$) (for, of course, a line of a pathopencil must not be supposed to rotate into the position left vacant by the absent $O$-line); and $R_2$ is to mean " congruent " in a sense to be given later.

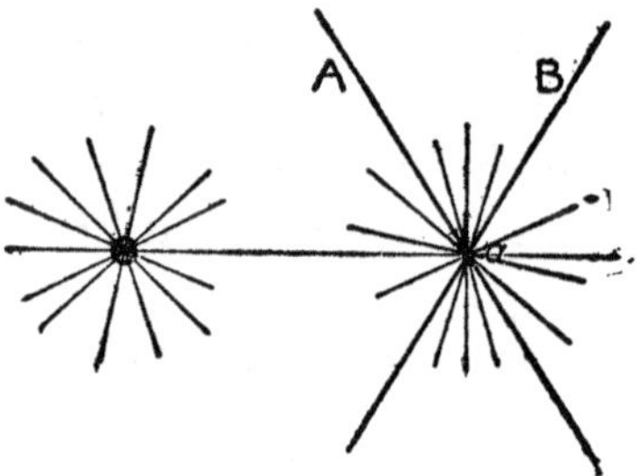


FIG. 2.

We have to show that the indicated meanings satisfy, or verify, the postulates of $H\Delta F$. That some of them are thus satisfied may be made evident by simple figures; and it will be interesting and enlightening to exhibit such evidence before giving the proof for all the postulates. At the same time, we will lay bare, by means of figures, the significance of one or two theorems of the new doctrine. I shall not here repeat the postulates, but will suppose you to have them in hand.

Postulate (1) is plainly satisfied, for any two lines $A$ and $B$ of $\Phi$ determine, as in Fig. 2, a pathopencil $a$, which consists of all the lines through $a$ except the $O$-line $Oa$.

Next consider postulate (8). That it is verified is evident in Fig. 3 where line *B* is clearly between *A* and *C* and between *C* and *A*. For another example, let us take

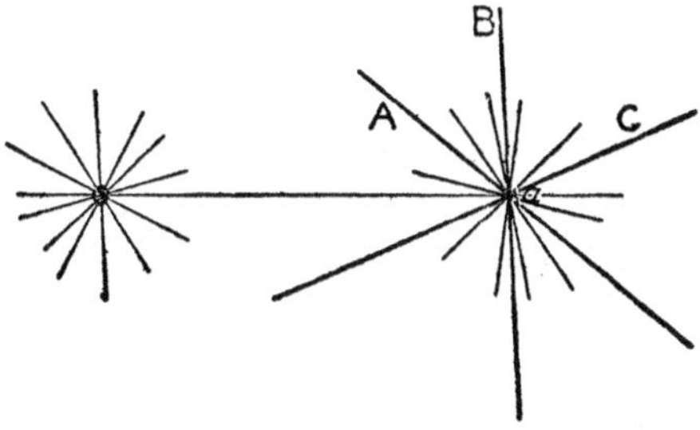


Fig. 3.

postulate (12), the famous postulate of Pasch. But first we must have some

Definitions.—A pair of lines, *A* and *B*, of a pathopencil, is a *segment AB* or *BA*; its *ends* are *A* and *B*; the lines between them are the *segment's lines.*

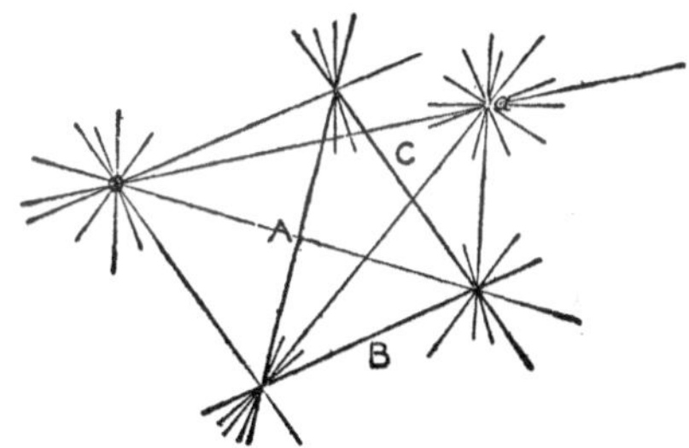


Fig. 4.

In the light of Fig. 4, it is obvious that postulate (12) is satisfied. Note that *A*, *B*, *C* are any three lines of Φ not belonging to a same pathopencil; that pathopencil *a* contains a line of segment *AB*, by hypothesis; and that *a* contains a line of segment *BC* but none of segment *AC*.

Before considering another postulate, let us illustrate the following *theorem* (a propositional function in the doctrinal function $H\Delta F$): *Any given* $v_2$ *separates the remaining* $v_1$*'s of the* $v_3$ *into two classes such that, if* $v_1'$ *and* $v_1''$ *are one of them in one of the classes and the other in the other, the segment* $v_1'\,v_1''$ *contains a* $v_1$ *of the* $v_2$*; and that, if* $v_1'$ *and* $v_1''$ *are both in one of the classes, the segment does not contain a* $v_1$ *of the* $v_2$. (It is theorem 5 of Hilbert's book.) A fairly careful examination of Fig. 5 will suffice to con-

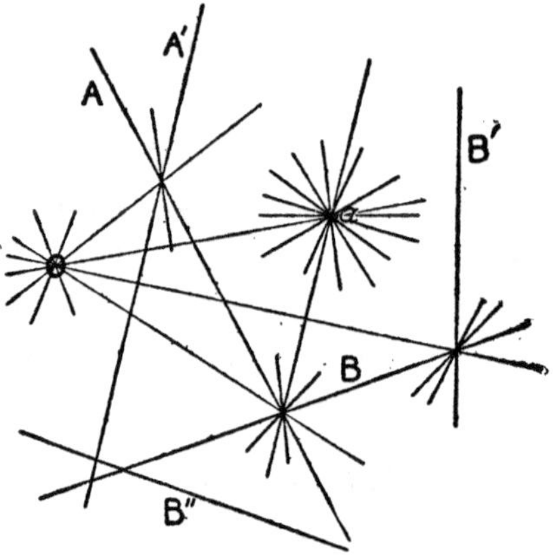


FIG. 5.

vince you that that theorem is verified in our new interpretation. One of the two classes of lines is composed of all the lines of Φ that go between *O* and *a*; all the other lines compose the other class. Segments *AA'* and *BB'* contain no line of the pathopencil *a*, but any such segment as *AB* contains a line of *a*.

You should not fail to compare Fig. 5 with Hilbert's figure for the corresponding proposition in doctrine $D_1$, the old familiar interpretation. The two figures are the same logically but very different psychologically: in the latter figure the truth of the proposition is perfectly and immediately evident to intuition, while in the former the truth of the proposition is very far from being thus evident. Why? The question, you observe, is one for psychologists, like hundreds of similar questions that arise here and elsewhere in mathematics, if only psychologists would learn enough mathematics even to *ask* the questions.

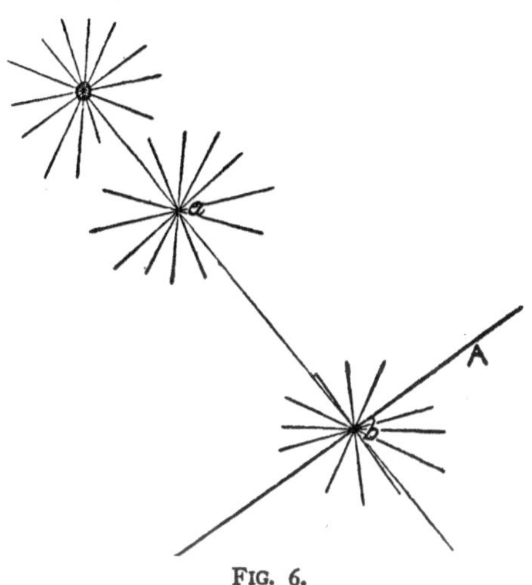


FIG. 6.

Let us now turn to postulate (13)—the postulate of parallels. Fig. 6 shows clearly that this famous Euclidean postulate is satisfied by our new interpretation. Here *a* is the given pathopencil; *A* is any given line not belonging to *a*; *b* is a pathopencil containing *A* but having no line in common with *a*, and there is plainly no other such pathopencil; in other words, *b* is parallel to *a* and there is no other such pathopencil containing *A*.

If, now, you attempt to show (and I advise you to make the attempt) by a figure that postulate (20), or other postulate involving *congruence*, is satisfied, using "congruent" in your figure in the sense it has in the old interpretation or doctrine $D_1$, you will quickly find yourselves in trouble. In the new interpretation, however, we are not going to employ "congruent" in that sense, but in a sense which I shall explain presently in the course of a simple argument designed to show, as by a

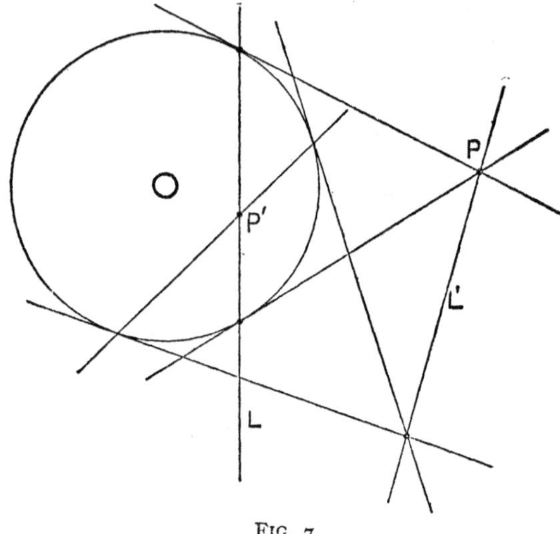


FIG. 7.

single stroke, that *all* of the postulates are satisfied by our new interpretation.

Before presenting that argument we must acquaint ourselves with what is called, in the projective geometry of a plane, the *Pole-Polar transformation with respect to a circle.* It is a very beautiful transformation, important, and easy to understand.

Let Fig. 7 be in a projective plane $\pi$. Tangents through $P$ are drawn to the circle. Line $L$ joining the points of tangency is called the *polar* of $P$, which is called the *pole* of $L$. You readily see that, if $P$ moves off, becoming an " Ideal " point of $\pi$, the polar $L$ goes through the center—is a line of the pencil vertexed at $O$; also, if $P$ moves up to the circle, $L$ becomes tangent at $P$. If $P$ is inside the circle, say at $P'$, $L$ is $L'$, whose construction is obvious; if, in particular, $P$ is at $O$, $L$ is $\pi$'s " ideal " line, or line at infinity. Thus you see that the given circle serves to set up a one-to-one correspondence between the points of $\pi$ as poles and the lines of $\pi$ as polars. This correspondence is called the Pole-Polar transformation of $\pi$ with respect to the circle. We say the transformation transforms or converts a point into its polar line, a line into its pole point, and each of these is called the *transform* of the other. If you will study the transformation a bit, playing with it, making a few figures, you will discover some of its important properties, such as these: it converts a *range* of points into a *pencil* of lines, and a pencil into a range; a segment of a range into a segment of a pencil, and a pencil segment into a range segment; if three points of a range or three lines of a pancil are in the order—$A$, $B$, $C$,—the transforms are in the same order.

And now for the argument showing that all the postulates in $H\Delta F$ are verified by our new interpretation. Imagine our field $\Phi$ laid down upon a Euclidean plane $\alpha$. Remember that the $O$-pencil is not in $\Phi$—I have put in a few of its lines merely to remind us that it is absent. Such a pencil is present in $\alpha$ just below. Remember also that $\Phi$ has an " ideal " line at infinity which $\alpha$ has not. Assume a definite circle $C$ about $O$ as center. Consider the pole-polar transformation as to $C$. Let the transforms of the points and lines of $\alpha$ be in $\Phi$; you readily see that, in a one-to-one way, the points of $\alpha$ are converted into the lines of $\Phi$ and the lines (ranges) of $\alpha$ into the pathopencils of $\Phi$; also that the *order* of the elements in $\alpha$ is carried over into their transforms in $\Phi$. But, as you readily see, congruence in $\alpha$,—that is, congruence as understood in interpretation $D_1$,—is not carried over. We, therefore, agree to give a new meaning to "congruent" for use in $\Phi$, and the meaning is this: if two segments or angles be congruent (in the old sense) in $\alpha$, then and only

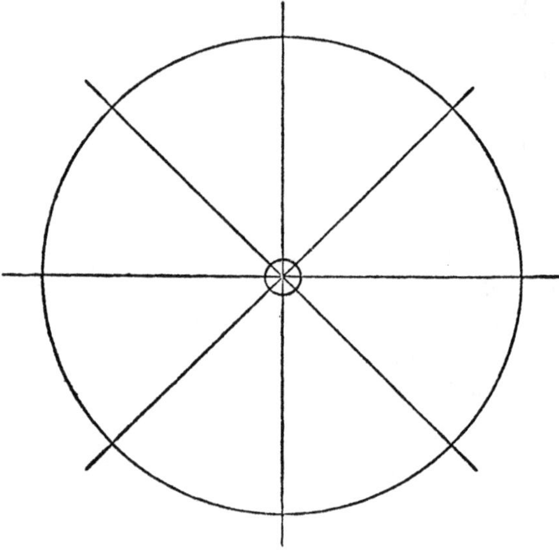

FIG. 8.

then their transforms shall be said to be congruent in $\Phi$. It is evident, without further talk, that all the postulates are satisfied and that we, accordingly, have a new interpretation of the doctrinal function $H\Delta F$. Let us denote this interpretation, or doctrine, by $D_2$. $D_2$ is evidently a two-dimensional geometry of the lines and pathopencils of $\Phi$ and is isomorphic with $D_1$, the ordinary geometry of the points and lines of a Euclidean plane.

I will close this lecture by indicating,—merely indicating,—the analogous new interpretation of the Doctrinal function $H\Delta F'$, which, you remember, includes the entire list of Hilbert postulates in their restated form. I shall denote the new doctrine, or interpretation, by $D_2'$. Let $S$ denote a projective space of three dimensions. We have already formed the concept of such a space. All the lines (or planes) of $S$ that have in common point $P$ are together called a *sheaf*, or *bundle*, of lines (or planes); all the planes having a common line constitute an *axial pencil* of planes. Let $O$ be a chosen point of $S$. Call the sheaf of lines (or planes) having $O$ for vertex the $O$-sheaf of lines (or planes). In thought remove from $S$ the $O$-sheaf of lines and the $O$-sheaf of planes. We thus remove from every other line sheaf one line, from every other plane sheaf an axial pencil and from every axial pencil (not contained in the $O$-sheaf of planes) one plane. The ensembles, thus rendered defective, may be respectively called a pathosheaf of lines, a pathosheaf of planes and a pathopencil of planes, or plane pathopencil. Analogous to the pole-polar transformation as to a circle,—which we have already explained and used,—there is for $S$ a pole-polar transformation with respect to any given *sphere* converting each point into a polar plane and each plane into pole point. Our field of operation—$\Phi'$—is $S$ bereft of the two $O$-sheaves. As you may have by this time surmised, our new interpretation, or doctrine $D_2'$, arises on giving the variable symbols in the postulates of $H\Delta F'$ meanings as follows: $v_1$ will mean a plane of $\Phi'$; $v_2$, a pathopencil of planes; $v_3$, a pathosheaf of planes; $v_4$, $\Phi'$; $R_1$, between in the sense that, if $A$, $B$, $C$ are planes of a pathopencil, $B$ will be said to be between $A$ and $C$ if either of the latter must rotate through the position of $B$ to coincide with the other; $R_1$ will mean congruent in the sense that segments, etc., in $\Phi'$ will be called congruent if they are transforms of segments, etc., congruent in $D_2$.

Obviously $D_2'$ is a three-dimensional geometry of planes, pathopencils of planes and pathosheaves of planes of $\Phi'$ and is isomorphic with $D_1'$, the familiar geometry of points, lines and planes of ordinary Euclidean space.

Note that $D_2$ and $D_2'$ are *logically* the same as $D_1$ and $D_1'$ but greatly differ from the latter *psychologically*.

## LECTURE V

### Another Geometric Interpretation

BRIEF INTRODUCTION TO THE METHOD OF DESCARTES AND FERMAT—INVERSION GEOMETRY AND INVERSION TRANSFORMATION—THE INFINITE REGION OF INVERSION SPACE A POINT—BUNDLES OF CIRCLES AND CLUSTERS OF SPHERES—PATHOCIRCLES AND PATHOSPHERES—ONE-TO-ONE CORRELATION.

In presenting a third interpretation of our two doctrinal functions, it will be convenient to borrow a few ideas from Cartesian Analytical Geometry and Inversion Geometry. It will be advantageous to explain them in advance.

The perpendicular lines $OX$ and $OY$, Fig. 9, are called *coordinate axes;* $O$ is the *origin* of distances, which, if measured upward or rightward, are regarded *positive*, but, if downward or leftward, *negative*. I am supposing the figure to be in a Euclidean plane. Choose some unit of length; then any point has a pair of numbers $(x, y)$, $P$'s distances from the axes and called its *coordinates*. Conversely, to any such a pair belongs a point. Let (1), Fig. 10, be any line through $O$; then (2), parallel to (1), is any line of the plane. Let $P(x, y)$ be any point of (1); let $m = \tan\theta$; then $y = mx$; this equation is the equation of (1); it is so called because to any pair $(x, y)$ satisfying it belongs a $P$ of (1) and any $P$ of (1) has a pair satisfying it. Plainly, the $y$ of $P'$ is equal to $P$'s $y+b$; hence the equation of (2), any line of the plane, is: $y=mx+b$.

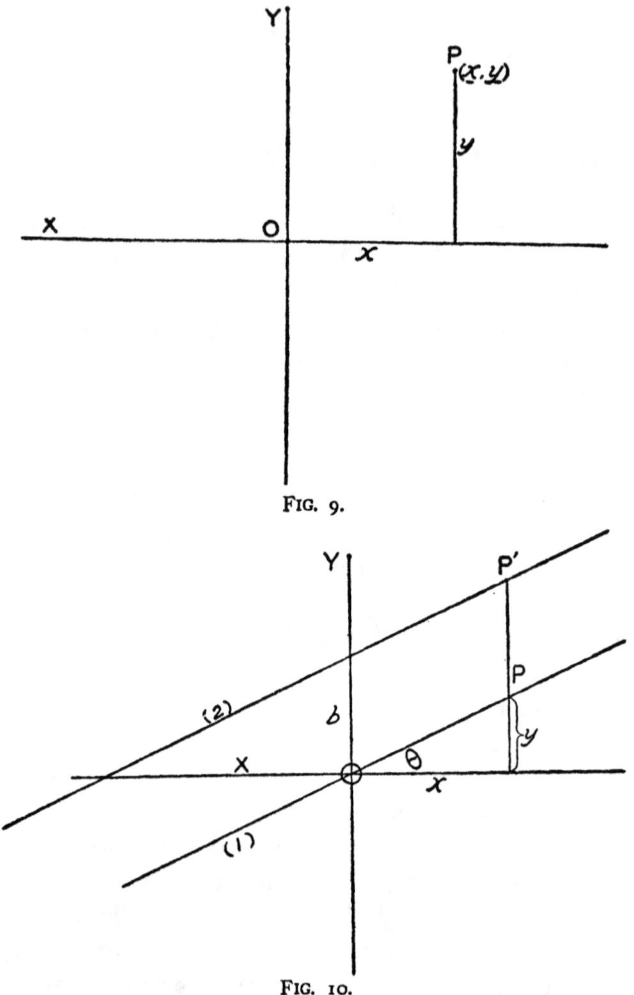


FIG. 9.

FIG. 10.

Conversely, any equation of first degree in $x$ and $y$ represents a line of the plane.

By Fig. 11 you see that, if $d$ is the distance between two points, $P_1(x_1, y_1)$ and $P_2(x_2, y_2)$, then $d^2 = (x_1 - x_2)^2 + (y_1 - y_2)^2$.

From the foregoing distance formula, you see that the *equation* of any circle, Fig. 12, of radius $r$ and center $(a, b)$ is $(x-a)^2 + (y-b)^2 = r^2$; that is, $x^2 + y^2 - 2ax - 2by + a^2 + b^2 - r^2 = 0$. Conversely, any equation of the form $x^2 + y^2 + 2Ax + 2By + C = 0$ represents a circle of center $(-A, -B)$ and squared radius, $A^2 + B^2 - C$.

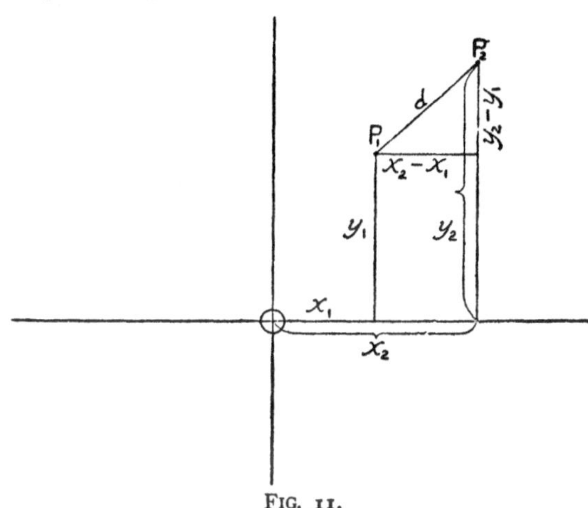


FIG. 11.

On any line through the center of a circle of radius $r$ let $P$, $P'$ be such that *distance OP times distance OP'* $= r^2$; the point $P$ (or $P'$) is called the *inverse* of $P'$ (or $P$); the circle and its center are called the inversion circle and center. Taking the circle's center for *origin*, Fig. 13, you will easily find that:

$$(1)\quad \begin{cases} x = \dfrac{x'r^2}{x'^2 + y'^2}, \\ y = \dfrac{y'r^2}{x'^2 + y'^2}; \end{cases} \qquad (2)\quad \begin{cases} x' = \dfrac{xr^2}{x^2 + y^2}, \\ y' = \dfrac{yr^2}{x^2 + y^2}. \end{cases}$$

Notice that to each point there corresponds one and but one point—except that the inversion center corresponds to no point (in the Euclidean plane). To remove this exception it is common to assume the existence of one and but one " ideal " point, or point at infinity, to serve

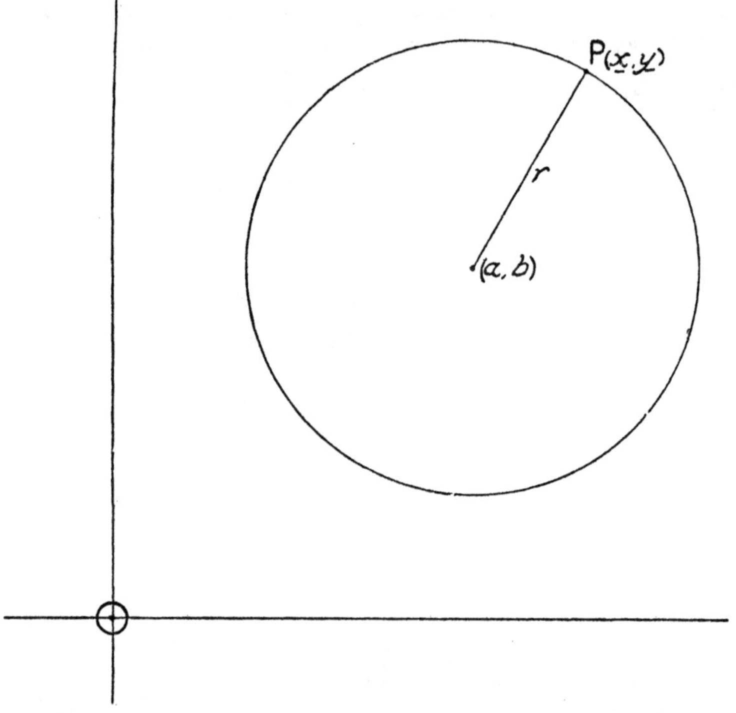


FIG. 12.

as the inverse of the center. The new sort of plane thus got is called the *Inversion Plane*. The foregoing point-to-point correspondence is called the *Inversion Transformation* of the plane with respect to the given circle. Clearly, any line through the center is converted into itself.

What is the transform, or inverse, of a line not through the center? Let $Ax+By+C=0$ be such a line; replace the coordinates $(x, y)$ of any point in it by their values taken from (1), simplify the result and (if you like) drop the primes; we thus get

$$(3)\ x^2+y^2+r^2\frac{A}{C}x+r^2\frac{B}{C}y=0.$$

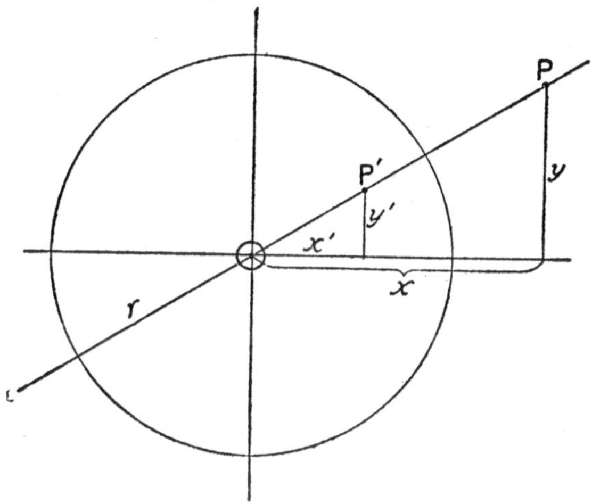


FIG. 13.

This, you note, is a circle through the inversion center, which is here the origin, for the coordinates (0, 0) of the origin satisfy the equation. Hence every line not through the center has for its *transform*, or inverse, a circle through the inversion center.

With these simple ideas held in reserve for use as we need them, let us proceed to our third geometric interpretation. It will be advantageous to deal first with $H\Delta F$. Denote by $\pi$ an *inversion* plane. Let $O$ be a chosen point of $\pi$. The ensemble of all circles through $O$ is called a bundle of circles. The bundle includes, as infinite circles (*i.e.*, circles of infinite radius), the straight lines through $O$. Now, in thought, let us, once for all, remove the point $O$ from $\pi$. Each circle of the bundle now lacks a point; we may call them *pathocircles*, and speak of the $O$-bundle of *pathocircles*. Our field of operation—which may be denoted by $K$—is composed of the pathocircles of the $O$-bundle and the points (except $O$, of course) of $\pi$. We are going to give the doctrinal function $H\Delta F$ an interpretation in the field $K$; it will be a geometry of certain elements of $K$. The interpretation arises from assigning to the variable-symbols in the postulates of $H\Delta F$ definite meanings as follows: $v_1$ will mean a point of $K$; $v_2$, a pathocircle; $R_1$ will mean *between* in the sense that, if $A$, $B$, $C$ be three points of a pathocircle, $B$ will be said to be between $A$ and $C$, if $A$ (or $C$) must go through $B$ in moving on the pathocircle to $C$ (or $A$); and $R_2$ will mean *congruent* in the sense that, if two segments or angles be congruent in the ordinary sense (interpretation $D_1$), their transforms, or inverses, with respect to a given circle with $O$ as center, will be called congruent in the field of $K$.

We have now to show that the postulates are verified by the meanings assigned. Before giving a proof valid for all of the postulates, it will be instructive to deal with a selected few of them singly by means of simple figures, as in the preceding lecture. Postulate (1) is evidently satisfied. In Fig. 14 the two points $A$ and $B$ determine the pathocircle $a$ of the $O$-bundle.

Fig. 15 exhibits the fact that postulate (8) is verified. Point $B$ is between $A$ and $C$ and between $C$ and $A$; neither $A$ nor $C$ is between the other two of the three points; of course, no point can move through the absent $O$.

Let us next have a look at postulate (12). But we must premise some

Definitions.—A pair of points, *A* and *B*, of a pathocircle is a *segment AB* or *BA; A* and *B* are its *ends;* the points between them are the *segment's points.*

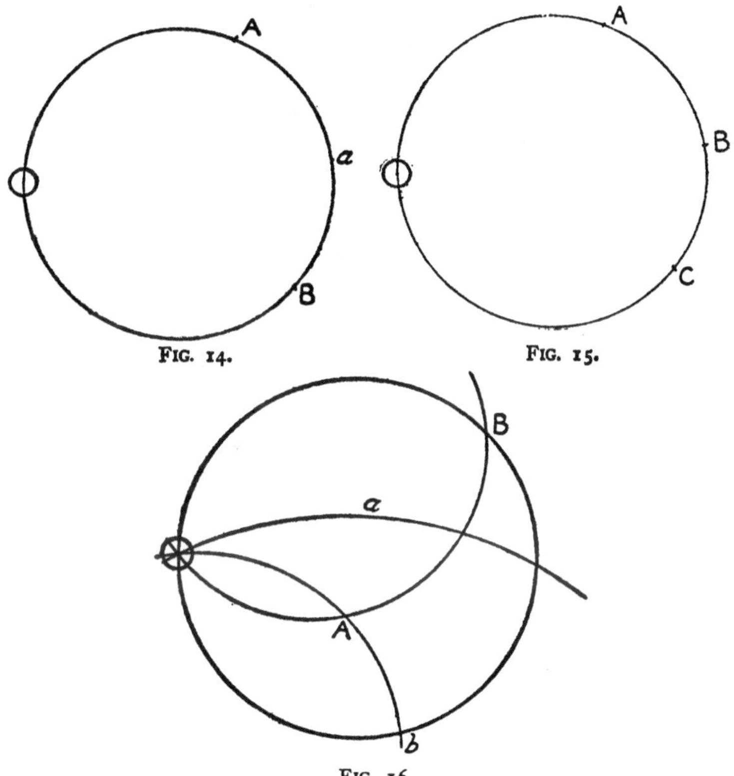


Fig. 14.

Fig. 15.

Fig. 16.

It is easy to see, Fig. 16, that the Pasch postulate (12) is verified. *A*, *B*, *C* are three points not on a same pathocircle; they determine three segments, *AB*, *BC*, *CA;* the pathocircle *a* going through *AB*, one of the three, goes through another, *BC*.

Let me suggest that, as an exercise, you make a figure illustrating that the *theorem* (corresponding to Hilbert's theorem 5) dealt with in the preceding lecture, is verified in the present interpretation.

Let us turn to the parallel postulate (13). That it is satisfied is clear in the light of Fig. 17. The given pathocircle is $a$; $A$ is a point not on $a$; through $A$ there is evidently one and but one pathocircle $b$ having no point in common with $a$; $a$ and $b$ are, of course, *parallel* to each other. This postulate, as you know, is *the* Euclidean

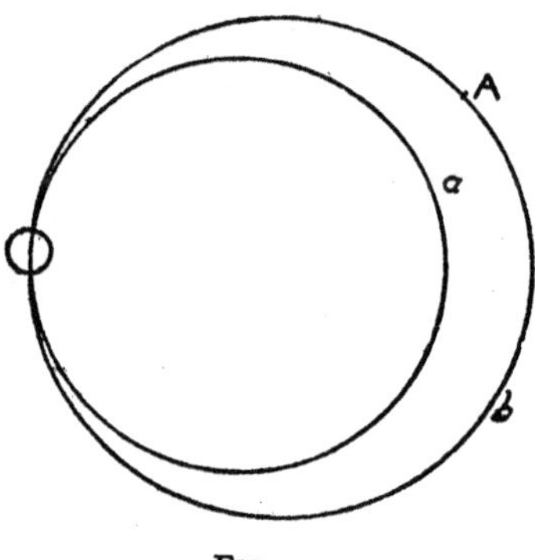


FIG. 17.

postulate *par excellence*—the one that mainly distinguishes Euclidean geometry from the famous non-Euclidean geometries of Lobachevski and Riemann. And so you see, in passing, that all interpretations of $H\Delta F$ or $H\Delta F'$ yield doctrines of Euclidean *type*—in the sense that in them the foregoing postulate of parallels is satisfied: they all of them contain some theorems whose proofs depend upon that postulate.

That all of the postulates of $H\Delta F$ are verified by the meanings we have assigned to their variables may be quickly made evident by help of the inversion transformation, explained a little while ago. Let us suppose our field $K$ to be laid down upon a Euclidean plane $\pi$. Remember that $O$ is absent from $K$ but that, below the vacant position, $\pi$ has a point, which we may call $O'$. In $K$ take a definite circle $I$ for inversion circle having $O$ for center. Regard the transformation as converting the points of $K$ (or $\pi$) into the points of $\pi$ (or $K$), noting that $O'$ of $\pi$ and the "ideal" point of $K$ are each the other's transform; that the lines of $\pi$ are converted into the pathocircles of $K$, and conversely; and that, if, in $\pi$, a point $B$ is between $A$ and $C$ on a line, then in $K$ the transform of $B$ is between the transforms of $A$ and $C$ on a pathocircle, the transform of the line. You see that there is thus established a one-to-one correspondence between the points and lines of $\pi$ and the points and pathocircles of $K$, in such a way that all postulated relations among the elements of $\pi$ hold equally among the corresponding elements of $K$.

Though logically superfluous, it will be instructive to illustrate the matter a little further by simple figures. In Fig. 18, $I$ is the inversion circle; $a$ is a line in $\pi$; pathocircle $a'$ is the transform of $a$; points $A'$, $B'$, $C'$ are the transforms of $A$, $B$, $C$; segments $AB$ and $BC$ are congruent in the familiar sense—in doctrine $D_1$; their transforms $A'B'$ and $B'C'$ are congruent in the new sense. You see that the postulate of Archimedes, postulate (20), is verified; for as congruent segments stretch upward in endless succession along $a$, their congruent transforms proceed on $a'$ in endless succession towards $O$, never reaching this vacant point-position.

Fig. 19 illustrates congruence of triangles in the new interpretation. Triangles $ABC$ and $A_1B_1C_1$ are congruent in $\pi$—in $D_1$; their transforms,—the new triangles

$A'B'C'$ and $A'_1B'_1C'_1$,—are congruent in $K$—in the new interpretation.

Let us denote the doctrine arising from the new interpretation of $H\Delta F$ by $D_3$. $D_3$ is, as you see, a two-dimensional geometry of the points and pathocircles of the field $K$ and is isomorphic with $D_1$ and $D_2$. We may say that $D_1$ is converted, element for element, figure for

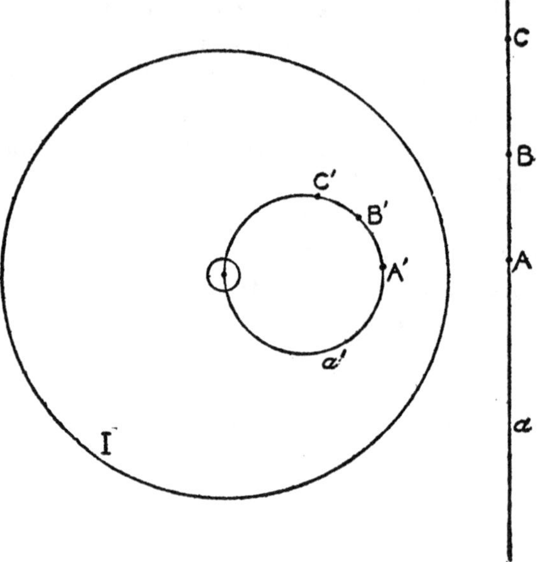


FIG. 18.

figure, proposition for proposition, into $D_3$ by the inversion transformation just as $D_1$ is completely converted into $D_2$ by the pole-polar transformation. You thus begin to glimpse the office and power of what mathematicians call transformation, which, at the close of the first lecture, I named, as you will remember, among the pillar-concepts of mathematics.

It remains to give $H\Delta F'$ an interpretation analogous

to that we have just given to $H\Delta F$. I will sketch it merely, inasmuch as you will find a fairly full account of it in Weber and Wellstein's *Elementare Geometrie*, which is the second volume of their *Encyklopädie der Elementar-Mathematik*—an excellent work handling in a maturely critical way the various elementary branches of mathe-

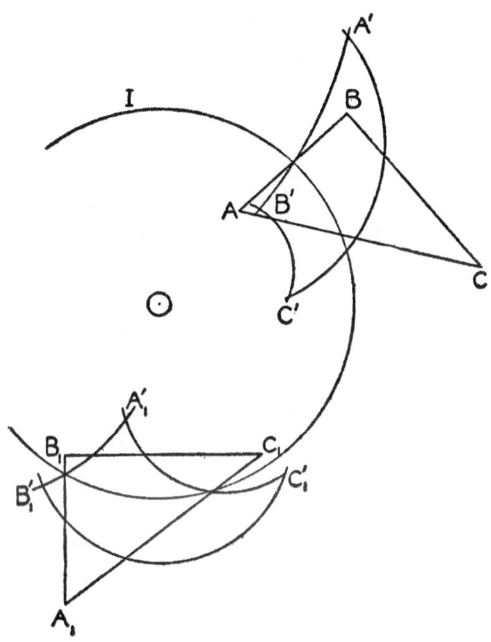


FIG. 19.

matics. You should not be misled by the adjective *Elementare*, for the discussions are designed for advanced students.

By way of a preliminary, I should say a word respecting inversion transformation of ordinary Euclidean space with respect to a given sphere. If the radius be $r$ and the center $C$, then two points, $P$ and $P'$, on a line through $C$, are inverses of each other provided the *distance* $CP$ *times the distance* $CP' = r^2$. You easily see that to each point there corresponds one and but one point, except that $C$ has no correspondent (in the Euclidean space). To annul the exception we assume one " ideal " point, or point at infinity, to serve as the transform of the inversion center $C$. The new space thus obtained is called *inversion space*. The lines and planes through $C$ are transformed into themselves. All lines and planes not through $C$ are converted respectively into circles and spheres through $C$.

And now for the field of our new interpretation. You probably guess what it is to be. Let $S$ be an inversion space; $O$ a chosen point in it. The ensemble of all the spheres (including planes as spheres of infinite radius) that go through $O$ may be called the $O$-cluster of spheres. Now remove the point $O$ from $S$; the cluster is now the $O$-cluster of pathospheres; and the cluster of circles bereft of $O$ will be called the $O$-cluster of pathocircles. Our field,—let us denote it by $K'$,—is composed of the points (except $O$) of $S$, the pathospheres and pathocircles of the $O$-clusters.

I need hardly say,—for you doubtless foresee,—that our new interpretation of $H\Delta F'$ springs from agreeing that $v_1$ shall mean a point of $K'$; $v_2$ shall mean a pathocircle; $v_3$ shall mean a pathosphere; $R_1$ shall mean between in the sense explained for the field $K$; and $R_2$ shall mean congruent in the sense that the transforms of segments or angles congruent in the familiar sense of $D_1'$ shall be congruent in the new sense.

Call the new doctrine thus arising $D_3'$. It evidently is a three-dimensional geometry of the points, pathocircles and pathospheres of the field $K'$ and matches

$D_1'$ or $D_2'$ proposition for proposition. Once more let me emphasize the fact that the differences—the very striking differences—of these three geometries are psychological; logically the three are one.

The next lecture will present a *non*-geometric interpretation of our two doctrinal functions.

# LECTURE VI

## Non-Geometric Interpretation

NOT ALL THAT GLITTERS IS GOLD—A DIAMOND TEST OF HARMONY—TWO-DIMENSIONAL DOCTRINE OF NUMBER DYADS AND SYSTEMS THEREOF—THE THREE-DIMENSIONAL ANALOGUE.

THE interpretations, or doctrines, which have hitherto concerned us—$D_1$, $D_2$, $D_3$ of $H\Delta F$ and $D_1'$, $D_2'$, $D_3'$ of $H\Delta F'$—ought to be called, and I have called them, *geometric* doctrines because their content or subject-matter,—that which the doctrines are doctrines of or about,—consists of things, whether sensible or purely conceptual, that are essentially and ultimately *spatial* in kind. The distinction is psychological; mathematicians, not being able to tell precisely what space is, and disdaining or affecting to disdain psychology, may ignore the distinction, if they like—such asininity not being penalized by municipal law in any land. Let us not be so uncandid or so dull as to ignore the essential distinction between spatial and non-spatial doctrines merely because they happen to have the same *form*. Not all that glitters is gold. Let us not so easily lose our common sense—a box of table sugar is not a box of table salt even if the two boxes are identical in size and form.

In the present lecture I invite your attention to a non-geometric interpretation of our doctrinal functions—to an interpretation, or doctrine, to be properly called non-geometric because, though the same in form as the foregoing geometries, it deals with non-spatial things and so has a non-spatial content. Some years ago I asked Mr. Wellington Koo, then a student at Columbia University and a pupil of mine, a brilliant pupil, in analytical geometry, to tell me what the Chinese word for geometry means as a word. He replied: " It means show by a figure." In the interpretation we are about to study we can have no figures, for figures are spatial affairs. This necessity of getting on without figures is, in a sense, fortunate—fortunate as an intellectual discipline—for, in the absence of sensuous representation by figures, we shall be driven to a kind of sheer thinking. And this warning, I hope, will prepare you for the needed effort.

As in the previous lecture, I will deal first with $H\Delta F$. At a later stage of our course, the nature of what is called the *system of real numbers* may be discussed. But for the purposes of the present lecture, I shall assume that you are sufficiently acquainted with the system, merely reminding you that it is composed of the positive and negative integers; the ordinary fractions; the irrationals, such as $\sqrt{2}$, $\sqrt[3]{7}$; and the transcendental numbers, like $e$ and $\pi$, for example. By the term number I shall mean a real number. In order to indicate the nature and the field of our new interpretation, it will be convenient to make use of this *definition*: If $a$, $b$, $c$ be three numbers, $b$ will be said to be *between* $a$ and $c$ (or $c$ and $a$) when and only when $a>b>c$ or $a<b<c$, where $>$ means *greater than* and $<$ means *less than*.

The new field of operation—which may be denoted by $N$—consists of all *dyads* $(x, y)$ of real numbers; that is, of all *ordered* pairs $(x, y)$, where by ordered I mean that $(x, y)$ will not be the same as $(y, x)$ unless $x=y$. It is, of course, understood that the dyads $(x_1, y_1)$ and $(x_2, y_2)$

are distinct unless $x_1 = x_2$ and $y_1 = y_2$. You see that the field is non-spatial, non-geometric, for numbers and number dyads have no essential reference to space and would continue to be perfectly good objects, or subjects, of thought if all spatial sense and all conception of space were to vanish; *symbols* for numbers and for dyads do indeed occupy *room*, but numbers themselves and dyads do not.

And now it is time to say that our non-geometric interpretation of $H\Delta F$ arises from assigning to the postulate variables constant values, or meanings, as follows: $v_1$ will mean a dyad of $N$; $v_2$ will mean a system of dyads, *i.e.*, the dyads satisfying an equation of the form $Ax + By + C = 0$, where either $A$ or $B$ is not zero, *i.e.*, $A \neq 0$ or $B \neq 0$; $R_1$ will mean between in the sense that, if $(x_1, y_1)$, $(x_2, y_2)$ and $(x_3, y_3)$ are three dyads of a same system, $(x_2, y_2)$ will be said to be between $(x_1, y_1)$ and $(x_3, y_3)$ if and only if $x_2$ is between $x_1$ and $x_3$ or $y_2$ is between $y_1$ and $y_3$; and $R_2$ will mean congruent in the sense that two dyadic pairs $(x_1, y_1)$, $(x_2, y_2)$ and $(x_3, y_3)$, $(x_4, y_4)$,—that is, two segments $(x_1, y_1,)(x_2, y_2)$, $(x_3, y_3)$ $(x_4, y_4)$,—will be said to be congruent when and only when $\sqrt{(x_1 - x_2)^2 + (y_1 - y_2)^2} = \sqrt{(x_3 - x_4)^2 + (y_2 - y_4)^2}$; with a like meaning for congruence of angles to be give later.

Are the postulates in $H\Delta F$ verified by the meanings thus assigned? It will be very instructive to examine the matter somewhat carefully.

Postulate (1).—Let $Ax + By + C = 0$ be an undetermined system $s$; $d_1$, $d_2$, any two dyads $(x_1, y_1)$, $(x_2, y_2)$ of field $N$; $d_1$ and $d_2$ will belong to $s$ when and only when

$$(1) \quad \begin{cases} Ax_1 + By_1 + C = 0, \\ Ax_2 + By_2 + C = 0; \end{cases}$$

three cases are possible and only tnree: ($\alpha$) $x_1=x_2$, $y_1\neq y_2$; ($\beta$) $x_1\neq x_2$, $y_1=y_2$; ($\gamma$) $x_1\neq x_2$, $y_1\neq y_2$. In ($\alpha$) $B=0$ and $\frac{C}{A}=-x_1=-x_2$; in ($\beta$) $A=0$, and $\frac{C}{B}=-y_1=-y_2$; in ($\gamma$) plainly $A\neq 0$, $B\neq 0$, and if $C=0$, then

$$\frac{A}{B}=-\frac{y_1}{x_1}=-\frac{y_2}{x_2},$$

but if $C\neq 0$, then

$$\frac{A}{C}=-\frac{\begin{vmatrix} 1 & y_1 \\ 1 & y_2 \end{vmatrix}}{\begin{vmatrix} x_1 & y_1 \\ x_2 & y_2 \end{vmatrix}} \quad \text{and} \quad \frac{B}{C}=-\frac{\begin{vmatrix} x_1 & 1 \\ x_2 & 1 \end{vmatrix}}{\begin{vmatrix} x_1 & y_1 \\ x_2 & y_2 \end{vmatrix}};$$

so that in all cases $s$ is determined, and postulate (1) is verified.

*Postulate* (2).—The question is: do every two dyads of a system determine it? Let $Ax+By+C=0$ be a system $s$; $(x_1, y_1)$ $(x_2, y_2)$, any two dyads, $d_1, d_2$, of $s$. These dyads, by postulate (1), do determine a system $s'$, say $A'x+B'y+C'=0$. We are to show that $s$ and $s'$ are the same system.

If $A=0$, then $A'=0$, for $d_1$ is $\left(x_1, -\frac{C}{B}\right)$ and $d_2$ is $\left(x_2, -\frac{C}{B}\right)$; hence

$$(1) \quad \begin{cases} BA'x_1-B'C+BC'=0, \\ BA'x_2-B'C+BC'=0, \end{cases}$$

since $d_1$ and $d_2$ belong to $s'$; as $x_1\neq x_2$, $A'=0$; so $s'$ is $B'y+C'=0$; hence $B'y_1+C'=0$, and as $d_1$ belongs to $s$,

$By_1+C=0$; so $\frac{C'}{B'}=\frac{C}{B}$, and hence $s'$ is the same system as $s$. If $B=0$, the same identity results.

If $A\neq 0$ and $B\neq 0$, then, if $C=0$, $C'=0$, for $d_1$ is $\left(x_1, -\frac{A}{B}x_1\right)$, $d_2$ is $\left(x_2, -\frac{A}{B}x_2\right)$, and $\frac{A}{B}=-\frac{y_1}{x_1}=-\frac{y_2}{x_2}$; from the last we see that $x_1\neq x_2$, $y_1\neq y_2$; from the equation of $s'$ we have

$$(2)\quad \begin{cases} A'Bx_1-AB'x_1+BC'=0, \\ A'Bx_2-AB'x_2+BC'=0; \end{cases}$$

if $x_1=0$ or $x_2=0$, then $C'=0$, as $B\neq 0$; if $x_1\neq 0$, $x_2\neq 0$, divide (2) by $x_1$ and $x_2$ respectively and then subtract; so it is seen that $C'=0$. Hence $s'$ is $A'x+B'y=0$, and, as $A'x_1+B'y_1=0$, $\frac{A'}{B'}=-\frac{y_1}{x_1}=\frac{A}{B}$ and so, again, $s'$ and $s$ are the same.

Finally, if $A\neq 0$, $B\neq 0$ and $C\neq 0$, then, by the foregoing reasoning, $A'\neq 0$, $B'\neq 0$ and $C'\neq 0$. Hence

$$\frac{A'}{B'}=-\frac{\begin{vmatrix} 1 & y_1 \\ 1 & y_2 \end{vmatrix}}{\begin{vmatrix} x_1 & y_1 \\ x_2 & y_2 \end{vmatrix}}=\frac{A}{B} \quad\text{and}\quad \frac{B'}{C'}=-\frac{\begin{vmatrix} x_1 & 1 \\ x_2 & 1 \end{vmatrix}}{\begin{vmatrix} x_1 & y_1 \\ x_2 & y_2 \end{vmatrix}}=\frac{B}{C};$$

and so in this, as in all other cases, $s$ and $s'$ are identical. And postulate (2) is verified.

*Postulate* (7).—This is the next postulate in $H\Delta F$. It is satisfied, for any system $Ax+By+C=0$ is evidently satisfied by infinitely many dyads, and it is evident that no system contains all the dyads of $N$.

*Postulate* (8).—If $d_1(x_1, y_1)$, $d_2(x_2, y_2)$, $d_3(x_3, y_3)$ belong to a same system and if $d_2$ is between $d_1$ and $d_3$, then $x_2$ is between $x_1$ and $x_3$, or $y_2$ is between $y_1$ and $y_3$; then, by definition, $x_2$ is between $x_3$ and $x_1$, or $y_2$ is between $y_3$ and $y_1$, and so $d_2$ is between $d_3$ and $d_1$. Hence the postulate is satisfied.

*Postulate* (9).—It is evident that in determining a dyad of any given system we can assign the $x$ (or $y$) at will. Now let $d_1(x_1, y_1)$ and $d_3(x_3, y_3)$ be two given dyads of any given system $s$; let $d_2(x_2, y_2)$ be a dyad of $s$ such that $x_1 < x_2 < x_3$; then $d_2$ is between $d_1$ and $d_3$; next let $d_4(x_4, y_4)$ be such that $x_1 < x_3 < x_4$; then $d_3$ is between $d_1$ and $d_4$. Hence the postulate is satisfied.

*Postulate* (10).—We need consider only four possibilities: ($\alpha$) $A=0$, and $s$ is $By+C=0$; ($\beta$) $B=0$, and $s$ is $Ax+C=0$; ($\gamma$) $A\neq 0$, $B\neq 0$, $C=0$, and $s$ is $Ax+By=0$; ($\delta$) $A\neq 0$, $B\neq 0$, $C\neq 0$, and $s$ is $Ax+By+C=0$.

You know that of three numbers one and only one is between the other two. In ($\alpha$) any three dyads of $s$ are of the form $\left(x_1, -\frac{C}{B}\right)$, $\left(x_2, -\frac{C}{B}\right)$, $\left(x_3, -\frac{C}{B}\right)$; hence one and only one of the $x$'s is between the other two, and so, too, of the dyads; in ($\beta$) like reasoning leads to the same conclusion; in ($\gamma$) let $d_1(x_1, y_1)$, $d_2(x_2, y_2)$, $d_3(x_3, y_3)$ be any three dyads of $s$; then $\frac{A}{B} = -\frac{y_1}{x_1} = -\frac{y_2}{x_2} = -\frac{y_3}{x_3}$; hence no two $x$'s (or $y$'s) are equal for, if they were, the corresponding $y$'s or ($x$'s) would be equal and we should not have three distinct dyads; hence one and only one of the $x$'s (and also one and only one of the $y$'s) is between the other two; hence so, too, the dyads; finally, in ($\delta$) we have

$$\frac{A}{C} = -\frac{\begin{vmatrix} 1 & y_1 \\ 1 & y_2 \end{vmatrix}}{\begin{vmatrix} x_1 & y_1 \\ x_2 & y_2 \end{vmatrix}} = -\frac{\begin{vmatrix} 1 & y_2 \\ 1 & y_3 \end{vmatrix}}{\begin{vmatrix} x_2 & y_2 \\ x_3 & y_3 \end{vmatrix}} = -\frac{\begin{vmatrix} 1 & y_3 \\ 1 & y_1 \end{vmatrix}}{\begin{vmatrix} x_3 & y_3 \\ x_1 & y_1 \end{vmatrix}},$$

since any two of the dyads determine $s$; if two of the $y$'s were equal, then $A=0$, contrary to hypothesis, unless the corresponding $x$'s were also equal, but then we should not have three distinct dyads. Hence one and only one of the $y$'s (or $x$'s) is between the other two, and, the same being consequently true of the dyads, the postulate is verified.

The next postulate in $H\Delta F$ is the beautiful postulate (12). First, however, we must have a

DEFINITION.—A pair of dyads, $d_1(x_1, y_1)$ and $d_2$ $(x_2, y_2)$ of an $s$, is a *segment* $d_1d_2$ or $d_2d_1$; $d_1$ and $d_2$ are its ends; all dyads between the ends are the *segment's dyads.*

*Postulate* (12).—Let us notice, in the first place, that, taken two at a time, three dyads, $d_1(x_1, y_1)$, $d_2(x_2, y_2)$, $d_3(x_3, y_3)$, not belonging to a same system, determine three systems, $s_1$, $s_2$, $s_3$, as follows:

$$s_1 : \frac{x-x_2}{x-x_3}=\frac{y-y_2}{y-y_3}=\lambda_1,$$

$$s_2 : \frac{x-x_3}{x-x_1}=\frac{y-y_3}{y-y_1}=\lambda_2,$$

$$s_3 : \frac{x-x_1}{x-x_2}=\frac{y-y_1}{y-y_2}=\lambda_3;$$

it is plain that there is but one restriction on the $\lambda$'s, namely, $\lambda_1 \neq 1$, $\lambda_2 \neq 1$, $\lambda_3 \neq 1$; for, except for the inequalities, the given dyads would not be distinct. Looking at $s_1$, for example, you see that, when the variable dyad $d(x, y)$ is between $d_2$ and $d_3$ (*i.e.*, when it belongs to the segment $d_2d_3$), $\lambda_1$ is *negative;* and that, if $\lambda_1$ is *negative* (and neither zero nor $\infty$), $d$ is in the segment $d_2d_3$. Clearly the same statement, *mutatis mutandis,* is valid for $\lambda_2$ and $\lambda_3$.

Solving the foregoing equations for $x$ and $y$, we get

$$\text{for } s_1: \begin{cases} x=\dfrac{x_2-\lambda_1 x_3}{1-\lambda_1}, \\ y=\dfrac{y_2-\lambda_1 y_3}{1-\lambda_1}; \end{cases}$$

$$\text{for } s_2: \begin{cases} x=\dfrac{x_3-\lambda_2 x_1}{1-\lambda_2}, \\ y=\dfrac{y_3-\lambda_2 y_1}{1-\lambda_2}; \end{cases}$$

$$\text{for } s_3: \begin{cases} x=\dfrac{x_1-\lambda_3 x_2}{1-\lambda_3}, \\ y=\dfrac{y_1-\lambda_3 y_2}{1-\lambda_3}. \end{cases}$$

Now let us suppose that $Ax+By+C=0$ is a system $s$ not containing any of the dyads $d_1$, $d_2$, $d_3$. The conditions that $s$ shall contain a dyad of each of the systems $s_1$, $s_2$, $s_3$, are respectively

$$\lambda_1=\frac{Ax_2+By_2+C}{Ax_3+By_3+C},$$

$$\lambda_2=\frac{Ax_3+By_3+C}{Ax_1+By_1+C},$$

$$\lambda_3=\frac{Ax_1+By_1+C}{Ax_2+By_2+C};$$

We have, as you see, $\lambda_1\lambda_2\lambda_3=1$; hence none of the $\lambda$'s is negative or else two (and only two) of them are negative. Now suppose that $s$ contains a $d$ in the segment $d_1d_2$; then $\lambda_3$ is negative; hence $\lambda_1$ or $\lambda_2$ is negative, and so $s$

contains a $d$ of segment $d_2d_3$ or segment $d_1d_3$. Hence, you see, our postulate is verified.

*Postulate* (13).—That this postulate of parallels is verified in our new interpretation may be quickly seen as follows. Let $Ax+By+C=0$ be any given system $s$, and let $d(x', y')$ be any dyad not belonging to $s$. Then *any* system $s'$ containing $d$ is $A'(x-x')+B'(y-y')=0$, or $A'x+By'+C'=0$ where $C'=-(A'x'+B'y')$. Solving $s$ and $s'$ for $x$ and $y$, we get

$$x=-\frac{\begin{vmatrix} C & B \\ C' & B' \end{vmatrix}}{\begin{vmatrix} A & B \\ A' & B' \end{vmatrix}},\ y=-\frac{\begin{vmatrix} A & C \\ A' & C' \end{vmatrix}}{\begin{vmatrix} A & B \\ A' & B' \end{vmatrix}};$$

the two terms of neither fraction can be *zero*, for, if they were, then $\frac{A}{A'}=\frac{B}{B'}=\frac{C}{C'}$, and $s$ and $s'$ would coincide, contrary to hypothesis; hence $x$ and $y$ have definite finite values and accordingly $s$ and $s'$ have a common dyad $(x, y)$, except when the denominator is zero, but this can happen when and only when $\frac{A'}{B'}=\frac{A}{B}$, and hence there is one and only one $s'$ having no dyad in common with $s$, this unique $s'$ being parallel to $s$. And, as you see, the postulate is satisfied.

Before examining postulate (14) we require a

Definition.—If $d'(x', y')$ be a given dyad of a system $s$, any dyad $d(x, y)$ will be said to be *on the one side or on the opposite side of* $d'$ according as $x>x'$ or $<x'$, except when $s$ is of the form $Ax+C=0$ and then the distinction of sides will depend on whether $y>y'$ or $y<y'$.

*Postulate* (14).—Let $d_1(x_1, y_1)$ and $d_2(x_2, y_2)$ be any segment $d_1d_2$; let $d'(x', y')$ be any given dyad of any given system $s$; it is clear that there is in $s$ at least one dyad $d''(x'', y'')$ such that $d_1d_2$ is congruent with $d'd''$, *i.e.*, such that

$$\sqrt{(x_1-x_2)^2+(y_1-y_2)^2}=\sqrt{(x'-x'')^2+(y'-y'')^2},$$

for $x''$ is at our disposal and $y''$ is a function of it. But is there in $s$ more than one such $d''$? We know that $s$ is $y=mx+b$ or else $x=m'y+b'$; let us use the former, for the reasoning will be the same as for the latter. If there be a second $d''$, denote it by $d'''(x''', y''')$, where $x'''=x''+\delta$; then since $d_1d_2$, $d'd''$, $d'd'''$ are congruent, we

$$\sqrt{(x'-x'')^2+(y'-y'')^2}=\sqrt{(x'-x'''^2+(y'-)y''')^2};$$

note that $y'=mx'+b$, $y''=mx''+b$, $y'''=m(x''+\delta)+b$; substituting these values in the last radical equation, and simplifying, we get $\delta^2+2(x''-x')\delta=0$; whence $\delta=0$ or $\delta=2(x'-x'')$; the former value of $\delta$ gives $x''=x'''$, and so does not give a second $d''$; the latter value of $\delta$ gives $x'''=x''-2(x''-x')$, and so there is one and but one other $d''$; now note that $x'''-x'=-(x''-x')$; hence if one $d''$ is on one side of $d'$, the other $d''$ is on the other side. And so the postulate is verified.

*Postulate* (15).—This postulate is so manifestly satisfied that we need not tarry to prove the fact.

*Postulate* (16).—That this postulate is verified may be readily proved as follows: Let $d_1(x_1, y_1)$, $d_2(x_2, y_2)$ and $d_3(x_3, y_3)$, three dyads of any given systems $s$, be such that the segments $d_1d_2$ and $d_2d_3$ have in common no dyad save $d_2$; let $d_1'$, $d_2'$, $d_3'$, three dyads of any given system $s'$, be such that $d_2'$ is the only dyad common to the segments $d_1'd_2'$ and $d_2'd_3'$. Let $d_1d_2$ be congruent with $d_1'd_2'$, and $d_2d_3$ with $d_2'd_3'$; we are to prove that $d_1d_3$ and $d_1'd_3'$ are congruent. We may take $s$ to be $y=mx+b$, and $s'$

to be $y=m'x+b'$; then, since $d_1d_2$ and $d_1'd_2'$ are congruent $\sqrt{1+m^2}(x_1-x_2)=\sqrt{1+m'^2}(x_1'-x_2')$; so, too, $\sqrt{1+m^2}(x_2-x_3)=\sqrt{1-m'^2}(x_2'-x_3')$; whence, by addition, $\sqrt{1+m^2}(x_1-x_3)=\sqrt{1-m'^2}(x_1'-x_3')$; but this last equation tells us that $d_1d_3$ and $d_1'd_3'$ are congruent; and so, we see, the postulate is verified.

Before attacking postulate (17) let us make due preparation for it in the way of a simple theorem, some definitions and a little acquaintance with a very fundamental kind of algebraic transformation.

*Theorem.*—Every system $s$ separates the remaining dyads of $N$ into two classes such that, if $d_1$ and $d_2$ be any two dyads the segment $d_1d_2$ contains or does not contain a dyad of $s$ according as the given dyads belong, one of them to the one class and one of them to the other, or both of the dyads belong to the same class. [The theorem is the correspondent of Hilbert's theorem 5.]

The proof is not difficult. The given system $s$ is of the form (1) $x=k$ or of the form (2) $y=mx+b$. If $s$ be of form (1), it is clear that the classes required are respectively composed of dyads for which $x>k$ and of those for which $x<k$. Next suppose $s$ to be $y=mx+b$. Let $d_1(x_1, y_1)$ and $d_2(x_2, y_2)$ be any two given dyads not belonging to $s$. It is plain that there is a system $s_1$, $y=mx+b_1$, containing $d_1$, and a system $s_2$, $y=mx+b_2$, containing $d_2$; so that $y_1=mx_1+b_1$ and $y_2=mx_2+b_2$. The dyads $d_1$ and $d_2$ determine a system $s'$, namely,

$$\frac{x-x_1}{x-x_2}=\frac{y-y_1}{y-y_2};$$

now let $d(x, y)$ be the dyad common to $s$ and $s'$; then

$$\frac{x-x_1}{x-x_2}=\frac{m(x-x_1)+b-b_1}{m(x-x_2)+b-b_2};$$

whence

$$x-x_1=\rho(b-b_1)\div(1-\rho m),$$

$$x-x_2=\rho(b-b_2)\div(1-\rho m),$$

where $\rho$ is a proportionality factor. You see that $x-x_1$ and $x-x_2$ are unlike or like in sign according as $b-b_1$ and $b-b_2$ are unlike or like in sign; that is, $d$ is between or not between $d_1$ and $d_2$ according as $b$ is between or not between $b_1$ and $b_2$; hence the theorem. We may agree to say that the dyad $d_1(x_1, y_1)$ is on the positive or the negative side of $s$ according as $b_1>$ or $<b$.

Definitions.—If $d$ be a dyad of a system $s$, the dyads of $s$ on a same side of $d$ constitute a *half-system emanating from d.* [So it is seen that any dyad of an $s$ separates the remaining dyads of $s$ into two opposite half-systems.] A pair of half-systems, $h$ and $k$, emanating from a dyad $d$ and not belonging to a same system, is an *angle* $(h, k)$; $d$ is the angle's *vertex*, and $h$ and $k$ its *sides;* its *interior* is the class of dyads such that, if $d_1$ and $d_2$ be any two of them, the segment $d_1d_2$ contains no dyad of $h$ or $k$; its *exterior* consists of all other dyads of $N$ except $d$ and the dyads of $h$ and $k$. Let $d$ and $d'$ be the vertices of two angles $(h, k)$ and $(h', k')$; let $dd_1$ and $dd_2$ be two segments of $h$ and $k$ respectively, and let $d'd_1'$ and $d'd_2'$ be segments of $h'$ and $k'$ respectively; suppose $dd_1$ is congruent with $d'd_1'$ and $dd_2$ with $d'd_2'$; then, if $d_1d_2$ is congruent with $d_1'd_2'$, we shall say that the given angles are congruent. [Note that we have defined congruence of angles in terms of congruence of segments. Note also and note carefully that, though we have for the sake of convenience used such terms as angle, vertex, side, and so on, which smell of geometry and suggest space, there is in such use no logically implicit geometric or spatial reference whatever.

The use of those terms here is purely metaphorical and, had we desired to dispense with their use, it would not have been difficult, as you no doubt see, to do so.]

Let me now explain briefly a simple but exceedingly important algebraic transformation which will be very helpful in dealing with postulate (17). Consider the pair of equations

$$(t) \quad \begin{cases} x = x' \cos \theta - y' \sin \theta + a, \\ y = x' \sin \theta + y' \cos \theta + b; \end{cases}$$

solving these for $x'$ and $y'$, we get the pair

$$(t') \quad \begin{cases} x' = x \cos \theta + y \sin \theta - a \cos \theta - b \sin \theta, \\ y' = -x \sin \theta + y \cos \theta - b \cos \theta + a \sin \theta; \end{cases}$$

you notice that $(t)$ and $(t')$ are equivalent, either pair being obtainable from the other. Either pair, say $(t)$ defines a *dyad-to-dyad* transformation; that is, given a dyad $(x', y')$, there corresponds to it, by virtue of $(t)$ a definite dyad $(x, y)$, and conversely. Of two dyads thus related, we say that each is the other's transform or that each is converted or transformed into the other. I wish to call your attention to four further properties of the transformation. One of them is that the dyads of a *system* are converted into the dyads of a *system*. To prove this proposition, take any system $Ax + By + C = 0$, replace $x$ and $y$ by their values from $(t)$, simplify, and note that you then have the equation of a system. Thus a dyad-to-dyad transformation is also a system-to-system transformation: the transform of a system is a system. Another property of the transformation, showing its power, is that, owing to the presence of three undetermined quantities, or parameters as they are called —$a$, $b$ and $\theta$—, we can convert any given system

(1) $Ax+By+C=0$ into any given system (2) $A'x+B'y+C'=0$ and, at the same time, any given dyad of the former into any given dyad of the latter; to show this possibility, transform (1) as above indicated, then equate the two ratios of the coefficients in the resulting equation to the corresponding ratios taken from (2); these two equations (two conditions on $a$, $b$ and $\theta$) insure that (1) has (2) for its transform; but our three parameters can satisfy a third condition; notice what it is; let $d_1(x_1, y_1)$ be the given dyad of (1), and $d_1'(x_1', y_1')$ the given dyad of (2); then $y_1 = -\frac{A}{B}x_1 - \frac{C}{B}$ and $y_1' = -\frac{A'}{B'}x_1' - \frac{C'}{B'}$; $d_1$ is to be converted into $d_1'$ and this gives the third condition, which is that $x_1 = x_1' \cos\theta + \left(\frac{A'}{B'}x_1' + \frac{C'}{B'}\right)\sin\theta + a$ or an equivalent one obtained from the second equation of ($t$). The writing out of the three conditions and solving them for $a$, $b$ and $\theta$ involves a little finger work but no logical difficulty. You may wish to perform the task as an exercise. Again, any one of our dyad-to-dyad transformations converts any given segment into a congruent segment. I say " any one of our dyad-to-dyad transformations," for we have many, infinitely many, of them, depending on the values we assign to the parameters $a$, $b$ and $\theta$. To prove the property in question let the segment be determined by $d_1(x_1, y_1)$ and $d_2(x_2, y_2)$; in $\sqrt{(x_1-x_2)^2+(y_1-y_2)^2}$ replace $x_1, y_1, x_2, y_2$ by their transforms $x_1\cos\theta - y_1\sin\theta + a$, $x_1\sin\theta + y_1\cos\theta + b$, $x_2\cos\theta - y_2\sin\theta + a$, $x_2\sin\theta + y_2\cos\theta + b$, simplify and then note that the radical expression has suffered no change. Finally, any one of our transformations leaves the *order* of dyads unchanged; that is, if $d_1$, $d_2$ and $d_3$ are converted respectively into $d_1'$, $d_2'$ and $d_3'$, then, if $d_2$

be between $d_1$ and $d_3$, $d_2'$ will be between $d_1'$ and $d_3'$. Let $d_1$, $d_2$, $d_3$ belong to $s$, $y=mx+c$; then $y_1=mx_1+c$, $y_2=mx_2+c$ and $y_3=mx_3+c$; then from ($t$) we have

$$x_1'=(\cos\theta+m\sin\theta)x_1+(c-b)\sin\theta-a\cos\theta,$$
$$x_2'=(\cos\theta+m\sin\theta)x_2+(c-b)\sin\theta-a\cos\theta,$$
$$x_3'=(\cos\theta+m\sin\theta)x_3+(c-b)\sin\theta-a\cos\theta;$$

hence

$$x_1'-x'_2=(\cos\theta+m\sin\theta)(x_1-x_2),$$
$$x_2'-x'_3=(\cos\theta+m\sin\theta)(x_2-x_3);$$

hence if $x_1>x_2>x_3$ or $x_1<x_2<x_3$, then $x_1'>x_2'>x_3'$ or $x_1'<x_2'<x_3'$; that is, if $d_2$ is between $d_1$ and $d_3$, $d_2'$ is between $d_1'$ and $d_3'$.

From the invariance of congruence and of order, or betweenness, it follows that, if the angle ($h'$, $k'$) be the transform of the angle ($h$, $k$), the interior of the former is the transform of the latter's interior and that the angles are congruent.

With the foregoing equipment we may proceed to the examination of

*Postulate* (17).—Let me ask you to read the postulate very attentively. It requires us to prove the following proposition: *Given an angle* ($h$, $k$), *a system* $s$, *a dyad* $d$ *of* $s$ *and a half-system* $h'$ *emanating from* $d$, *there is one and but one half-system* $k'$ (*emanating from* $d$) *such that the angle* ($h'$, $k'$) *is congruent with the angle* ($h$, $k$) *and that the interior of* ($h$, $k$) *is on a given side of* $s$.

In virtue of our dyad-to-dyad transformation it is evident that, without loss of generality, we may take the sides $h$ and $k$ to be half-systems belonging respectively to the systems $s_1$, $y=m_1x$ and $s_2$, $y=m_2x+b$, and emanating from their common dyad, say, $d'(x', y')$; that we may take $s$ to be $y=0$, $d$ to be the dyad $(0, 0)$ of $s$, and $h'$

to be a half-system belonging to $s$ and emanating from $(0, 0)$. Let us now choose, as we evidently may choose, $d_1(x_1, y_1)$ on $s_1$, $d_2(x_2, y_2)$ on $s_2$, $d_3(x_3, 0)$ on $s$ and $d_4(x_4, y_4)$ on $y = mx$ so that the segments $d'd_1$, $d'd_2$, shall be congruent respectively to $dd_3$ and $dd_4$; note that $dd_4$ is part of the side $k'$ of the angle whose existence is to be established. We have to show that $m$ may be so chosen that $d_1d_2$ and $d_3d_4$ shall be congruent. By virtue of the given congruences, the condition that $d_1d_2$ and $d_3d_4$ shall be congruent and that consequently the angles $(h, k)$ and $(h', k')$ shall be congruent is readily found to be

$$m^2 = \frac{(m_1 - m_2)^2}{(1 + m_1m_2)^2};$$

there are, you see, two real values of $m$, of opposite signs, corresponding to the two sides of $s$ (or $h'$); and the postulate is, accordingly, satisfied.

*Postulate* (18).—So plainly satisfied as not to detain us.

*Postulate* (19).—That this one is satisfied follows at once from our definition of congruence of angles and the fact that postulate (17) is satisfied.

We now come, finally, to the Archimedean postulate of continuity.

*Postulate* (20).—By reason of the properties of our dyad-to-dyad transformation we may, without loss of generality, choose the system, $y = 0$, for system $s$, and for given dyads of $s$ the dyads $d(0, 0)$ and $d'(x', 0)$. Let $d_1(x_1, 0)$, $d_2(x_2, 0)$, $d_3(x_3, 0)$, . . . be such that $d_1$ is between $d$ and $d'$ and between $d$ and $d_2$, that $d_2$ is between $d_1$ and $d_3$ . . ., and that the segments $dd_1$, $d_1d_2$, $d_2d_3$, . . . are mutually congruent. We are to prove that in the dyad series there is a dyad $d_n(x_n, 0)$ such that $d'$ is between $d$ and $d_n$. The series of $x$'s is an increasing or decreasing series, say increasing. Then $x_1=x_2-x_1=x_3-x_2= \ldots =x_n-x_{n-1}$. The sum $nx_1=x_n$; we choose $n$ so that $nx_1>x'$, then $x_n>x'$, but $x'>0$; hence $x'$ is between 0 and $x_n$ and hence $d'$ is between $d$ and $d_n$; which was to be proved.

To prove the compatibility of postulates we have to find a set of things regarding which the postulates make true statements when the things are put in place of the variables. The better the things are known, the better is the test. Now number dyads and systems thereof are the best known of things; and so, in showing that they verify the Hilbert postulates, we have established their compatability by the diamond test.

Let us denote the doctrine arising from the foregoing interpretation of $H\Delta F$ by $D_4$. $D_4$, as I have said and as you must now plainly see, is in all strictness non-geometric, having no spatial content. It is purely algebraic or numerical—a two-dimensional theory of dyads and systems of dyads of real numbers. In point of *form* it is Euclidean, having the same form as Euclidean plane geometry; but to say that is to say that Euclidean geometry has the same form as the Dyad doctrine. If the latter had happened, as it might have happened, to be developed prior to Euclidean geometry and received a name proper to it, there would be precisely as much sense and propriety in calling Euclidean geometry by that name as there now is in calling the Dyad doctrine Euclidean geometry.

This lecture has grown, I fear, to a wearisome length. Yet I must ask your permission to continue long enough to indicate very briefly an interpretation of $H\Delta F'$ analogous to the foregoing interpretation of $H\Delta F$. The field $N'$ of the interpretation in question is composed of all the *triads* $(x, y, z)$ of the real numbers. The interpretation arises from assigning to the variables in $H\Delta F'$ the following values, or meanings: $v_1$ is to denote a triad of $N'$; $v_2$, the system of triads common to a pair of equations, $Ax+By+Cz+D=0$, $A'x+B'y+C'z+D=0$; $v_3$, the system of triads satisfying one such equation; $v_4$ is to mean $N'$; $R_1$ is to mean between in the sense that if the triads, $t_1(x_1, y_1, z_1)$, $t_2(x_2, y_2, z_2)$, $t_3(x_3, y_3, z_3)$, belong to a system of the former kind, then $t_2$ will be between $t_1$ and $t_3$ when and only when $x_2$ is between $x_1$ and $x_3$ or $y_2$ is between $y_1$ and $y_3$ or $z_2$ is between $z_1$ and $z_3$; and $R_2$ is to mean congruent, or equal, in the sense that the segment $t_1t_2$ will be congruent to segment $t_3t_4$ when and only when,

$$\sqrt{(x_1-x_2)^2+(y_1-y_2)^2+(z_1-z_2)^2} = \sqrt{(x_3-x_4)^2+(y_3-y_4)^2+(z_3-z_4)^2},$$

and, for angles, in the sense analogous to that given in the preceding interpretation.

The interpretation is worked out with some detail in a very interesting and enlightening way in the *Elementare Geometrie* cited in the preceding lecture. The doctrine, $D_4'$, arising thus from $H\Delta F'$ is, as you see, non-spatial and non-geometric; it is a purely algebraic three-dimensional theory of triads and systems of triads of real numbers and is, of course, isomorphic with ordinary Euclidean geometry of space.

## LECTURE VII

### Essential Discriminations

DISTINCTION OF DOCTRINE AND METHOD—ANALYTIC GEOMETRY AND GEOMETRIC ANALYSIS—THE TWAIN BEGOTTEN OF CONVERSE TRANSFORMATIONS—AN INFINITE FAMILY OF SISTERS — ALL HERITORS OF THEIR MOTHER'S FORM—THEIR COMMON CHARACTER AND INDIVIDUALITIES—EXCESSIVE MEANING OF CONTENT — GENERIC AND SPECIFIC MEANINGS OF EUCLIDEAN AND NON-EUCLIDEAN—THREE PROPERTIES COMMON TO POSTULATE SYSTEMS—FERTILITY AND COMPENDENCE AND COMPATIBILITY.

IN this lecture and the next one, I invite you to join me in considering a variety of kindred matters closely connected with the preceding lectures. Some of these matters are suggested in the foregoing lengthy title. In course of the discussion some of the unanswered questions you have asked and some others that you are no doubt prepared to ask will, I hope, receive suitable answers. I say " some " of them, for I trust we are not so stupid as to be able to answer all the questions we are able to ask. Let us begin with one of the questions that must be asked and can be answered satisfactorily.

*A Word about Analytic Geometry and Geometric Analysis.*—In Lecture V, I gave a little introduction to what is called the analytic or algebraic geometry of the

Euclidean plane. We saw that, a pair of axes and a distance unit being chosen, to any point $P$ of the plane there belongs a pair $(x, y)$ of real numbers, and conversely; and that to each line there belongs an equation $Ax+By+C=0$, and conversely. Now such a pair and such an equation are respectively what we called in Lecture VI a dyad and a system of dyads. The question is: Is not the dyad doctrine $D_4$ simply ordinary Euclidean geometry $D_1$ in disguise? I might answer, quite justly, that $D_4$ is no more and no less $D_1$ in disguise than $D_1$ is $D_4$ in disguise. You may now wish to say: very well, are not $D_1$ and $D_4$ identical? The answer is *no*, for $D_1$ is a doctrine about spatial things—points and lines—while $D_4$ is a doctrine about non-spatial things—dyads and systems of dyads of pure real numbers. Perhaps you would rejoin, saying: Is not $D_4$ simply the analytic, or *algebraic, geometry* of the Euclidean plane? It is evidently just to answer: $D_4$ is that, no more and no less than $D_1$ is the *geometric* algebra of $N$, which is the field of number dyads and systems thereof just as the plane is the field of points and lines. And you know that $D_1$—the ordinary plane geometry of Euclid—is not an algebra. The fact is that, unless we are content to confound things that are essentially different, we must here distinguish *four* different things: namely, $D_1$, $D_4$, and *two converse aspects* of what is in usage somewhat indiscriminately called analytic, or algebraic, geometry of the Euclidean plane. One of these aspects ought to be called analytic, or algebraic, geometry; and the other, geometric analysis or geometric algebra. "Ought," I mean, for the sake of philosophic clarity, not necessarily in common every-day parlance or practice. Let us be quite clear in this business. What is commonly called the analytic, or algebraic, geometry of the Euclidean plane has its birth in a certain transformation—a point-to-dyad transformation—which consists in the fact that a one-to-one correspondence subsists between the points of the plane and the number dyads $(x, y)$ of $N$. By virtue of this transformation, to any given relation among points in doctrine $D_1$ there corresponds a definite relation among dyads in doctrine $D_4$; and conversely, for the correspondence runs both ways. Do not fail to note now very carefully, for this is the crux of the matter, that, owing to the mentioned correspondence, we can translate a problem respecting points into a problem respecting dyads, then solve the latter (algebraically) and finally translate the result in terms of points, thus getting a proposition in $D_1$; and, conversely, we can translate a problem respecting dyads into a problem respecting points, then solve the latter (geometrically) and finally translate the result in terms of dyads, thus getting a proposition in $D_4$: in other words, we can investigate algebraically the point relations making up $D_1$ and, conversely, we can investigate geometrically the dyad relations making up $D_4$. It is now obvious that, instead of calling both of these converse procedures analytic, or algebraic, geometry, the former ought to be called analytic, or algebraic, geometry; and the latter geometric analysis or geometric algebra. Observe that neither of them yields a new doctrine; each of them is simply a new *method* of establishing an old doctrine; and the fundamental distinction between the two doctrines, $D_1$ and $D_4$, remains in undisturbed serenity.

You perceive at once that the foregoing discussion applies, *mutatis mutandis,* to $D_1'$ and $D_4'$.

*The Possibility of Yet Other Interpretations of $H\Delta F$ and $H\Delta F'$.*—To each of these doctrinal functions have now been given three geometric interpretations but only one non-geometric one, and the latter is algebraic. It is natural to ask: are there other algebraic interpretations? The answer is, there are. I shall not tarry to present them, for we have many other things to consider, but we may pause a moment to convince ourselves of their existence. Let us recall our third interpretation of $H\Delta F$, for example, giving rise to the geometric doctrine $D_3$. It is plain that in it we may replace point by dyad and pathocircle by an equation determining a perfectly corresponding system of dyads, and thus obtain a new algebraic interpretation of $H\Delta F$ and therewith a new two-dimensional theory of dyads and dyad systems. And so on—an algebraic interpretation for each geometric one and conversely.

How many geometric and how many algebraic interpretations of $H\Delta F$ or of $H\Delta F'$ are possible? Is the number finite or infinite? I will state—without giving the proof—that each of the two functions admits of an infinitude of interpretations of either sort. And I may add,—again omitting the proof, which is easy,—that from any given interpretation, whether geometric or algebraic, one can derive an endless series of different interpretations, correspondingly geometric or algebraic, drawing them, each out of its predecessor, unceasingly as the successive joints of an infinitely-many-jointed telescope. Most of the interpretations thus obtainable and the corresponding doctrines are devoid of interest for us human beings, but that statement is a commentary upon our supersimian curiosity and not upon the intrinsic merits of the doctrines.

Do $H\Delta F$ and $H\Delta F'$ admit of interpretations that are both non-geometric and non-algebraic? Yes: each of the functions admits of an infinite variety of such interpretations. It is very easy to prove it. To do so, consider any one of the interpretations we have encountered, say the fourth one of $H\Delta F$—the doctrine $D_4$. Any other one would do as well. You know that we can, if we choose, associate in our thought any two given objects, $O_1$ and $O_2$, thus obtaining a third object, $O_3$ (which is simply $O_1$ and $O_2$ associated together). Now let $O$ denote some given object of thought, say the center of gravity of the Milky Way or Cæsar's love for Cleopatra or the taste of good whiskey—any specific thing, no matter what. Now associate $O$ with *each* dyad involved in $D_4$; association of $O$ with $(x_1, y_1)$ gives a new object $O_1'$; association of $O$ with $(x_2, y_2)$ gives us another new object $O_2'$; and so on. Never mind how arbitrary or artificial or uninteresting the new objects may be, for that is of no *logical* importance at all. Observe that we have a one-to-one correspondence between the dyads in field $N$ and the objects $O'$, which we may think of as constituting a field $M$. You see that to a given system of dyads there now corresponds a definite class of the $O'$'s, which class we may, if we like, call a system of $O'$'s. Let us next agree,—evidently we *may* agree,—to say that two or more $O'$'s *satisfy* a relation when and only when the corresponding dyads satisfy the relation. You see immediately that, in virtue of our agreement, or convention, the $O'$'s of $M$ and the $O'$-*systems* verify the postulates of $H\Delta F$ just as well as do the dyads and dyad systems of $N$, that is, perfectly. And you see that there thus arises a new interpretation of $H\Delta F$ and a new doctrine whose content differs from that of $D_4$ as an $O'$ differs from a dyad. If you choose a different $O$ you obtain a new kind of object $O'$ and a new doctrine. You thus get as many doctrines as there are objects $O$ to use. If God has not made an infinite number of $O$'s for you, you doubtless see that you can make them yourselves. I grant that the vast majority of doctrines that are constructible in the way indicated are trivial—mere weeds of the doctrinal garden; it was, however, not our task to estimate their worth, but to demonstrate their infinite multiplicity.

*Sense in which All Doctrines Derivable from $H\Delta F$ and $H\Delta F'$ Are Like in Form, or Structure.*—Let me request you to remind yourselves vividly of the fact that each of the doctrinal Functions consists of a system of propositional functions, called postulates, and a set of propositional functions logically deducible from the postulates and called theorems. Be good enough to recall also the fact that, if we replace the variables in the postulates of one of the doctrinal functions by admissible constants—a term already explained—we thereby obtain a doctrine, which is true, and then called a value of the function, or false, according as the substituted constants verify or do not verify all of the postulates. Because the doctrine, whether true or false, matches the doctrinal function, statement for statement, and because the statements (propositions) composing the doctrine and the corresponding statements (propositional functions) composing the doctrinal function are identical in respect of form, we say that the *doctrine* and the *function* are themselves like in *form*, or *structure*. You see that, therefore, the infinitude of values of either one of our doctrinal functions and the infinitude of false doctrines derivable from it are all of them like in form, or structure, for each of them is like in form, or structure, to the function from which all of them are derivable.

*Senses in which All Doctrines Derivable from $H\Delta F$ and $H\Delta F'$ Are Like and Unlike in Content, or Subject-matter.*—

The doctrinal functions, we have seen, have no specific content, no definite subject-matter, and are neither true nor false. On the other hand, each of the derivable doctrines has specific content, or subject-matter, and is true or else false. Being, as we have seen, like in form, or structure, the doctrines are discriminated among themselves solely by differences of content, or subject-matter. Now, their contents, or subject-matters, differ in respect of what we may call their *meanings*. *Query*: Is there any respect in which the contents of the various doctrines are identical? The answer is that the contents of all of the *true* doctrines,—of all of the *values* of the doctrinal function concerned,—are identical *in the respect* that the various contents equally verify, or satisfy, the postulates of the function; but such partial identity of content can not be affirmed of two of the false doctrines. Let us now confine our attention to the true doctrines for it is these that we value. It is perfectly clear that the meaning of the content, or subject-matter, of such a doctrine,—the meaning, that is, of the things which the doctrine is a doctrine of or *about*,—is *not exhausted* by the requirement that the things shall be verifiers of the postulates. Ordinary points and lines, for example, or number dyads and dyad systems, have countless uses and significances over and above the service they render by satisfying the postulates of $H\Delta F$. The meaning that the content of a true doctrine has *beyond* that it must have to verify the postulates of the function of which the doctrine is a value may be called the content's, or subject-matter's, *excessive meaning*. Thus you see that the infinitely many diverse doctrines having a given doctrinal function for their common matrix are discriminated from each other by diversities in the excessive meanings of their contents, and each doctrine is *identified* by something peculiar in the excessive meaning of its content. The statement just made holds good for *true* doctrines only, for it is evident that to false doctrines the notion of excessive meaning does not apply. How are the false doctrines having a common functional matrix discriminated and identified? I leave the question to such of you as may be interested to consider it.

We have seen that all doctrines, whether true or false, that have a same doctrinal function for matrix are like in form—they have, that is, the same logical frame or structure. It is pretty obvious and very noteworthy that, on the other hand, what we have called the excessive meaning of a true doctrine's content is thus not logical, but is purely *psychological.* A point and a number dyad, for example, or a line and a pathocircle, or a plane and a pathosphere, though they fit into the same logical scheme, performing the same (logical) office in relation to the postulates, are discriminated not only by their differences as concepts,—which are psychological phenomena—but also and especially by the exceedingly different imageries or intuitions with which they and the doctrines they figure in crowd the mind. You are students of philosophy. As such you ought to be interested in psychology and I trust none of you is afflicted with *psychological blindness.* Not long ago, a professional philosopher in good standing, told me that he saw no " psychological " difference between a point and a straight line. I cannot understand how any student with a feeling for psychology can fail to have a little quickening of pulse when he sees clearly for the first time the fact now staring us in the face: namely, that we are living in a world where it is possible to have an infinitude of true doctrines and an infinitude of false ones which, though differing among themselves *psychologically* in an endless variety of ways, are yet but *one* in point of form, absolutely identical in *logical* frame or structure. I know of no other equal revelation of the truly amazing economic power of Logic in our world. Think of having to live in a world where no two doctrines, no two theories, could own an identity of logical constitution. I suspect that in such a world there could be no logic, no science, no philosophy, no genuine life of intellect, no civilization.

*Sense in which All Doctrines having $H\Delta F$ or $H\Delta F'$ for Their Matrix Are Euclidean.*—The adjectives, Euclidean and non-Euclidean, are, as you are aware, customarily employed to designate certain types of *geometry*. In this use each of the adjectives has two different meanings —one of them very specific and common, the other generic and less common. In order to avoid confusion in reading geometric literature it is important to know what the two meanings are. In its generic and less common meaning the adjective "Euclidean" is used to designate the kind of geometry that is, in all important or essential respects, identical with the kind found in Euclid's *Elements*. Having that meaning of Euclidean in mind, we should say that a given geometry is non-Euclidean if, for example, it is algebraic (or analytic) in method, for the method of the *Elements* is that of so-called pure (non-algebraic) geometry; or if it is a geometry of four or more dimensions, for that of Euclid is three-dimensional; or if, like projective geometry, for example, or inversion geometry or the so-called hyperbolic geometry of Lobachevski or the so-called elliptic geometry of Riemann, it uses one or more postulates inconsistent with Euclid's postulates; or if, like the endless series of geometries (actual or potential) initiated by Julius Plücker's great creation of Line Geometry, it employs some spatial entity or entities other than the point, line and plane (of Euclid's *Elements*) for primary element or elements, or subject-matter. For all such distinctions are sufficiently important. On the other hand, as I need hardly say, a merely idiomatic or expressional difference,—such, for example, as the Greek's saying, " a straight line can be drawn from *every* point to *every* point " whereas we say " from *any* point to *any* point,"—is no warrant for calling the latter non-Euclidean. So much for the *generic* meanings of the adjectives—Euclidean and non-Euclidean as applied in geometry. And now let us be very clear as to what the specific and more common meaning of each term is. One of Euclid's postulates—his postulate 5—had the fortune to be an epoch-making statement—perhaps the most famous single utterance in the history of science. It is this:

> *If a straight line falling on two straight lines make the interior angles on the same side less than two right angles, the two straight lines, if produced indefinitely, meet on that side on which are the angles less than two right angles.*

Apparently convinced that this proposition could not be deduced as a theorem from his other postulates and axioms, or common notions, Euclid assumed it. It was for him an assumption, an hypothesis, a primitive proposition, a postulate—a *basal* proposition of the *Elements.* It is commonly known as Euclid's parallel-postulate because it is equivalent to the postulate that, if $P$ be a point and $L$ a line, there is but one line through $P$ parallel to $L$. Unlike Euclid, his successors for two thousand years, like his predecessors, were not convinced that the postulate was incapable of demonstration and many of the best of them devoted their genius to vain attempts at proving it. At length, however, mathematicians learned better and began to produce geometries on the basis of postulate systems in which Euclid's parallel postulate was contradicted. That such geometries are just as logically possible as Euclid's I will show in a subsequent lecture. What I wish now to say is that any geometry built upon a postulate system containing Euclid's parallel-postulate, or its equivalent, is called Euclidean, however widely it may differ in other respects from Euclid's *Elements;* and, correspondingly, any geometry, like that of Lobachevski or that of Riemann, whose postulate system contains a contradictory of Euclid's parallel-postulate, is said to be non-Euclidean, no matter how much it may be like Euclid's *Elements* in other respects. Such are the specific and more usual senses in which these familiar adjectives are employed in the literature of geometry.

It must occur to you at once that there is no good reason for confining the use of the terms, in the sense just indicated, to *geometry*. For the Hilbert postulate (13) being in agreement with the parallel-postulate of Euclid, it is evident that we may with evident and perfect propriety call Euclidean all of the infinitely many doctrines having $H\Delta F$ or $H\Delta F'$ for matrix, whether the doctrines be true or false, and whether they be geometric or algebraic or neither the one nor the other.

*What Are the Properties that a Collection of Propositional Functions Must Have in Order to be a Postulate System?*—I will name three properties: pregnance, or productiveness, or fertility; compendence, or connectedness; and compatibility, or consistency. In discussions of the question, the last property is always specified; the first two, seldom or never, though they are evidently essential, as you will presently see.

In saying that the collection of propositional functions must be pregnant, or fertile, I mean that it must be such that one or more consequences, or theorems, can be logically deduced from its component functions. In other words, it must be capable of giving rise to a doctrinal function containing one or more propositional functions besides those serving as postulates. No one, I imagine, would deliberately call a barren collection of propositional functions a postulate system.

In saying that a collection of propositional functions, if it is to be a postulate system, must be compendent, or connected, I mean something sufficiently easy to grasp, once it is perceived, but not very easy to state precisely and clearly. I will try to be intelligible. You know that owing to the presence of variables in the postulates of a postulate system, the latter has no specific subject matter; we may say, however, that, since the postulates talk about the variables as about subject-matter, the system has " apparent " subject-matter, or, better, we may say the system has undetermined subject-matter represented by the variable-symbols or variable-names. Now, if you will examine the postulates of some postulate system, say those of the Hilbert system, you will discover—what you may not have before noticed consciously—that the variable-symbols are each *connected* with every other; that is to say, the $v_1$'s are in $v_2$'s, which are in $v_3$'s, which are in $v_4$, $R_1$ is a relation of $v_1$'s, $R_2$ a relation of segments composed of $v_1$'s and of angles composed of half-rays (or half-$v_2$'s), and so on. This connectedness gives the undetermined subject-matter of the system unity, makes it hang together, gives it and the systems compendence. Suppose you had, as one might have, a collection of propositional functions some of which talked of the variables $x$, $y$, $z$, . . . , and some of which talked of the variables $x'$, $y'$, $z'$, . . . , in such a way that the former symbols were not connected with the latter, then the undetermined subject-matter of the collection of functions would lack unity; the collection would not be compendent and would, therefore, not be a postulate system. If the subcollection involving $x$, $y$, $z$, . . . connected them and if the same were true of the subcollection involving $x'$, $y'$, $z'$, . . ., and each of the two subcollections were moreover, pregnant and compatible, then the original collection would constitute *two* postulate systems, but these would *not* together constitute *one*. They would be independent; neither of them being included in the other, they would not be related like the systems involved respectively in $H \Delta F$ and $H \Delta F'$, for the former of these is a part of the latter.

In saying that a collection of propositional functions, if it is to be a postulate system, must be compatible, or consistent, we mean that its functions must be such as not to involve contradiction among themselves—they will be compatible unless at least two of them contradict each other explicitly or implicitly. The reason for this requirement is obvious. For, if the collection contained two mutually contradictory functions, the functions of the collection would admit of no verifiers; whatever set of admissible constants we might substitute for the variables in the functions would yield a set of propositions of which at least two would be mutually contradictory; hence, if we called the collection a postulate system and then consistently called the system together with the theorems derivable therefrom a doctrinal function, it would be one having no values, admitting of no interpretation, giving rise to no true doctrine. For this reason an incompatible collection of propositional functions is not called a postulate system.

If a collection of propositional functions be compatible, how may the fact be ascertained? There is but one known test: if we find a set of constants that we are convinced verify the functions, then and only then we say the collection is compatible. If you will examine Hilbert's book, you will find that he showed, or rather indicated very briefly how to show, that his postulate system is compatible by indicating how to show that the postulates are verified by the pairs and certain systems of pairs of a specified set of algebraic numbers. In Lecture VI, I have shown in detail the compatibility of the postulates of $H\Delta F$ by showing that they are verified by the set of dyads and certain systems of dyads of the real numbers, and have indicated the analogous procedure for the postulates of $H\Delta F'$. That is one of the reasons why Lecture VI is so detailed—I desired it should incidentally serve to exemplify what we may call compatibility proof.

## LECTURE VIII

### Postulate Properties

SCIENTIFIC PLATFORMS—THEIR FERTILITY, COMPENDENCE AND COMPATIBILITY—DIFFERENCES OF EQUIVALENT PLATFORMS—VARIETIES OF PLATFORMS AND FUNCTIONS—MEANINGS OF INDEPENDENCE AND CATEGORICALNESS—THEORETICAL AND PRACTICAL DOUBT.

UNLESS I am mistaken, you, as students of philosophy, should find no little interest in certain questions connected with the properties I have mentioned as essential to a genuine postulate system and as therefore common to all such systems. I desire to draw your attention to some of the questions without thereby promising or pretending to answer all of them, for I can not do so satisfactorily.

If a collection of propositional functions be fertile or infertile, how may we ascertain the fact? Of course, if we actually deduce one or more consequences from them, we then know that the collection is fertile. But the question I desire to ask and to commit to you for future consideration is this: Is there a criterion for deciding *a priori* whether a given collection of propositional functions is fertile or not, and if there is, what is the criterion?

Another question—which I believe to be important and difficult—is this: What is the essential nature of the rôle of variables in propositional functions by virtue of which a collection of such functions has an ambiguous, or undetermined, subject-matter that may or may not be compendent? May I leave this subtle matter to your reflection?

As to compatibility, suppose we have a collection of functions such that we have not been able either to verify them or to prove them incompatible. Doubtless, we must say, in such a case, that we do not know whether the collection is compatible or not. Might not the collection be incompatible without our being able ever to discover the fact? Might it not be compatible, though we should never be able to know it? Another question—very different from the preceding one—is this: Is there, conceivably, a compatible or a not incompatible collection of propositional functions having no verifiers in our world?

What, essentially, is logical compatibility? Must we be content with mere examples of it or with what seems at all events to be such examples? Whatever logical compatibility may be, it evidently is such that compatibility and incompatibility are related somewhat as pleasure and pain, as cosmos and chaos, as music and noise, as health and disease, as harmony and discord, as beauty and ugliness—so that Logic and Science are no less under the empire of the muses than are the Arts.

Is compatibility, then, an emotion, a feeling, a mere sentiment? If it be, it is not one of ideas, but is a sentiment of forms—propositional forms. What is propositional form? The question arose before and I said we should return to it. Well, here it is. I can not answer it. I know, in a sense, and so do you, what such form is, but I cannot define it abstractly,—not satisfactorily. Possibly you can—sometime; and if you do, you will thereby make a great contribution to the science of Logic. Whatever the thing may be, it is something in respect of which, for example, the statements "$x$ is $y$" and "Socrates is a man" are identical; something in respect of which the statement "If $x$ has the character $y$ and whatever has the character $y$ has the character $z$, then $x$ has the character $z$" is identical with the statement "If Socrates is human and whatever is human is mortal, then Socrates is mortal"; something in respect of which either statement of the first pair and either one of the second pair are, as specimens of logical material, radically unlike, irreducible to the same type.

One word more regarding compatibility and I shall quit the theme. I know you desire to ask,—for in discussing the matter of testing for compatibility, students never fail to ask,—how we may be certain that we have found a set of verifiers for a given collection of propositional functions. The answer required is that one disciplined in the fine art of doubting never can be *absolutely* certain. Absolute certainty is a privilege of uneducated minds—and fanatics. It is, for scientific folk, an unattainable ideal. Perhaps we can in no case reach a higher degree of certainty than that the dyads and systems thereof, employed in Lecture VI, satisfy the postulates there concerned. Yet a capable doubter may doubt whether we sufficiently understand the nature of the real numbers to be absolutely certain even in that case. Even less, but only slightly less, difficult is it to doubt the adequacy, as verifiers, of the point and the straight line, though these have been used as such verifiers since the memory of man runneth not to the contrary. But, though we can never attain absolute certainty in the premises, we can reach a certainty so nearly absolute that one who, having such a certainty respecting any *practical* affair, yet refrained, for lack of certainty, from action where action was called for, would be rightly judged, not necessarily stupid, but foolish or morbid or insane.

*The Hilbert Postulate System Not Intrinsically Superior to Others.*—In our discussions this particular system has been the subject of so many commentaries that, despite the precautions explicitly stated in Lecture II, it may seem to you to be the subject of our study instead of being merely one of the instruments employed. Perhaps I need not remind you that what we have been mainly examining is the nature of the important general conception denoted by the term "postulate system," and that, instead of beginning with an abstract definition of the concept, we have preferred to study it by means of a specific representative, or typical, example. For such an example, we have chosen the Hilbert system because of its familiarity, accessibility and fame. Our purpose had been served equally well, however, had we employed some other system, whether logically equivalent or non-equivalent to that of Hilbert. Systems of both kinds abound, and I shall presently refer you to some of them.

*Equivalence of Postulate Systems and Identity of Their Doctrinal Functions.*—Two postulate systems, $S$ and $S'$, are said to be equivalent if, and only if, every postulate in $S$ is in $S'$ or is deducible as a theorem from those in $S'$ and every postulate in $S'$ is in $S$ or is deducible from those in $S$. The same conception may be approached and viewed as follows: $S$, we know, gives rise to a doctrinal function, say, $\Delta F$, composed of the postulates in $S$ and the theorems deducible therefrom. Similarly, $S'$ yields a doctrinal function, say, $\Delta F'$. Let us agree to call $\Delta F$ and $\Delta F'$ identical if, and only if, every propositional function in $\Delta F$ or $\Delta F'$ is in the other, $\Delta F'$ or $\Delta F$, it being, of course, understood that some mere rewording may be required to show that a statement in one is in the other. You see, at once, that two postulate systems are equivalent or non-equivalent according as the corresponding doctrinal functions are identical or non-identical; and conversely. It is a fact of no little scientific and philosophic interest—for it is far from "self-evident"—that, within limits, the postulates and the theorems in a doctrinal function may interchange their respective rôles without destroying the function's identity. Some questions arise here which, so far as I know, no one has asked, and which I am unable to answer. One of them is: what are the "limits" within which the mentioned interchange of rôles may occur?

The only way to know that two equivalent systems are equivalent is to prove them equivalent. It would be very enlightening and a lot of fun to illustrate the process, but it would delay our course too much. Perhaps you will try your hand at the game. Two extremely interesting systems which, I believe, though I have not proved it in full detail, are equivalent to Hilbert's system and consequently to one another are the systems devised respectively by Professor O. Veblen and Professor Mario Pieri. Veblen's system, called "A System of Axioms for Geometry," is found in Volume V of *The Transactions of the American Mathematical Society* (1904). This system, in a modified form, was subsequently presented by its author as the initial monograph in the *Monographs on Topics of Modern Mathematics* (edited by Professor J. W. A. Young)—a volume which, though its articles differ widely in aim, spirit and excellence and though it attempts, pretty successfully, to avoid philosophic questions, may yet be recommended to philosophical students as a collection of essays affording an introduction to a variety of important elementary topics of modern mathematics: namely, Veblen's *The Foundations of Geometry*—which does not deal with the foundations of geometry in general, but gives the reader a finely histological view of the ultimate tissues and minute logical structure of the first parts of Euclidean metric geometry; *Non-Euclidean Geometry*, by Professor F. S. Woods—resembling Veblen's essay in spirit and method, hardly surpassed as an introduction, for beginners, to the geometries of Lobachevski and Riemann; *Modern Pure Geometry*, by Professor T. F. Holgate—not postulational or rigoristic like the articles just now mentioned, and not concerned with pure geometry in general, but giving such an acquaintance with pure projective geometry as one gains of an immense city by riding about in it on the top of a comfortable 'bus; an exceedingly enlightening essay by Professor E. V. Huntington, dealing postulationally and very refreshingly with *The Fundamental Propositions of Algebra;* an interesting and instructive article by Professor G. A. Miller treating *The Algebraic Equation* in part historically, in part critically, in a manner a little too mature, perhaps, and a bit sketchy for beginners; *The Function Concept and the Fundamental Notions of the Calculus*, by Professor G. A. Bliss—an essay chiefly notable, I think, as showing how swiftly and quickly a competent reader may be conducted into the presence of the cardinal concepts of the calculus and be given some sense of their power; and three instructive and stimulating essays by Professors J. W. A. Young, L. E. Dickson and D. E. Smith concerned, respectively, with *The Theory of Numbers*, *Con-*

*structions with Ruler and Compasses* and *The History and Transcendence of* $\pi$.

A moment ago I referred to a remarkable postulate system devised by the Italian mathematician, Pieri. It was published in 1899 in *Memorie della R. Academia delle Scienze di Torino* under the title "*Della Geometria elementare come sistema ipotetico-detuttivo; monografia del punto e del mote*"; an excellent abstract of it was published in 1905 by the late Louis Couturat in his *Les Principes des Mathématiques* and was partly reproduced in 1911 by Professor J. W. Young in his admirable *Lectures on Fundamental Concepts of Algebra and Geometry*. I desire, in passing, to recommend these books of Couturat and Young as well worth your attention, provided you will really read them—pondering what is said in them—and not be content with merely glancing through them. They handle, in excellent style, some important matters which these lectures touch but lightly or not at all.

You can not fail to observe, if you will examine and compare them—as I hope you will—that Veblen's system and that of Pieri differ from Hilbert's in various ways. For example, Veblen's system contains 12 postulates; Hilbert's, 21; Pieri's, 20; again, while Hilbert's system contains, as we have seen, five undefined terms, or variables, Veblen's has but two—"*point*" and "*between*"—and Pieri's also has but two—"*point*" and "*motion.*" In studying the Italian's beautiful system, your understanding of it will be much facilitated by noticing that the undefined term "motion" is used in the sense of a unique and reciprocal correspondence—a one-to-one transformation—between points, and not in the man-in-the-street's sense of a physical time-consuming change of place.

*Other Varieties of Postulate Systems and Doctrinal Functions.*—By " other " varieties I mean such as are not equivalent to the foregoing systems. Before citing them, or rather some of them—for I am far from intending to list all that have been devised—a word of caution seems desirable. In inventing a postulate system the inventor is never, or almost never, aiming at the establishment of what we have been calling a doctrinal function. He is aiming at establishing autonomously a particular one of the many doctrines which, as we have seen, a doctrinal function has for its values. Such special interest of the inventor, guiding and controlling him, is nearly always betrayed by the air or color of his speech: often by his giving his system a kind of name indicating that the system has a specific subject-matter, which it has not; nearly always by calling the postulates propositions, which they are not, instead of propositional functions, which they are; and usually by denoting the undefined terms by names instead of variable-symbols as if the undefined terms were constants instead of variables. This precaution will, I trust, help to keep you from gaining a false impression from the following citations. The systems in the list are almost random selections, and the list is far from exhaustive but, by help of the numerous references in the systems cited, it will afford you a clue to all or nearly all extant systems.

*The Axioms of Projective Geometry*, by A. N. Whitehead, Cambridge University Press, 1906.

*The Axioms of Descriptive Geometry*, by A. N. Whitehead, Cambridge University Press, 1907.

" A Set of Axioms for Line Geometry," by E. R. Hedrick and Louis Ingold, *Transactions of the American Mathematical Society*, Vol. XV, 1914.

"On a Set of Postulates Which Suffice to Define a Number-Plane," by R. L. Moore, *Trans. Amer. Math. Soc.*, Vol. XVI, 1915.

"A Set of Postulates for Real Algebra, Comprising Postulates for a One-Dimensional Continuum and for the Theory of Groups," by E. V. Huntington, *Trans. Amer. Math. Soc.*, Vol. VI, 1905.

"Set of Independent Postulates for Betweenness," by E. V. Huntington and J. R. Kline, *Trans. Amer. Math. Soc.*, Vol. XVIII, 1917.

"Complete Existental Theory of the Postulates for Serial Order," by E. V. Huntington, *Bull. Amer. Math Soc.*, Vol. XXIII, 1917.

"*Sui principi fondamentali della Geometria della Retta*," by G. Vailati, *Revista di Matematica*, Vol. II, 1892.

"A Set of Five Independent Postulates for Boolean Algebras, with Application to Logical Constants," by H. M. Sheffer, *Trans. Amer. Math. Soc.*, Vol. XIV, 1913.

"*Sulle ipotesi che permettono l'introduzione delle coordinati in una varietà a piri dimensioni*," by F. Enriques, *Rendiconti del Circolo matematico di Palermo*, Vol. XII, 1898.

Finally, I will add to the foregoing short list a reference to the famous postulate system by which G. Peano—founder and leader of the important Italian school of workers in the foundations of mathematics, owning such names as Pieri, Padoa, Vailati and others,—sought to characterize the class of finite integers. It is found in various editions of the *Formulaire de Mathématiques* (as 1899, 1901) and has been often quoted and critically discussed—especially by Bertrand Russell in his *Principles of Mathematics*, by Couturat in his *Les Principes des Mathématiques*, already cited, and again, very recently and illuminatingly, by Russell in his *Introduction to Mathematical Philosophy*—a work which no student of philosophy can afford to neglect.

*Independence of Postulates.*—You observe that in some of the foregoing titles the word "independent" occurs: What does it signify? It is used in a technical sense and means that the postulates of the system in question are such that none of them is a logical consequence of the rest. As students primarily of philosophy and therewith of epistomology, you should, I believe, be specially interested in the way in which postulates are tested for independence. What is the way? It is this: if we find a set of constants verifying all the postulates but one, then and only then we say that this one is independent of the others, for if it were not, it could be deduced from them but, in such case, it would be verified by their verifiers. If the system contain $n$ postulates, then, to test the entire system, it is obviously necessary to make $n$ partial tests—one for each postulate. If a collection of propositional functions is to be a postulate system, is it essential that the functions be independent? The answer is *no;* independence is desirable but not essential; it would be essential if the doctrinal function corresponding to the system were required to have a minimum of assumption and a maximum of deduction, and that is indeed a genuine ideal. Custom, however, does not require that the postulates of a system be independent; those of Peano's system, for example, are independent; but those of Hilbert's are not, as you can readily see by comparing the last one with the next to the last. You will note, however, that Hilbert did not neglect the question of independence. If you disregard postulate (21), his postulates compose five sets. He proved that those of any set are independent of each other and that each set is independent of the other sets.

*Categoricalness, or Sufficiency, or Completeness of a Postulate System.*—A postulate system is said to be categorical, or sufficient, or complete, when and only when it has a certain property to be stated presently. The first of the adjectives was introduced into the literature of the subject by Veblen, who received the suggestion from Professor John Dewey; the second one had been previously employed by Huntington, and the third by Hilbert. The property in question, which a postulate system may or may not have, is very interesting, sometimes important, and a bit subtle—not easy to make quite clear. The term "category," as you know, is a Greek word denoting a class, and we shall see that it has that meaning here. What is meant by categoricalness of a postulate system? Let me remind you that some of the undefined terms, or variables—say, $v_1, v_2, v_3, \ldots, v_n$—in the postulates of a system $S$ denote *elements*, or *substantives*, and the others—say, $R_1, R_2, R_3, \ldots, R_k$—denote *relations*, or *connections* (among the elements). Let me further remind you that, accordingly, a set of verifiers of $S$—a set of constants, or meanings, verifying $S$—is composed in part of *element*-constants—the values or meanings assigned to the $v$'s—and in part of *connection*-constants, or *relation*-constants—the values, or meanings, assigned to the $R$'s. Now, in *any* given set of verifiers of $S$, let $c_1, c_2, c_3, \ldots$ be the element-constants (representing the $v$'s), and $r_1, r_2, r_3, \ldots$, the relation-constants (representing the $R$'s); and in *any* other set of verifiers, let the element-constants be $c_1', c_2', c_3', \ldots$, and let the relation-constants be $r_1', r_2', r_3', \ldots$. If it be possible to set up a one-to-one reciprocal correspondence between the $c$'s and the $c_1''$s, the $c_2$'s and $c_2''$s, . . . , in such a way that, if two or more $c$'s be related by some $r$, the corresponding $c''$s are related by the corresponding $r'$, then and only then we say that the system $S$ is *categorical*, or sufficient. Two sets of verifiers that are transformable into each other in the manner indicated are said to be of the *same type*. It is easy to see the dictional propriety of the adjectives "categorical" and "sufficient" as thus used. For if $S$ be categorical, or sufficient, it determines a category, or class, of sets of verifiers, which sets are all of them of the same type, and it ($S$) is "sufficient" to do that. The Hilbert system is categorical, as are some of the other systems above listed. For an interesting discussion of the term categorical, of the advantages and disadvantages of categoricalness, together with detailed proofs that certain systems are, and certain others are not, categorical, I may refer you to the previously mentioned *Fundamental Concepts* of Young and to Huntington's article in the above-cited *Monographs*.

*Euclid's Postulate System Defective.*—In a previous lecture I stated that Euclid's postulate system is defective —mainly by omission—and promised to prove the fact at a later stage. Owing, however, to time limitations and to the insistence of many other topics which remain to be considered, I have decided to omit the proof and to be content with referring you to Young's *Fundamental Concepts* (pp. 12, 143) where the defectiveness in question is demonstrated simply and clearly.

## LECTURE IX

### Truth and the Critic's Art

MATHEMATICAL PHILOSOPHY IN THE RÔLE OF CRITIC—A WORLD UNCRITICISED, THE GARDEN OF THE DEVIL—"SUPERSIMIAN" WISDOM—AUTONOMOUS TRUTH AND AUTONOMOUS FALSEHOOD—OTHER VARIETIES OF TRUTH AND UNTRUTH—MATHEMATICS AS THE STUDY OF FATE AND FREEDOM—ITS PURE BRANCHES AS DOCTRINAL FUNCTIONS—ITS APPLIED BRANCHES AS DOCTRINES—THE PROTOTYPE OF REASONED DISCOURSE OFTEN DISGUISED AS IN THE DECLARATION OF INDEPENDENCE, THE CONSTITUTION OF THE UNITED STATES, THE ORIGIN OF SPECIES, THE SERMON ON THE MOUNT.

We have seen that a doctrinal function is composed of two sets of propositional functions: an assumed set,—fertile, compendent, compatible, sometimes independent, sometimes categorical,—called a system of postulates; and a set logically deducible from the postulates, and called theorems.

We have seen that an autonomous doctrine,—a doctrine derivable from a doctrinal function by replacing the variables in the postulates with admissible constants,—is composed of two sets of propositions: a set derived from the postulates—one for each postulate; and a set similarly matching the theorems.

We have seen that a doctrinal function is, like the propositional functions composing it, neither true nor false; and that a doctrine derived from it, is, like each of its component propositions, either true or else false.

We have seen that a doctrinal function gives rise to an infinitude of true doctrines,—values of the function,— and an infinitude of false ones.

We have seen that a doctrinal function, owing to the presence of the variables in its propositional functions, has no specific or definite, but only an ambiguous or undetermined, subject-matter; and that, on the other hand, a doctrine, owing to the presence of the "substituted" constants in its propositions, has a specific, or definite, kind of subject-matter.

We have seen that, in respect to *structure* or *form,* a doctrinal function and all of the derivable doctrines are identical, while, in respect to *content,* or subject-matter, no two of them are identical.

We have seen that, in the case of a doctrinal function, the theorems, (which are *forms*) are logically deducible from the postulates (which are *forms*)—the deduction being purely *formal;* and that, in the case of a derived doctrine, the propositions matching the theorems can not be logically deduced *as* propositions from the other propositions *as* propositions but only *as forms,* in which respect, however, the propositions and the corresponding propositional functions are, as we have seen, *identical;* so that, in any and all cases, it is the *form* of the premises, and never their subject-matter, that determines their logical consequences.

Hereupon, there supervenes an important critical question: Given a doctrinal function and one of the doctrines derivable from it, which of the two things ought to be called a branch of pure mathematics and which one a branch of applied mathematics? It is evidently not merely a question of taste, for the two things are not on the same level, they are not coordinate: the doctrinal function is a matrix, the doctrine is one of the things it moulds; the former *is* form, the latter *has* form, and subject-matter besides. In the light of the foregoing conspectus of their differences and similitudes, it is obvious, I think, what the answer must be: the doctrinal function is a branch (or part) of pure mathematics; the doctrine, a branch (or part) of applied mathematics—of what Lord Bacon called "mixed" mathematics, the mixture consisting in the mingling or union of form and subject-matter—of structure and something having it or conforming to it—of a prototype, model, mould, or pattern, and material owning the impress thereof. Are we, then, to say that the various kinds of geometric doctrine,—ordinary Euclidean metric geometry, for example,—and the various kinds of algebraic doctrine,—the algebra of the real numbers, for example,—are all of them so many branches of applied mathematics? From that conclusion there is, I believe, no escape. They are quite as *genuinely,* though not quite so *obviously,* applied mathematics as are, for example, rational mechanics, mathematical statistics, mathematical physics, and mathematical astronomy, for the things which geometries and algebras are doctrines about are just as genuinely, though less evidently, kinds of subject-matter (as distinct from pure form) as are the things which the other mentioned branches are doctrines about.

In the view thus presented, pure mathematics appears as a large (potentially infinite) ensemble of doctrinal functions and applied mathematics as the ensemble of doctrines derivable from them, having them for matrices, and owning their forms. The view, as you will readily see upon a little reflection, is recommended and confirmed by its harmony with many an insight similarly or otherwise gained and usually justified in other terms.

It accords, for example, with the splendid *mot* of Bertrand Russell that "mathematics is the science in which one never knows what one is talking about nor whether what one says is true"; for a doctrinal function, as we have said so often, has no determinate subject-matter and, without losing its integrity as a function, might conceivably be not even verifiable by any of the subject-matters in our world.

It accords with another just saying (before quoted) of the same author that "pure logic, and pure mathematics (which is the same thing), aims at being true, in Leibnizian phraseology, in all possible worlds, and not only in this higgledy-piggledy job-lot of a world in which chance has imprisoned us"; for the connection of the theorems of a doctrinal function with its postulates,—the logical *lien* binding the former to the latter as conclusions to premises indissolubly, forever,—depends in no manner or degree upon the content, the accidents, or the vicissitudes of the "big buzzing blooming confusion" which we call *our* universe.

It accords perfectly with the critical judgment, elsewhere[1] expressed, that "it is in implications and not in applications that (pure) mathematics has its lair"; for the very essence of a doctrinal function,—constituting of its elements a single indestructible Form of forms,—is that its postulates logically *imply* its theorems.

It accords with the often quoted definition of pure

[1] *Human Worth of Rigorous Thinking*, p. 303.

mathematics given by Benjamin Peirce as "the science which draws necessary conclusions"; for the theorems of a doctrinal function are necessary consequences of its postulates in the sense that the former just *are* the implicates of the latter.

It accords with the judgment of Pieri that pure mathematics is a "hypothetico-deductive" science; for the postulates of a doctrinal function appear in the rôle of *hypotheses* and the theorems in that of conclusions logically *deduced.*

It accords with the exquisite penetrating saying of William Benjamin Smith that pure mathematics is "the universal art apodictic"; for the logical validity of a propositional function as such is completely independent of any and all particular subject-matters, whether of our world or of any other that may be conceivable or possible, and the logical coherence of the theorems and postulates of such a function is apodictically certain.

It accords with the seemingly shallow but really profound saying of Henri Poincaré that mathematics is "the giving of the same name to different things"; for, despite the confusion thus arising, a doctrinal function and its various values are commonly given a single name, which is usually that of a specially important or familiar one of the values.

It accords well with the saying of an eminent jurist that "mathematics is the attempt to seize hold of God where the hair is shortest"; for the pure *forms* of thought present clean-shaven aspects—they are "bald as the bare mountain tops are bald, with a baldness that is sublime," and the discourse of a Gauss or a Lagrange is naturally less "woolly" than that of a Cicero or a Justinian or a Coke or a Montesquieu or a Blackstone.

It does *not,* however, accord,—and that, too, is confirmatory,—with such definitions,—no longer current among competent critics,—as held mathematics to be the science of "number" and "space" or the science of "quantity" or the science of "measurement" or the "science of indirect[1] measurement"; for, as you clearly see, a propositional function as such has no essential concern with such particulars as number or space or quantity or measurement direct or indirect.

But, as you see, it does accord with the often expressed view that mathematics is the science of *form* and with the view that it is the *normative* science par excellence.

Finally, it accords perfectly with the saying—reiterated many times and in may forms since the golden days of Plato—that mathematics contemplates Being under the aspect of Eternity; for it is perfectly clear that doctrinal functions, though their discovery by man is a temporal event, are themselves timeless—"older than the Sun or the Sky" and destined to survive all things that are under the law of change and the doom of death.

We have seen that one doctrinal function may, without losing its proper autonomy, be a part of another one, —one autonomous Form of forms being thus an integral constituent of another such Form more inclusive,—in which case any doctrine derivable from the former is similarly a part of the corresponding doctrine derivable from the latter. For example, $H \Delta F$ and its values are thus related, as we saw, to $H \Delta F'$ and its values. Whether all doctrinal functions,—both those that are known and those that remain to be discovered,—are somehow logic-

[1] Auguste Compte: *The Positive Philosophy* (translation by Harriet Martineau).

ally connected together as an immense hierarchical gang of subordinates to one supreme Function, or Form, which embraces the whole of pure mathematics, is a question most worthy of your best attention as students of philosophy. It has been answered affirmatively, as I said in the introductory lecture, by Whitehead and Russell in the *Principia.* I shall not here attempt to justify the answer but, for information regarding the manner of the answer and the evidence supporting it, I again refer you to that monumental work, which, as it is a composite of the most scientific philosophy and the most rigorous science, you will find a little harder to read than philosophical works of the usual rhetorical type.

The view I have been presenting, in which pure mathematics appears as a vast array of doctrinal functions, gives the science, from one point of view, a pretty severe aspect. For a doctrinal function is not only timeless, as said, and indestructible, but,—and the fact merits our most pensive meditation,—when once the principles, or postulates, are chosen, the die is cast—all else follows with a necessity, a compulsion, an inevitability that are absolute—we are at once subject to a destiny of consequences which no man nor any hero nor Zeus nor Yahweh nor any god can halt, annul or circumvent. Mathematics is, in a word, the study of Fate. Let me hasten to say that the Fate is not physical, it is spiritual—the unbreakable binding thread of destiny runs through the universum of rigorous Thought: the fate is logical Fate. Is it a tyrant? And the intellect, then, a slave? A tryant has whims but Logic is lawful. Where, then, is the intellect's freedom? What do you love? Poetry? Painting? Architecture? Statuary? Music? The muses are *their* fates. If you love them, you are free. Logic is the muse of *thought*. When I violate it, I am erratic; if I hate it, I am licentious or dissolute; if I love it, I am free —the highest blessing the austerest muse can give.

The remainder of this lecture consists of a brief discussion of a very much neglected subject of very great importance; and its importance is not only of the theoretical kind but, as I trust you will be able to see, of the most effectively practical kind also; practical, that is, for such as have the talent and training,—the gumption and discipline,—to employ effectively the most delicate and most powerful of intellectual instruments. I may call the subject

*The Rôle of Postulate Systems and Doctrinal Functions in the Structure and Criticism of Thought.*—Here, as generally in these lectures, I use the term Thought in a very comprehensive sense: not in a sense so inclusive as it has sometimes—in William James's *Principles of Psychology,* for example, where it often signifies or covers "mental states at large, irrespective of their kind"; but rather in the sense it has in Theodore Merz's great *History of European Thought in the Nineteenth Century* where the term embraces both what we ordinarily mean by "Science" and what by "Philosophy"; in other words, I am using the term Thought to signify that sort of discourse which deliberately owns *allegiance,* even though it often fails in *loyalty,* to the authority of Logic. The subject is, you see, immense, penetrating all the sciences and all the philosophies, natural, or social, or speculative—all fields, in short, where men have sought by means of *reasoned* discourse to gain or to give wisdom and light for the guidance of humankind. To treat it as it deserves to be treated,—both in full generality and in detail,—would require the writing of a large volume. Since every doctrinal function includes a postulate system as its logical base and since pure mathematics, as we have seen, consists of doctrinal functions, such a volume might be appropriately entitled: *The Rôle of Pure Mathematics in the Criticism of Thought;* or, better perhaps, *Pure Mathesis in the Rôle of Critic.* Possibly, one of you will one day undertake the production of such a work. The entire present course of lectures evidently bears upon the task but the bearing is, in the main, implicit. In what remains of the hour, I desire to discuss the subject, very sketchily indeed, but explicitly and in terms. And I will begin with a word regarding

*Autonomous Truths and Autonomous Falsehoods.*—We have repeatedly spoken of the logically organic body of propositional functions constituting a doctrinal function as being an *autonomous* form. We have done so, as you know, because the thing presents a certain aspect of self-sufficience or independence. If such a function be included in another one, it does not owe its existence, its unity or its integrity to that relation. It stands alone, erect, eternal, holding its principles, its base,—the postulates,—within itself, as it contains within itself the logical *lien* binding its elements into one solitary, self-sufficing, indestructible whole. A doctrine derived from it (in the way now familiar) is not so pure as the function whence it was derived; it is, so to speak, the doctrinal function *dipt*—dipt or immersed in subject-matter, in a kind of material giving each of the propositional functions *significance,* each of them *thus loaded* being a proposition and, as such, true or false; hence, we cannot say that the doctrine *is* form but, as we have seen, it *has* form, and the form it *has* is precisely that which the doctrinal function *is;* and so we say that such a doctrine, too, is autonomous. A doctrine is the off-spring of a marriage—the marriage of subject-matter and pure form; the latter is the mother and transmits its own autonomy to all its children. If the doctrine be true, we may call it an autonomous truth,—the most beautiful and most precious thing in the world,—for it has the doctrinal function's beauty of form; it has the beauty of truth; and is, besides, tinged with the warmth and living colors of some species of subject-matter in which our practical life is immersed and finds its interests and its sustenance; if, on the other hand, the doctrine be untrue, then it is not a falsehood merely, but is an autonomous falsehood; this is indeed not a precious thing, it is the very opposite; and yet, strange to say,—for so pervasive is beauty in our world,—an autonomous falsehood, despite its having the ugliness of untruth, has all the beauty of perfect form—the form of the doctrinal function whence it was derived.

An autonomous falsehood's perfection of form is both a great advantage, and a minor disadvantage, in the quest of truth, for it makes it in one respect much easier, and in one respect somewhat harder, to detect the falsity. It makes it easier, for, as you know, an autonomous doctrine consists of a set of propositions [$p$], derived from the doctrinal function's postulates, and a set [$p'$], derived from the function's theorems; and hence two ways,—a *direct* way and an *indirect* way,—are open in which to try whether the postulates are satisfied: the direct way, by comparing the known facts in the field of the doctrine's subject-matter with [$p$]; the indirect way, by comparing them with [$p'$]. It makes it harder, for formal perfection is in itself a thing so impressive, so fascinating, so pleasing, that it tends to camouflage a defect of content and thus to deceive by a kind of agreeable dazzling of the mind. The disadvantage in question, though always inferior to the mentioned advantage, is naturally more serious in cases where the question of postulate verification is especially difficult to answer with perfect certitude. Such cases are not merely supposable; they are in fact of very frequent occurrence in the history of science. Just at present, we have indeed a living illustration in the world-wide discussion of relativity theories, wherein the satisfiedness, or verifiedness, of certain famous postulates (or deductions therefrom)—once regarded as established, long so regarded in the case of some of them—has been called in question and is now held in doubt or denied. For a presentation of the great matter of these theories, I have real pleasure in referring you to C. D. Broad's article, "Euclid, Newton, and Einstein," in *The Hilbert Journal, Vol. XVIII., April, 1920;* the article, which is easily the best I have seen on the subject, is quite notable as a sound, intelligible, semi-popular exposition of an exceedingly recondite scientific development.[1] The art of such exposition, let me say in passing, is difficult and important—quite as difficult and, in its service, quite as important as research itself; a high degree of skill in it is, I think, not less rare than a high degree of research ability; once in a great while, the two things are united in one personality, as in W. K. Clifford, for example, in Thomas Huxley, in Helmholz and Ernst Mach, but not in Henri Poincaré who, though he repeatedly essayed the task of popular exposition and indeed produced many a lightning flash

[1] Since this was written many attempts have been made to explain the doctrine in popular terms. Among the best attempts may be mentioned Bolton's *Introduction to the Theory of Relativity* [E. P. Dutton & Co.] and W. B. Smith's article, *Relativity and Its Philosophic Implications* [Monist, Dec., 1921].

in the layman's sky, yet lacked the requisite patience for continuous clarity.

*Heteronomous, or Anautonomous, Doctrines, True and False.*—Everyone knows that each of the great subjects belonging to the domain of Thought has a more or less reasoned literature,—often an immense literature,—of its own. Everyone knows that any such literature,—the literature of any such subject,—is composed of a number of more or less logically organized bodies of propositions. It is common and convenient, as everyone knows, to speak of such a body of propositions,—no matter what the subject,—as a theory or a philosophy or a science or a doctrine. Let us here employ the term last mentioned. Everyone knows that in every great subject such doctrines are not only numerous but that, by modification of old ones and addition of new ones, the number is constantly increasing. Together they constitute our more or less reasoned wisdom,—what Clarence Day would call our *supersimian* wisdom,—about the world.

I desire to draw your attention to the fairly obvious fact that most doctrines,—the vast majority of doctrines whether true or false,—are *not* autonomous. Autonomy,—the quality of being autonomous,—is an ideal; it is an ideal to which doctrines in every subject, or the builders of them, do indeed more or less consciously aspire and to which they slowly, for the most part very slowly, approximate but which they seldom even nearly attain. When a doctrine does reach (or *nearly* reach, for it can not *quite* reach) the ideal, when it attains close approximation to autonomy, then and, strictly speaking, only then it has become mathematical; the immense majority of doctrines are, then, non-mathematical, lacking autonomy. They are heteronomous, or *anautonomous,* doctrines. These are, I grant you, terrifying words. Why not say mathematical and non-mathematical, and have done with it? Because the other words serve to direct and fix attention upon what is precisely characteristic of mathematical doctrine, on the one hand, and of the non-mathematical, on the other.

Why is it that nearly all doctrines in the world, even those which deal with the most familiar subjects, not excluding some doctrines that are currently called mathematical, have been and are *an*autonomous? The causes are evidently many. There is the general feebleness, the logical meagreness, of the human intellect; there are the strong unruly passions of men driving them in uncharted courses as rudderless vessels in a storm; there are their lusts and greeds aiming at the gratification of propensities infinitely beneath and commonly hostile to the craving for truth; there are laziness, fickleness, and impatience; there is the marvelous copiousness and prodigality of mother Nature enabling her children to get on somehow even though they have but meagre care for wisdom; and, finally, there is the inherent intractableness of the great subject-matters with which most doctrines deal.

Hence a rough general answer to our question evidently is that the building of an autonomous doctrine regarding any great matter is, for us humans, constituted and circumstanced as we are, exceedingly difficult, while the making of the other kind is *easy*: there are so many, many ways in which a doctrine may fail of autonomy—so many possibilities, so many opportunities, so many solicitations from within ourselves and from without, for going wrong in the business and incurring delay. Do but reflect a little upon the matter. An autonomous doctrine, we have seen, is one derivable from a doctrinal function and inheriting its form. Today indeed we are familiar with the general conception of such functions and have numerous examples of it; and we are in some danger of inferring or supposing that the formation of that conception and the discovery of the known functions exemplifying it have been accomplished easily. But such an inference or supposition would be very erroneous. Those whom we conventionally call the authors of the known doctrinal functions are not, strictly speaking, their discoverers. Far from it. The discovery of them and of the general concept they exemplify is not the achievement of an individual but of a very, very long series of individuals; it is, like all other forms of wealth, like all other elements of civilization, a *racial* achievement—the slowly accumulated fruit of many generations of dead men's toil. And clear consciousness of the outcome,—of the fact and nature of the fruit,—is of very recent date. To realize vividly that such is the case, you need only reflect that doctrinal functions are composed of propositional functions and that, as we saw in a previous lecture, the supremely important notion of propositional function came to recognition and received a name only a few years ago. Compared with the vast backward stretch of human time,—say, a quarter or a half million years,—the interval from Euclid's day to ours is indeed very short; virtually we are among Euclid's contemporaries; yesterday he was here; yet his *Elements* is our human race's earliest example of a doctrinal function and even it is an imperfect example, failing, as we have seen, to state certain of the postulates explicitly, and being in appearance, as he probably conceived it to be in fact, a specific doctrine instead of a doctrinal function.

But the sheer difficulty of attaining doctrinal autonomy, great as the difficulty is, is by no means solely responsible for the fact that so few of the doctrines in the world are autonomous and that nearly all the rest of them are very remote from that estate. Part, a very large part, of the explanation is found in the fact,—abundantly manifest in the history of thought and for most of us strongly confirmed, I fear, by our introspective knowledge of ourselves,—that we humans in our doctrinal constructions and preachments are but seldom *much* concerned to make them even approximately perfect in respect to logical *form;* in their *content* our interest is, in general, far greater; but even as to content, we have, I think, to own that we are, in general, much less concerned to have our doctrines *ultimately true* than to have them *instantly effective.* I trust I am not sufficiently depraved to believe in the total depravity of man; for many of his supersimian traits and for some of his simian qualities, I have profound admiration; but in candor we must own, I believe, that *wholly disinterested* pursuit of truth is very rare. We humans desire indeed to be regarded devoted lovers of truth and we flatter ourselves that we are such in fact; sometimes we are, but, in general, we are not; in general, we prefer something else; we often boast that we are not theoreticians, and the boast has its basis in fact; we are not theoreticians, we are practicians, though we dislike the word and call ourselves practical instead. Being primarily and predominately practical in our interests, when we are building doctrines, though we always pretend to be thus endeavoring to set forth truth, we are, with rare exceptions, animated by a very different motive; we are not trying to formulate something that, by painstaking research, we have found to be true; we are, instead, though we do not confess and may not even know our real motive, trying to make an instrument that will "work," that will be an *effective* means to some practical end; we are not,—however much we pretend to be,—endeavoring to *enlighten* our fellow men—we are endeavoring to *influence* them: our aim is not the advancement of wisdom; it is, in current slang, to put something over or across. And, as already intimated, the dominance of this motive is entirely consistent with sincerity. The builder or the advocate of a doctrine ostensibly aiming at truth but really aiming at some practical achievement, may be entirely sincere—he may indeed be, like Mahomet, for example, like Deacon Paris or Lenin, a *fanatic,* incapable of doubt, incapable (that is) of doubting the validity or justice of his central thesis, and hence incapable of *scientific* devotion to truth.

Now, it is evident that one making or advocating a doctrine, if he be animated, not by the genuine philosopher's love of truth, but by the spirit of the partisan and propagandist, if he have not the disinterestedness of the genuinely scientific worker but have instead the interest of one bent on driving through to the goal of some practical purpose by any and every available means thereto—it is evident, I say, that such a one will *not desire* to bring his doctrine to the perfection of logical form but will often indeed desire the very opposite; and the reasons are plain: to make a doctrine autonomous requires much patience and time, but the practician, the partisan, the propagandist, is by nature impatient—he is eager for results; in trying to make a doctrine autonomous, we usually discover that the doctrine is false (for most doctrines are false), but such a discovery, which tends to dampen ardor, is just what your partisan or your fanatic most desires to avoid; an autonomous doctrine, because its elements are arranged in order and their logical connections are bared, is thereby prepared for relatively easy examination by others, so that, if the doctrine be false, the fact is specially liable to detection, but it is not the aim of your propagandist to make such detection easy; a doctrine, once it is made autonomous, though it has thus gained in *light*, has lost its *heat*, it is lacking in punch, as we say, or "pep," it is prepared for the service of mere enlightenment; but your propagandist, your fanatic and your partisan do not seek to enlighten, they seek to influence,—to get action,—and so they keep their doctrines amorphous, malleable, and charged with emotion for emotional utterance and emotional effect.

Well, you may say, what is to be done about it? What is the remedy? The remedy is—Criticism—the Gadfly: patient, unsparing logical criticism of one's own work in doctrine building; and, in all subjects, keen, merciless, stinging, gadfly criticism of any and all half-baked, logically amorphous, flabby doctrines pretending to be important embodiments of truth or vessels of wisdom. Men must be driven by art,—the art of criticism,—to levels of excellence higher than those to which they are drawn by unenlightened nature.

I am, I hope, not misunderstood in this matter. I am far from contending,—no one can be so foolish as to contend,—that in *every* field of thought workers can be constrained by criticism to put their results in the logically perfect form of an autonomous doctrine; man can not be constrained to perform the impossible nor to do instantly what has at best a very remote possibility of being done at all: what I do contend is that in *all* departments of thought men can be constrained by criticism to have constant regard to the principles and the spirit of mathematics—the spirit of dispassionate thought—to estimate the logical cogency of their thinking in accordance with mathematical standards, to employ the postulational method in many instances where it has never been employed nor even attempted, to hold it as a model in all cases, and in all their work to own the *authority,* even though they can not attain the *perfection,* of the doctrinal function as the highest and purest of logical ideals; and I contend that, if this were done, both the *logical* quality and the *truth*-quality of what I have called "the more or less reasoned literature" of the world would be thereby rapidly, constantly and immeasurably improved.

Finally, I would direct your best attention to the fact that everywhere in that literature,—the literature of Thought,—there are to be found certain phenomena, certain common characters, which invite us to the indicated type of criticism as to a great and hopeful enterprise. What I mean is this: if you will select any well-known doctrine, no matter how amorphous, belonging to any field, no matter how remote or seemingly remote from mathematics—it may be in natural science or in philosophy or in theology or in ethics or in law or in education or in politics or in economics or in history or in sociology or in education—if I say, you will select from any such field a doctrine worthy of attention and examine it, you will find that the author has more or less consciously recognized, in at least some small measure, the necessity of working with *principles* which he may not have explicitly stated as such either in whole or even in part; you will find that he has consciously or unconsciously made use of certain (or uncertain) *primitive* propositions or propositional functions,—certain assumptions, that is, or postulates,—which he may or may not have regarded, and may or may not have recorded, as such; you will probably find that he has tried to define certain of his terms and will certainly find that other terms ostensibly defined or not, have in fact been virtually employed, deliberately so or not, as *undefined* terms (primitives, or variables); you will find that he has stated a series of propositions which he has made some effort to prove, to demonstrate, to deduce, by a process of reasoning, from something or other; in a word, you will find that within the doctrine, however formless, however ill ordered its parts, however loosely knit its texture, there is shadowed forth more or less clearly, very dimly it may be, something of the figure and frame of the logical prototype called doctrinal function, as if this thing were so built into the very constitution of intellect as in some measure to guide and shape its activity whether we will or no.

It would amply compensate us for the toil involved, had we the time for it, to devote one or more lectures of this course to illustrating the truth of what I have just said by critically examining one or more outstanding doctrines of the non-mathematical, or anautonomous, sort with a view to discovering the presence in them of the mentioned phenomena. But, except for a few hints to be presently given, I must leave the task for you, commending it as being in my judgment the best possible discipline in the great art of doctrinal criticism, for which the present condition of the world calls more loudly than ever before and which it is your supreme privilege and supreme duty as philosophers to master, foster, and practise.

You have the clue and the material abounds on every hand. "I do not frame hypotheses" (*Hypotheses non fingo*) said Newton and he accordingly called his principles of dynamics "*Axioms or Laws of Motion*" (*Axiomata sive leges motus*). Today, however, even we, who are hardly Newtons, know that his "axioms" are not absolute certitudes, that they are not self-evident propositions, that they are indeed not perfectly clear; we know that they *are* "hypotheses," pure assumptions, postulates, genuine propositional functions in which the *variables* are, in Euclidean fashion, denoted by *names,* which names or some of them are, again in Euclidean fashion, "defined"—defined by definitions, or descriptions, serving merely to indicate one specific interpretation of the functions. For a good approximation to the sort of criticism I am recommending, let me refer you to an examination of Newton's doctrine of dynamics by the late Ernst Mach in his masterful *Science of Mechanics.* Do you wish to say that *this* doctrine is mathematical? Very well, it is mathematical but it is not purely such and I have cited it partly on that account and partly because of its fame. Let us, however, take a glance in other directions. Consider, for example, that most significant of all American political documents—*The Declaration of Independence.* It is, or contains, in epitome, a political doctrine of the highest importance. In saying, "We hold these truths to be self-evident," its authors virtually said, *We lay down the following postulates;* and the list they give of "self-evident truths" is clearly a list of their political postulates. These are propositional functions; a little scrutiny will enable you to detect the undefined, or variable, terms, which the authors of course assumed would be understood in some specific sense. The postulates, you observe, are swiftly followed by important *deductions.* I can not here further elaborate the matter, but you can not fail to detect in it the outlines or rudimentary presence of a singularly impressive doctrinal function of political type and to feel invited to examine the great document and perhaps to elaborate it in accordance with the conception and method of such functions.

For another example, consider the *Constitution of the United States.* It may be regarded in the same light, only to do so requires a little more ingenuity. Omit, "We the people of the United States do ordain and establish this Constitution for the United States of America"; what is left embodies a doctrine—a doctrine in the field of government. What are its postulates? Everything from article I to the end of the document—of course, the provisions are not stated in the *manner* of postulates but they can be so stated. Where are the theorems? These are not stated at all but are *involved* in the meanings of the great phrases respecting justice, tranquillity, and so on of the heavily laden preamble. I wonder if what I have said is a sufficient hint. The doctrine in question is, in a word, this: the provisions in the Constitution,—that is, the postulates,—*imply* the body of unstated propositions involved in the great terms of the preamble. Such, in a nut-shell, was the thought, the doctrine, of the fathers. Let me offer a similar hint, a mere hint, regarding the philosophy of Descartes. You know how strenuously he sought for a basis of indubitable fact. He was seeking, though he did not so conceive the task, for the indubitable verification of certain propositional functions, which he did not indeed formulate nor evoke from the shadowy background of his thought. One of the verifications he found or thought he found is, as you know, of world wide and immortal fame. Consider the propositional function: if $x$ performs a kind $y$ of activity, then $x$ has the property $z$. Let $x$ denote *I,* let $y$ denote *thinking,* let $z$ denote *being,* drop the hypothetical form, and you have Descartes's *Cogito, ergo sum; Je pense, donc je suis; I think, therefore I am.* Enough of hints. The suggested type of analysis is evidently applicable on every hand—to the *Sermon on the Mount,* to the *Republic* of Plato, to Darwin's *Origin of Species,* to the *League of Nations Covenant,* to Marxian *Socialism,* to the *Soviet Constitution of Russia,* to the *Constitution of the German Republic,* to the Einstein *Doctrine of Relativity,* to the Bryanistic *Ethics of Prohibition*—to all manner of doctrinistic contentions of wise men, knaves, fanatics and fools.

The type of criticism I am here advocating and urging as supremely important shapes itself, as you see, very simply. Confronted by a doctrine in any department of thought, Criticism demands answers to these questions: What is *assumed*—what are the *postulates?* What are the *undefined,* or *variable,* terms? What are the *theorems* or proved propositions and what the *defined,* or *constant,* terms? *How* have the theorems been *deduced,* and the defined terms *defined?* What *meanings* have been *assigned* to the *variable* terms, and *how?* Upon these questions, criticism, if it is to be criticism of Thought, is bound to insist—there is no alternative. Such criticism is a *civilizing* agency—the guardian of the principles of freedom. Without it, the world becomes a wilderness of error and lust—the garden of the Devil.

Easy to ask, the questions are, in general, not easy to answer, and the difficulty of answering rightly is usually greatest just where it is most important to compel an answer—in the case, that is, of amorphous, emotion-charged "dynamic" doctrines that pretend to aim at enlightenment but really aim at victory and win it by appealing, not to love of truth, but to lust for power or gain. If the author be unable to answer, criticism must drive him back to the silence of the cloister for further study. If he contend, as sometimes he will contend, that he has defined all his terms and proved all his propositions, then either he is a performer of logical miracles or he is an ass; and, as you know, logical miracles are impossible.

Allow me, in closing, an additional word to guard against a possible misapprehension. We have seen and said that a doctrine, in becoming autonomous, though it thus gains in light, loses in heat—it tends to become static. Is it not true, however, that to do its work in the world a doctrine must be dynamic? The answer is, it must; and nothing has been said to the contrary. It is necessary to distinguish: to test for truth is one thing; to utter is another. Of these two things, the former is the duty to which men must be driven by criticism if they be not drawn to it by love of truth. Once the test is made and the doctrine found not wanting, then, and not before, it may be legitimately urged home with full ardor by all the arts of utterance even though the truth be thus made to burst upon us like the thunder of Wagnerian music, making the mountains tremble, the seas vibrate, and seeming to shake the very rafters of the sky.

## LECTURE X

### Transformation

NATURE OF MATHEMATICAL TRANSFORMATION—NO TRANSFORMATION, NO THINKING—TRANSFORMATION LAW ESSENTIALLY PSYCHOLOGICAL—RELATION AND FUNCTION AND TRANSFORMATION AS THREE ASPECTS OF ONE THING—ITS STUDY THE COMMON ENTERPRISE OF SCIENCE—THE ART OF MATHEMATICAL RHETORIC—THE STATIC AND THE DYNAMIC WORLDS—THE PROBLEM OF TIME AND KINDRED PROBLEMS—IMPORTATION OF TIME AND SUPPRESSION OF TIME AS THE CLASSIC DEVICES OF SCIENCES

LOOKING back to the days of my youth, I see pretty clearly and a little sadly that in this good land of ours secondary and collegiate mathematical instruction was, with little exception, then remarkable for two things: its emphasis and its silence. It was very diligent and very emphatic about small matters; about great ones it was dumb.[1] I am led to this reflection by recalling my first and second introductions to the term, "transformation." The first was in algebra; there was a chapter on the cubic and the biquadratic equations, which we were to learn

[1] See in this connection Professor J. C. Fields' brilliant address—*Universities, Research and Brain Waste*—published by the University of Toronto Press, 1920.

how to solve; in the chapter was a section headed, *Transformation of Equations;* in it the $x$ of our equations was replaced by something else; so we got new equations; these were managed so that, after a few "stunts," we had the roots sought; just what part of the proceeding was dubbed transformation and what not, we were left to conjecture, but, in those days, even conjecture,—guessing, —was a kind of sin, for mathematics was "the exact science," it was just "pure reasoning." I wonder if you, who are of a later generation, were more fortunate. Well, my second introduction to the term in question occurred in analytical geometry, for the book contained, over toward the middle of it, an arid little chapter entitled *Transformation of Coordinates*—a meagre, dull, stupid, stupefying parched little desert discussion, of which no use was made there and very little in the subsequent part of the course. In neither of the "introductions" was there an illuminating word by text-book or by instructor to signalise the significance of the matter in hand—no insight, no outlook, apparently no sense of being in the presence of a great matter, at once a powerful instrument and a subject of first-rate importance. What I have said of the term "transformation" might be said with equal truth of other great terms,—of Function, for example, which is at length happily winning its way to due recognition in elementary instruction,—of Invariant and Group, of which I hope to speak at a later stage—and especially of Relation, which, though long current in mathematical literature as a convenient term used in a sense semi-scientific (or semi-technical) and semi-literary, is at length coming to be recognized, owing to recent work in the logical foundations of mathematics, as denoting better than any other term the ultimate tissue of mathematical science. For, while this science is, as we have seen, composed of doctrinal functions, these forms are themselves woven of abstract, or formal, relations.

In this lecture, I purpose to deal with the mathematical *idea* denoted by the term "transformation." I need not say that an hour's lecture can neither impart much knowledge of transformation theory nor give skill in the use of transformations as instruments; for the former requires prolonged study and the latter is the slow-maturing fruit of practice. What I hope to do in the hour is to make clear what mathematicians mean by a transformation and by a law of transformation; to show how fundamental and omnipresent the process of transformation is in *all* our thinking; to give an inkling of the endless number and endless variety of existing transformations; to show how transformations appear now as powerful tools and now as interesting themes; to disclose the intimate connection of the notion of transformation with that of relation and that of function; and briefly to indicate how the phenomena of transformation in the *abstract static* world of mathematics correspond to the phenomena of change in the *concrete dynamic* world of sense.

Let me begin as simply as I can—so simply, indeed, as possibly to suggest that transformation is a trivial term, which it is infinitely far from being. The notion of transformation has its *root* in the power we have, when given any two objects of thought, to *associate* either of them with the other. If $a$ and $a'$ be two such objects, we can, in thought, associate: (1) $a$ with $a'(a \rightarrow a')$; or, conversely, (2) $a'$ with $a(a \leftarrow a')$; or (3) each with the other $(a \longleftrightarrow a')$. If we do (1), we say we have *trans-*

*formed* (or converted) $a$ into $a'$ and that $a'$ is the *transform* of $a$; if we do (2), we transform $a'$ into $a$, which is then the transform of $a'$; if we do (3), we transform each into the other and each is the other's transform.

Let us take another step: suppose $C$ is a class composed of $a$, $b$, $c$, and that $C'$ is a class composed of $a'$, $b'$, $c'$. We may transform in the manner,—according to the *law*,—shown in the table (1): $(a \rightarrow a')$, $(b \rightarrow b')$, $(c \rightarrow c')$; or by the law (2): $(a \rightarrow a')$, $b \rightarrow c')$, $(c \rightarrow b')$; or by the law (3): $(a \rightarrow b')$, $(b \rightarrow a')$, $(c \rightarrow c')$; or by the law (4): $(a \rightarrow b')$, $(b \rightarrow c')$, $(c \rightarrow a')$; or by the law (5): $(a \rightarrow c')$, $(b \rightarrow a')$, $(c \rightarrow b')$; or by the law (6): $(a \rightarrow c')$, $(b \rightarrow c')$, $(c \rightarrow a')$. In each case we have, we say, transformed the *class* $C$ into the *class* $C'$, and $C'$, we say, is the transform of $C'$. You see, too, that we could conversely transform $C'$ into $C$ in six corresponding ways—by any one of six corresponding laws, which we could express by tables as above. You see, too, that we could transform the classes $C$ and $C'$ each into the other, now by one law, now by another, as, for example by the law (7): $(a \longleftrightarrow a')$, $(b \longleftrightarrow b')$, $(c \longleftrightarrow c')$; or by (8): $(a \longleftrightarrow b')$, $(b \longleftrightarrow a')$, $(c \longleftrightarrow c')$; and so on. Notice that in each of the transformations (1), . . . , (6), each element of $C$ has only *one* transform and that each element of $C'$ is a transform; each of those transformations is a *one*-to-*one* transformation, and has a direction, or *sense*, namely, from $C$ to $C'$; not from $C'$ to $C$; observe that in (7) or (8) the transformation is again one-to-one but runs *both* ways. In each of the cases considered, one of the classes is transformed into the *whole* of the other, not merely into a part of it. But such need not be so; we can transform two or more elements of $C$ into a same one of $C'$ or one of $C$ into more than one of $C'$ and thus transform $C$ into a *part*, or sub-class, of $C'$, as by the law (9): $(a \rightarrow a')$, $(b \rightarrow a')$, $(c \rightarrow b')$, or into the *whole* of it, as by the law (10): $(a \rightarrow a')$, $(b \rightarrow a')$, $(c \rightarrow b')$, $(c \rightarrow c')$. You will note that neither (9) nor (10) is a one-to-one transformation. And by a little work, or a little play, you will readily discover that the possibilities of transformation increase or decrease with the sizes of the classes and vary with other circumstances such as whether the classes are of the same size (containing the same number of elements), whether the classes have elements in common,—overlap or intersect, as we say,—possibly coincide, and especially whether the classes are *finite* (like those we have considered) or *infinite* (like the class, for example, of all the integers, 1, 2, 3, . . . and so on *endlessly*).

Let us think a bit about the effect of such circumstances.

Suppose $C'$ has more elements than $C$ has; it is plain that no one-to-one transformation of $C$ into $C'$ can cover the whole of $C'$, but it can be covered by transformations that are not one-to-one.

Suppose $a$ of $C$ is the same as $a$ of $C'$,—$C$ and $C'$ thus intersecting; then in the above transformations (1), . . . , (6), $C$ itself contains the transform of one of its own elements, $a$ being the transform of $a$ in (1) and (2), of $b$ in (3) and (5) and of $c$ in (4) and (6).

Suppose $C$ and $C'$ coincide,—the elements of either being those of the other,—an important special case of intersection; then, you see, each of the transformations (1), . . . , (8) converts (transforms) the class $C$ into *itself*.

Next suppose $C$ to be composed of *all* the integers 1, 2, 3, . . . $n$, . . . ; and suppose $C'$ to be composed of all the *even* integers 2, 4, 6 . . . , $2n$, . . . ; you notice that $C'$ is a *part*, or subclass, of $C$; now let us transform $C$ by the simple *law* which requires us to *associate*, in thought, *each element of C with its double;* symbolically expressed the law is: $(1 \rightarrow 2)$, $(2 \rightarrow 4)$, $(3 \rightarrow 6)$, . . . , $(n \rightarrow 2n)$, . . . ; you observe that the class $C$ is thus transformed into the class $C'$ by a one-to-one transformation; now, $C'$ is, as said, a part of $C$, so that $C$ has been, you see, transformed, by a one-to-one transformation, into a *part* of itself; and that is remarkable, for it can not, as you know, be done with just any class—it can not, for example, be done with a class composed of one thing, or two, or three, or a dozen, or a dozen million. If a class be such that it *can* be thus transformed into some part of itself, it is said to be *infinite*—an *infinite class.* The concept denoted by the term, " infinite class," is one of the most important of our modern mathematical and philosophical concepts. In noticing that it is *defined* by means of transformation, you get a glimpse of the latter's fundamental importance.

May I here relate a bit of relevant personal experience? Some years ago a student of philosophy and I undertook to read together a pioneer work in the mathematical doctrine of infinity—the *Paradoxien des Unendlichen* by Bernhard Bolzano, mathematician, philosopher and theologian. We came to a passage where Bolzano shows that the *class* of points composing a straight line segment is an infinite class. He does it very simply and very clearly by showing, about as follows, that the segment

$0 \qquad \tfrac{1}{2} \qquad 1$

FIG. 20.

can be transformed, in a *point*-to-*point* fashion, into a part of itself. Let $y = \frac{1}{2}x$; use the equation as a *law* of transformation converting a point whose distance from

*O* is *x* into the point whose distance is *y*, or $\frac{1}{2}x$; the point *O*, its distance being zero, is transformed into itself, for the half of zero is zero; the point 1 is transformed into the point $\frac{1}{2}$; and the points *between* *O* and 1 have for their transforms the points *between* *O* and $\frac{1}{2}$; nothing, as you see, could be clearer; yet our plan of joint reading had to be here abandoned, for my fellow student, prophet of philosophy, would not follow Bolzano's reasoning and remained invincible to the bitter end. Can you beat that?

For a simple example illustrating both the concept of transformation and that of infinite class *geometrically*, consider Fig. 21. We are going to transform the class composed of the points of segment *AD* into the *subclass* composed of the points of segment *A''D''*. Let all the points of *AD* be joined to *P*; any such join, say, *PB*, contains a point, *B'* of *A'D'*; associate *B* with *B'*; in this way segment *AD* is transformed, point for point, into segment *A'D'*; next join *Q* to all the points of *A'B'*; any such join, say, *QB'*, contains a point, *B''* of *A''D''*; associate *B'* with *B''*; in this way segment *A'D'* is transformed, point for point, into segment *A''D''*; now observe that, starting with any point *B* of *AD*, the first transformation gives us *B'* of *A'D'*, and the second leads to *B''* of *A''D''*; finally, associate each such *initial* point *B* with the *final* point *B''*; the result, as you see, is a point-to-point transformation of the entire segment *AD* into one of its parts, the segment *A''D''*; and so we see, in the light of the last transformation, that segment *AD* is an *infinite* class of points; the same is, of course, true of any other segment, however short, for, in the foregoing argument, *AD* is any segment you please. In passing, we may notice also that we can choose *Q* so that segment *A''D''* shall be any part we please of *AD*, and we have the

*Theorem:* that *any segment, however long, can be converted, by a point-to-point transformation, into any one of its parts, however short.*

Before further illustrating the concept of mathematical transformation, let us ask what we mean by the *law* of such a transformation. The answer is pretty evident. It is that the law of a transformation is any rule, formula, scheme or device which, given any one of the elements or

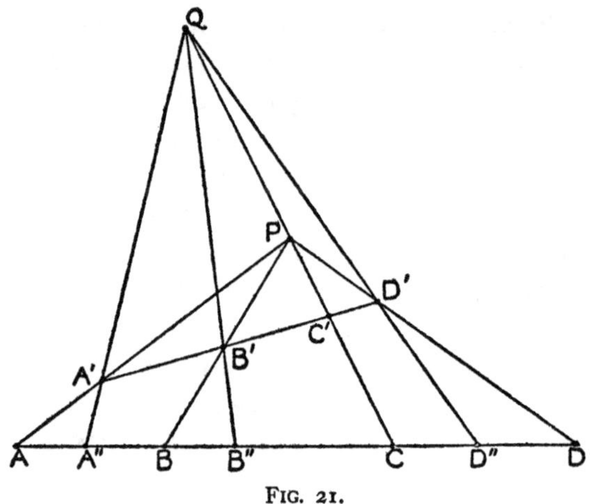


FIG. 21.

objects we are dealing with, determines its transform (or transforms, if it have more than one). It is, you should note, a *psychological* affair, the law being a device for guiding the transfer of attention from a given object to a definite other object (or objects). Such a law may be variously expressed: if the class to be transformed be small, it is practicable to express the law by tabulation, as by the foregoing tables (1), (2), . . . , (10); this is theoretically possible for *any finite* class, but is impracticable if the class, though finite, be very large; to express a law of transformation for an *infinite* class, by tabulation, is not even theoretically possible, but in such cases we must use an *incomplete* table or ordinary speech, as in the foregoing example where each integer is associated with (transformed into) its double and as in the example of Fig. 21; or, finally, we must express the law, as in Bolzano's example, by means of an equation or system of equations. Of all methods of expression, the equational method is the most common, and is usually the most satisfactory when it is possible, which it is sometimes not.

We have seen that the meaning of mathematical transformation has its root in the power we have to associate any idea or thing with any other, however like or unlike the former. As this power is fundamental and is continually exercised by *all* human beings in *every* kind of matter, we find, as we should expect to find, not only that mathematical transformation pervades mathematical thinking, but that such transformation is only a refinement of a process present in *all* our human thinking: a fact clearly illustrating the general truth that mathematical activity, instead of being remote from common life, merely consists in doing, with a peculiar finesse and ideality, what all human beings, when they think about the ordinary affairs of life and the world, are doing in a fashion relatively rough and crude. An ordinary dictionary, for example, is a good illustration of a kind of transformation that would be genuinely mathematical were it more precise; for, by definition, the *class* of *words* is transformed into the *class* of verbal *meanings*, and, conversely, the latter class is transformed into the former; the transformation runs both ways, but it is not one-to-one, since a given word commonly has two or more meanings and to a given meaning may correspond two or more words (synonyms). A telephone directory is a similar example, more nearly mathematical than the other one. Indeed, you will find, by a little looking about, that such illustrations abound on every hand. A perfect example of a genuinely mathematical one-to-one transformation running both ways is afforded by the vulgar *process of counting the objects of a class*, a class of count-words (one, two, . . .) being transformed, in a certain order, into the class (of objects to be counted), and conversely.

A moment ago I told you that, when a boy, my second puny introduction to the concept of mathematical transformation occurred in a dry little chapter (called transformation of coordinates) located near the middle of a beginners' course in analytical geometry. You will recall that in a previous lecture of the present course I gave a very brief introduction to the analytical geometry of the plane. As you will recall, it was shown how to transform (in one-to-one fashion) the *class* of the plane's *points* into the *class* of the real number *pairs* $(x, y)$, and, conversely, the latter class into the former; we saw that the former transformation gives birth to the *method* of analytical geometry; and the converse transformation, to the converse method—that of geometric analysis. We may think of the two transformations as one having two converse aspects, and, following usage, may speak of the two methods as one—called analytical geometry. Now observe carefully that the analytical geometry (of the plane), instead of merely using transformations among its processes, actually *springs* out of—owes its very existence to—a transformation, that of points into number pairs and of such pairs into points. Nay, the whole of analytical geometry viewed (properly) as a method is simply a vast transformation based upon the one just stated. To envisage the matter in a large way, conceive two immense canvases suspended parallel to each other each of them bisecting the universe of space. Imagine yourself comfortably seated between them; fancy that on the face of one of them are marked and drawn the points of a plane, its point loci,—curves and sets thereof limitless in number and variety,—and that on the face of the other canvas are recorded the real number pairs, pair systems—$(x, y)$-equations and sets thereof, more numerous than the sands of the sea; choose a unit of length and in the former face a pair of axes. What happens? You behold the phenomena on either sheet transformed into those of the other—and this infinitely multitudinous transformation is the method of plane analytical geometry. The like is true of analytical geometry of three or more dimensions. You see that instead of transfomation being a chapter in analytical geometry, the latter is itself only a huge chapter in the infinitely more embracing theory of Transformation.

And now do you ask what transformations are good for? That is very much like asking what Thinking is good for; for without transformations, thinking could not go on. We have just seen that analytical geometry is born of transformation and does its work thereby; we have seen that the Olympian concept of Infinity owes its birth to transformation; we have seen that, except for transformation, we could not even count the cattle in a field; we have seen that transformation pervades the practical thinking of the workaday world; in previous lectures, as you will recall, we saw that certain simple transformations,—the inversion transformation and others,—enable us to establish divers verifications of the doctrinal functions of Hilbert, doctrines being thus derived from doctrines and compared with one another, element for element and proposition for proposition, giving rise in this way to what might be appropriately called the Comparative Anatomy of Doctrines. These examples of the use of transformation, though they must suffice for the present, are only as pebbles picked up at random on the ocean beach.

"The hour contracts" and we have yet to speak of

*The Connection of the Concept of Transformation with that of Relation and that of Function.*—We shall see that the three concepts are very similar—but three aspects indeed of one and the same thing seen from different points of view. Let us, in the first place, try to understand clearly what a *relation* is. This is necessary because, though countless hosts of relations are present everywhere in the world and are used by everybody all the time, even in their dreams, yet the scientific conception of what a relation precisely is, is not familiar; even the great majority of logicians and mathematicians are not familiar with it; it seems a little strange that such is the fact, for the logical theory of relations,—the logical theory having the nature and the properties of relations as subject-matter,—goes back to the logical work of J. H. Lambert (1728–77),—mathematician, physicist, astronomer,—and especially to that of Augustus DeMorgan (1806–78),—mathematician, logician and wit; was advanced by important researches of our fellow countryman, the late C. S. Peirce, who called it the logic of *relatives;* and has now reached a high state of development in the *Principia* of Whitehead and Russell, who have made the theory of abstract relations supreme in logical doctrine. This great theory—fundamental alike in philosophy and in mathematics—has not yet become in most universities a subject of regular instruction, but it is, I believe, destined by its intrinsic importance to win such recognition.

Relations, as we are presently to see, are determined by propositional functions of *two* or more variables, and are accordingly described as dyadic, triadic (3-cornered), tetradic (4-cornered), . . . , $n$-adic ($n$-cornered), and so on. The most important ones are the *dyadic* relations.

What is meant by a dyadic relation? I will answer as clearly and simply as I can and will do so by the help of two familiar examples. Consider the two propositional functions: (1) $2x+3y-1=0$; (2) $x$ is a parent of $y$. Each of these is said to determine a dyadic relation. What *is* the relation determined by (1)? We see that (1) will be satisfied if, for example, we replace $x$ by 1 and $y$ by $-\frac{1}{3}$ and so we say that the ordered pair $(1, -\frac{1}{3})$ is a *couple* of *verifiers* of (1); another such couple is $(0, \frac{1}{3})$; there are, you see, infinitely many such couples; the *set*, or *class*, of all the couples of verifiers of (1) is said to *be* the *relation* determined by (1). Each of the couples may be called an *element* or *constituent* of the relation. What is the relation determined by (2)? Suppose John Smith is the father of Bill Smith, then the ordered pair (John Smith, Bill Smith) is a couple of verifiers of (2); the class of all such verifying couples is the relation determined by (2). In the light of these simple illustrations you will rightly understand that a dyadic relation is the class of all the couples ($x$, y) that verify (satisfy) some propositional function $F(x, y)$ containing two (and only two) variables, say, $x$ and $y$.

It is necessary to note carefully the following distinction in usage: in ordinary function-theory—say, in algebra or in analytical geometry—it makes no essential difference whether the $x$-terms in a propositional function come first or the $y$-terms first; that is, for example, no essential difference between (1) $2x+3y-1=0$, or $x=\frac{1}{2}(1-3y)$ and (1′) $3y+2x-1=0$ or $y=\frac{1}{3}(1-2x)$; (1) and (1′), for example, represent the same straight line; but in relation-theory it is essential to take account of the order in which the variables occur; the relations determined respectively by (1) and (1′) are not the same; for example, the former relation contains the couple $(0, \frac{1}{3})$ but the latter does not; on the other hand, the latter contains the couple $(\frac{1}{3}, 0)$ but the former does not; if we denote the former relation by $R$ and the latter one by $R'$, we then write $0R\frac{1}{3}$ to say that "0 has the relation $R$ to $\frac{1}{3}$" and write $\frac{1}{3}R'0$ to say "$\frac{1}{3}$ has the relation $R'$ to 0"; but we have neither $0R'\frac{1}{3}$ nor $\frac{1}{3}R0$, for both of these propositions are false; in general, as you see, if we have $x_1Ry_1$, then we have $y_1R'x_1$, and conversely, but not $y_1Rx_1$ nor $x_1R'y_1$ (except in very special cases); here we encounter the important notion of the *converse* of a relation—two relations, $R$ and $R'$, are each the other's converse if they are such that, whenever one of them holds between the terms of a couple $(t_1, t_2)$, the other holds between the terms of the *inverse* couple $(t_2, t_1)$; thus the relations determined by the propositional functions (1) and (1′) are each the other's converse. It is sufficiently obvious that every relation has a converse. (In the case of some important relations,—such as equality, for example, or similarity or diversity or identity,—the relation and its converse are the same.) The converse of the relation determined by (2) is that determined by the propositional function (2′), $y$ is a child of $x$, so that, if the couple (John Jones, Mary Jones) be a constituent of the former relation, the couple (Mary J, John J) is a constituent of the latter.

You are now in a fairly good position to see that the

concept of mathematical transformation, the concept of relation and the (ordinary) concept of mathematical function are, as I have said, virtually but three aspects of one and the same thing seen from different points of view. For look again at the propositional functions: (1) $x=\frac{1}{2}(1-3y)$, (1′) $y=\frac{1}{3}(1-2x)$. Observe that (1) is: (*a*) a law of *transformation* by which a class of numbers $y$ is converted into a class of numbers $x$; (*b*) a determiner of a *relation*, namely, that composed of the couples $(x, y)$ verifying (1); (*c*) a determiner of $x$ as a function (in ordinary sense) of $y$, namely, the function $\frac{1}{2}(1-3y)$. Observe that (1′) determines at once the *converse transformation*, the *converse relation* and the *converse*,—commonly called the *inverse*,—*function*. Observe that if $x_1$ and $y_1$ verify (1), then the pair $(x_1, y_1)$ is: (*d*) composed of a *thing transformed* and *its transform;* (*e*) a *pair* of *values* of the *function;* (*f*) a *couple*, or *constituent*, of the *relation*. Of course, the like is true of (1′).

Look again at the propositional functions: (2) $x$ is a parent of $y$; (2′) $y$ is a child of $x$. You see that (2) is at once: (*g*) the determiner, or law, of a *transformation*, associating any given $x$ (a parent) with some $y$ or $y$'s (child or children of the $x$), the $y$ or $y$'s being the $x$'s transform or transforms; (*h*) the determiner of $x$ as a *function* (in ordinary sense) of $y$, for to any value of $y$ (some child) there corresponds a value or values of $x$ (some parents); (*i*) the determiner of a *relation*, composed of the couples of verifiers of (2). It is plain that (2′) yields the respective converses of the foregoing transformation, function and relation.

The connections shown by these particular examples hold in general: given a transformation, you have a function and a relation; given a function, you have a relation and a transformation; given a relation, you have a transformation and a function: *one* thing—*three* aspects; and the fact is exceedingly interesting and weighty. Impressed by the immeasurable scope of the *ordinary* function concept, some thinkers have said, with a striking approximation to truth, that mathematics and indeed the whole of science is just the study of functions. It can, you see, be said, with the same approximation to truth, that the whole of science, including mathematics, consists in the study of transformations or in the study of relations.

Time is lacking for extensive pursuit of the matter here. Before leaving it, however, I should like to signalize the parallelism in another way. A *relation* $R$ has what is called a *domain*,—the class of all the terms such that each of them has the relation to something or other,—and also a *codomain*—the class of all the terms such that, given any one of them, something has the relation to it; a *transformation* $T$ proceeds from a *class*,—that of the things transformed,—to a *class*—that of the transforms; the independent variable (or argument, as it is called) of a *function* $F$ has a *range*,—the class of values the argument may take,—and the function has a *range*,—the class of values the function may take. Note the matching of the foregoing things; it is easiest to do it by an example. Consider the simple propositional function: $x=2y$. It determines a relation $R$, a transformation $T$ and a function $F$ (*i.e.*, $x$, or $2y$). Let $K$ denote the class of real numbers and $K'$ the class of their doubles. You see that $K$ is at once the *codomain* of $R$, the *class* transformed by $T$, and the *range* of $F$'s argument $y$; also that $K'$ is at once the domain of $R$, the *class* of transforms (under $T$), and the *range* of $F$. In this particular example we happen to have identity, or coincidence, of domain and codomain, of transform class and transformed class, of function-range and argument-range; but this is, in general, not so; and I recommend that you do the matching by some other propositional function, say, $x$ is the husband of $y$, or $x$ is the specific gravity of $y$, or the integer $x$ is greater than the integer $y$, or $x$ is ethically so sublime that he should not allow $y$ to make a glass of beer or " turn water into wine."

I can not refrain from tarrying here long enough to illustrate, by just one example, the now evident fact that any problem, process or operation having to do with (ordinary) functions is a problem, process or operation having to do with relations or with transformations, and conversely. The example is as follows: If $R_1$ and $R_2$ be two relations such that it is significant (true or false and not merely nonsensical) to say that $R_1$'s codomain and $R_2$'s domain intersect, then there is a relation $R'$—called the *relative product* of $R_1$ by $R_2$—such that, if $xR_1y$ and $yR_2z$, then $xR'z$; respecting functions the corresponding fact is this—if $F_1$ and $F_2$ be two functions such that it is significant to say that the range of $F_1$'s argument intersects the range of $F_2$, then there is a function $F'$—which might be called the functional product of $F_1$ by $F_2$—such that, if $x=F_1(y)$ and $y=F_2(z)$, then $x=F'(z)$; finally, as to transformations, the corresponding fact is this—if $T_1$ and $T_2$ be two transformations such that it is significant to say that the class of transforms (under $T_1$) and the class of things transformed by $T_2$ intersect, then there is a transformation $T'$—called the product of $T_1$ by $T_2$—such that, if $T_1$ converts $x$ into $y$ and $T_2$ converts $y$ into $z$, then $T'$ converts $x$ into $z$. I hope that what I have now said is sufficient to make clear the exceedingly important fact that the meanings of the great terms —transformation, relation and function (in ordinary sense)—are essentially identical.

*The Rhetoric of Mathematics.*—Before closing this lecture I wish to say something about the *psychology* of the mathematician's use of the *word* transformation and in connection therewith to speak briefly of what may be called the rhetoric of mathematics, a subject worthy of much fuller treatment than we have time to give it here. Are mathematicians rhetoricians? Rhetorician? "That is, of all things"—the mathematician will say—"exactly what I most certainly am not." And he should not be harshly blamed for disowning the character; for, by empty-headed advocates of good causes and by full-headed advocates of bad ones, the art of rhetoric has been so much abused in the world that "rhetorician" has come to be, oftener than not, a term of reproach. Nevertheless Rhetoric is a perfectly good name for the greatest of all the arts—the art of expression by speech. "Thought," said Henri Poincaré, "is only a flash of light between two eternities of darkness, but thought is all there is." How much poorer we should be, had the great thinker not expressed *this* thought, so beautiful and so poignant, all will know who have worthily meditated upon life and the world. Thought unexpressed is thought concealed, and concealed thought—light hid under a bushel—fades and perishes with the thinker. Expressed, however, it lives and grows, engendering its kind, adding its flame to the flame of other thought, and so that radiance which is "all there is" increases and tends to abide: it is expression, and especially expression in speech,—expression by the art of rhetoric,—that gives increase and perpetuity of light to the narrow vale between the dark eternities.

And, now, rightly using the term "rhetoric" to denote the art of expression by speech, my thesis is that mathematicians are all of them devoted rhetoricians and the best of them masters of the art. The thesis is not difficult to maintain. For what does the art demand? What are the first qualities of Style? Clarity? Energy? Order? Unity? Convincingness? Restraint? Beauty? In respect to these things no literature surpasses the literature of mathematics. It may not indeed be easy to understand, for the understanding of it requires a fair measure of mind,—imagination, especially, and logical sense,—but the difficulty inheres in the subject and not in the manner of handling it, for the latter is clear—clear in its definitions, clear in its enunciations, clear in its demonstrations; its energy may not be easy to feel, for the feeling of it requires a certain order of sensibility, but energy is always present in a high degree—indeed the whole vast symbolism of mathematics, invented with a view to the effective use of intellectual energy, is charged therewith beyond the measure of common words; its order may not be easy to appreciate, for it is the order of logic, beginning with principles and pursuing their destined consequences under the subtle rule of fate; its unity may not be easy to grasp, for it is the unity of a whole owing its integrity to the inner bond of implication; its convincingness may not be easy to sense for it is disinterested, dispassionate, purely intellectual, ideal; its restraint is the restraint of direct achievement by the simplest means; and its—Beauty? Its beauty is two-fold: the exquisite austere beauty of sheer form; and a unique kind of *dictional* beauty, due to the union, in mathematical nomenclature, of two qualities not elsewhere united. I mean a certain literary quality, not essential to mathematics as such, and a certain perfection of logical quality which neither "the literature of power" nor (outside of mathematics) "the literature of knowledge" attains. Pray do not fear that, in saying this, I am speaking as a partisan. Why should I? Mathematics and literature are, both of them, ineffably precious. I am merely endeavoring to state an interesting fact. And if the meaning and the truth of what I have said respecting mathematical diction be not yet sufficiently evident to you, they will become so if, when you have the opportunity, you will examine the matter attentively. It would be sufficient to examine fifty or a hundred representative mathematical terms, such, for example, as the following, taken quite at random from a vast multitude: Real — ideal — imaginary — transcendental — elliptic — parabolic — hyperbolic — value — range — field — domain — harmonic — anharmonic — symmetric — asymmetric — golden section — degrees of freedom — determination—necessary— sufficient — discriminant — determinant — variable — constant — invariant — covariant — calculus — congruent — divergent — oscillating— maximal — minimal — sheaf (of lines) — pencil (of planes) — family (of curves) — cluster (of spheres) — asympotic contact or approach—point of osculation—conjugate (elements or figures or forms)—interval—neighborhood—correlation — dependent — independent — closed — open — boundary — inside — outside — on — slope — continuity — discreteness — finite — infinite — infinitesimal — limit — chance — law. The literary significance of such representative terms—the wealth and variety of their general meanings, the warmth of some of them, their colors, the imageries awakened by them, the associations they carry—all that is evident. In addition to that, each of them denotes, as you may ascertain, a sharply defined mathematical concept, which in every instance is due to selecting and refining some feature or aspect of the term's *general* meaning. We have, then, as you see, in each of the terms two distinct qualities—the literary quality of its general meaning and the logical quality of perfect precision of specific meaning; but that is not all; not only are the two qualities present in the terms, but they are connected in them—they are there joined in a spiritual union not to be found beyond the borders of mathematical speech.

I had not intended to speak at so great length of mathematical rhetoric and can offer no plea in mitigation except the fascination of the theme and a growing sense of its importance. I must now hasten to say in connection therewith, what I have so long delayed saying with respect to the psychology of the mathematician's use of the *word* "transformation."

Functions, propositional functions, doctrinal functions, propositions, classes, points and point configurations, numbers and systems thereof—mathematical entities in general, simple or complex, elemental or composite,—are, all of them, stable things; immobile and immutable; they neither come nor go; they are not born and they do not perish; they have neither origin nor destiny, neither past nor future; they are timeless—inhabitants of eternity; they *are*: the world of mathematical entities is a static world; it owes its unity and integrity to the presence within it of an infinite system of interlocking relations; and those mathematical relations, too, like the entities constituting them and related by them, are static. And, now, what term do mathematicians employ to denote these static things? They employ, as we have seen, the dynamic term Transformation—as if they fancied themselves to be dealing with temporal things, with actual vicissitudes, with transmutations, with the changeful phenomena of the fluctuant world of sense. Why? Partly, no doubt, because they enjoy the illusion, for it stimulates their minds, enveloping the train of their abstract thought in a beautiful mist of sensuous imagery, and does so without diverting it from its true course. But is the illusion really an illusion? In a sense it is—in the sense indicated; but in a deeper sense it is not. For in dealing with the static world of immobile, unchanging, eternal things,—the world of concepts,—we are in fact dealing with the dynamic world of mobile, changing, temporal things,—the streaming world of sense,—in the only way in which the latter can be dealt with by logical thought. What is that way? Thought *arrests* the chaotic stream and gives it *order*—arrests it, I mean, and gives it order, in the sense of carving and shaping its confused and formless content into permanent *kinds*, *classes*, *concepts*—unchanging, immobile conceptual entities constituting a static world. For Thought this world of static elements and static relations represents, under the aspect of eternity, the temporal elements and temporal transformations of the dynamic world of Sense. And so we have a kind of provisional answer to our question: mathematicians discourse in dynamic terms about static things because they are constrained to think in static terms when they think about dynamic things. The real problem, however, is not thus solved—it is merely pointed at; for what is the secret of the mentioned constraint and the consequent compromise? I can hardly state the problem adequately, much less am I able to solve it; it is a problem for you and the future —a momentous problem for science and for philosophy,

It is evident that the nut to be cracked, or one of the nuts, is *Time*. We have seen that in the world of logic things and their relations are timeless, they *are*—all are present at once; but the things of the other world and their transformations are temporal, they are not all present at once, but occur in temporal order—each thing becomes its own successor and, in so becoming, ceases to be, so that there is a Past (which is empty) and a Future (never filled)—only a mobile Now, sole field and vehicle of change and transformation. How can either of these sharply contrasted worlds represent the other—the things that *are*, standing for the things that *happen*, the permanent for the changeful, rest for motion, relations for transformations, the beginningless and everlasting for the momentary children of birth and decay—the *timeless* for the *temporal?* It is evident, as said before, that one of the troublesome factors is *Time*. In endeavoring to manage this factor, science has tried two and only two ways—the way of *importation* and the way of *suppression*. We are going to see what they are.

The former is the way of Newton, the way of his *Fluxions and Fluents*. From the objective dynamic world of sense and physics time is imported into the subjective static world of conception and logic—it is *smuggled* in with motion: points are not immobile, they move; lines and curves do not really exist,—they are not unbegotten inhabitants of eternity,—they are engendered *in time* by motion of points; the same is held respecting surfaces, which are but the paths of moving lines and curves; and respecting solids, produced by moving suffaces; $x$, $y$ and $z$ are viewed as varying actually, they *grow*, their increments are *fluents;* and the static world is invaded by velocities and accelerations. The Newtonian method of dealing with the problem,—dynamicising the static world,—flooding the realm of eternal things with the waters of time,—has had a great vogue, has produced inestimable results and is still dominant; but it is not ultimately satisfactory; for *Gefühl ist alles*, and we can not rid ourselves of the feeling that points do not move, that numbers do not change, that relations are not transmutations, and that, in general, logical and mathematical entities are immutable.

And so, in the recent literature of science are to be found increasing tokens of dissatisfaction and reaction. The troublesome factor of Time is to be *suppressed;* instead of dynamicising the static world of conception and logic, we are to staticise the dynamic world of sensation and physics. I have alluded to tokens. The atmosphere of present-day " relativity " discussion is charged with them. Let me direct your attention to a striking one. I refer to Minkowski's famous interpretation of what is known as the Lorentz transformation. My present concern is with a single feature of the interpretation. It may be set in light as follows. Think of a " substantial " particle $p$ of our physical world; we are accustomed to saying that, at a time-instant $(t)$, $p$ is at a space-point $(x, y, z)$; that, at instant $(t')$, $p$ is at the point $(x', y', z')$; and so on; thus, to give account of $p$ we must give both its *when* and its *where*—its $t$ and its $x$, $y$, and $z$; we have thus a tetrad $(x, y, z, t)$; now let us, says Minkowski, view the matter in another way; let us regard this tetrad as one thing and name it *Weltpunkt*—world-point; such a world-point has *four* coordinates, $x$, $y$, $z$, $t$, and the world constituted by such points is a 4-dimensional world; the points of this world—of which there are $\infty^4$—all exist at once, they coexist; the fluxion called time is abolished; motion, as a change of place during a flow of time, is gone; in the new world, *where* (if the term be used) has a *new* meaning—it has absorbed both the old where and the old when. Where *is* the particle $p$? Where, that is, in the new world? At the point $(x, y, z, t)$. Where is the particle $p'$? At $(x', y', z', t')$. The particles $p$ and $p'$ are never the same; there are relations, but no transformations; no history in ordinary sense—no past—no future; child, youth, man *coexist* as phases of one individual; the same is true of morning, noon, night and so on: all is *static*—as a " painted ship on a painted ocean."

You see what has happened here and how. By suppressing the fluxional character of time along with its implicates,—motion, transformation, change,—and by regarding time as simply a cosmic dimension to be joined with the familiar dimensions of space, the Dynamics of our spatially 3-dimensional world has been made to appear as a Statics of a 4-dimensional world. I need not say that this way of handling time, however beautiful and helpful, is, like Newton's way (of which it is the antithesis), not ultimately satisfactory. I should add that Minkowski was far from regarding it as a final solution.

And so science and philosophy are still confronted and to-day confronted afresh by the age-old problem of Time. No one has been able to tell satisfactorily what is meant, or should be meant, by *when*. From time immemorial, human beings have talked of " instants," but no one has discovered what an instant *is*. It is important to observe that the time problem is not solitary; it is but one of a class of kindred problems or is perhaps an aspect or a fragment of a larger problem embracing them all. For what is meant, or should be meant, by *where?* By *here*? By *there*? By a *point?* And we talk of matter as of time and space. But what *is* an *atom* or an *electron?* And what is *ether?* No one has been able to answer these and like questions satisfactorily. It is fair to ask what *sort* of answer would *be* satisfactory. By help of an analogy this question can, it appears, be answered approximately with a good deal of confidence. Respecting the nature of Number there are questions analogous to the foregoing questions respecting Time, Space, Matter. As we ask for the meanings of—instant—when—now—then—point — where — here — there — electron — atom — ether — and the like, so in the domain of Number we ask such questions as these: What is a cardinal number? A positive integer? A negative integer? A rational number? An irrational number? A real number? A complex number? And so on. Now, in recent years, by workers in the logical foundations of mathematics, especially by the researches of Peano, Frege, Russell and Whitehead, these latter questions—analogues of the former kind—have at length been answered with a pretty high degree of satisfaction. Answered how? Answered in terms of a small number of *logical* data (or concepts or constants) more fundamental and more embracing than the terms defined: answered, that is, in terms of such familiar logical notions as *class*, *relation*, *symmetric relation*, *asymmetric relation*, *serial relation*, and a few other varieties. In this procedure in the Number field we have probably a model of what to seek in the other fields and a clue to it: the ultimate constituents of time, space and matter are to be conceived in terms of *logical* data. When this great task is accomplished, will the results be *entirely* satisfactory? I suspect not; the problem of defining the various kinds of number in terms of *logical* constants has, as I have said, been pretty satisfactorily solved; but, unless I am mistaken, there remains a *psychological* problem,—an immense and difficult one,—the task of discovering the connections of the number concepts with the data of sensation and sense-perception. Is a class, for example, or a relation, a percept or a concept or both? So, too, respecting time, space and matter, if the problem of defining their elements in logical terms were solved, there might still remain to be solved a psychological problem; or it may be that the solutions of the two problems will be advanced simultaneously. Duration, for example, seems to be a datum of sense, and so, too, as William James long ago pointed out, voluminousness, or bulk, appears to be a datum of sense; it may be that an instant and time itself will be logically and psychologically defined in terms of sense-given durations; and that a point and space itself will reach similar definition in terms of sense-given bulks. And similarly for similar things.

You are to be congratulated on the date of your generation when these kindred problems, or these kindred phases of the one great problem constituted by them, are pressing for solution; for the problem is indeed immense, embracing, not merely the now exciting question of " relativity," but—what is infinitely more—the nature of the ultimate data and ultimate structure of Knowledge. Let me, in closing, refer you to some of the works of some of the pioneers—to Russell's *Scientific Method in Philosophy*,[1] to his *Analysis of Mind*, to certain parts of Whitehead's *Organization of Knowledge*,[2] to his *Concept of Nature*, and especially to his truly momentous book, *The Principles of Natural Knowledge*. Regarding the last

[1] Reviewed by C. J. Keyser in *The Bulletin of the American Mathematical Society*.

[2] Reviewed by C. J. Keyser in *Science*.

it must be said that it contains obscure passages. The obscurity is to be ascribed partly to the very great difficulties of the subject and partly to the new ideas thronging the author's mind and impatiently pressing for utterance. The ideas will gradually win their way to greater clarity of exposition by Whitehead himself or by his collaborators and successors in the work; for his book both makes and marks the beginning of an epoch, and, when it perishes, it will "perish by supersession." In the same connection, you should examine Professor Eddington's *Space, Time, and Gravitation.*

# LECTURE XI

## Invariance

THE AGES-OLD PROBLEM OF PERMANENCE AND CHANGE—THE QUEST OF WHAT ABIDES IN A FLUCTUANT WORLD—THE BINDING THREAD OF HUMAN HISTORY—THE TIE OF COMRADESHIP AMONG THE ENTERPRISES OF THE HUMAN SPIRIT—NEED OF CRITICAL HISTORY OF THOUGHT.

INVARIANCE is, in all strictness, a subject of universal interest: it penetrates, as we shall see, not only all the sciences and all the arts, but also the common life of mankind everywhere and always. And no wonder. For the most obvious, the most embracing, the most poignant and the most tragic fact in the pageant we call the world is the fact of Change; in the world of sights and sounds, in the world of *sense*, nothing abides. "The life of man," said the Spirit of the Ocean, "passes by like a galloping horse, changing at every turn, at every hour." And so the sovereign fact in the life of *reason* is the quest of things Eternal. The mathematical theory of transformations,—dealt with in the preceding lecture,—is the logic of change; the mathematical theory of invariance,—the principal theme of the present hour,—is the logic of eternal things, the logic of permanence. The latter theory, like the former, with which it has the closest connections, is immense, manifold, technical and intricate; extensive knowledge of it can be gained only by pursuing it through many months with the tireless energy of a sleuth-hound. It is my aim to give you just a little introduction to the matter, a clue to it, a good grasp of its central idea, a very slight acquaintance with its methods, and a fair sense of its general significance and its bearings as a prototype for that quest of abiding reality which has dominated all the great truth-seeking activities of man and has served to unite them,—religion, philosophy, art, science,—as but different aspects of one supreme enterprise: emancipation from the tyranny of change—discovery of a stable world—a haven of refuge from the raging tempests of the sea.

Let us begin as we began in the case of transformation —as simply as we can; indeed, to begin aright we must return and begin our new study just where we began to study the meaning of mathematical transformation; for we may say at once that, in general sense, an invariant, as the word indicates, is to signify something which, when other things connected with it suffer change, remains itself unchanged; and now change, as we have seen, is represented in logic (in mathematics) by means of relations which, as we have also seen, mathematicians call transformations; so that the mathematical term " invariant " or " invariance " would be unintelligible or meaningless save for its connection with the mathematical notion of transformation.

We will accordingly suppose, as in the preceding lecture, that $a$ and $a'$ denote two objects of thought and that by a transformation—which we may denote by $T$—$a$ has been transformed, or converted, into $a'$, $(a \rightarrow a')$. Now, an object of thought has what we call *properties*, some (at least one) of which are peculiar to it and some belong to one or more other objects as well. Let us suppose that $a$

has the peculiar properties $p_1, p_2, \ldots$ ; that $a'$ has the peculiar properties $p_1', p_2', \ldots$ ; and that the properties $\pi_1, \pi_2, \ldots$ belong to *both* $a$ and $a'$; we say that the properties $\pi_1, \pi_2, \ldots$, since they belong both to $a$ and to its transform $a'$, are *invariant under* $T$,—have suffered no change,—are *preserved;* and that $p_1, p_2, \ldots$, as they belong to $a$ but not to its transform, are *variant under* $T$,—properties lost under the transformation,—not carried over by it. It is plain that under the converse (commonly called the inverse) transformation $T'$, $(a \leftarrow a')$, $p_1', p_2', \ldots$ are *variant* while $\pi_1, \pi_2, \ldots$ are *invariant* as before. It is evident that, if a property be invariant under some transformation, it will be invariant under the converse transformation. I am aware that what I have now said is so general, abstract and simple as to make the concept dealt with seem unreal—tasteless, pallid, thin, intangible. But the seeming is seeming only. The idea in question, far from being detached from reality, literally pervades it—pervades our thinking about it and our handling of it. How may we convince ourselves that this is true? We may do it by looking about us a little and by a little reflection—by considering a few specific concrete examples and observing that such examples abound in countless multitudes on every hand.

For a few examples that everyone can understand, consider the following. Let $I$ denote the class of familiar integers: 1, 2, 3, . . . ; and suppose these to be transformed in accordance with the law: (1) $y = 2x$. The transforms constitute the class of even integers: 2, 4, 6, . . . . We note that integers are converted into integers, and so the property of being an integer is preserved—it is an invariant under transformation (1); the value of an integer, however, is a property not preserved—it is doubled; neither is the product of integers preserved, for, if $x_1$ and $x_2$ be any two numbers in $I$, and their transforms be $y_1$ and $y_2$, then the product $x_1x_2$ is transformed into $y_1y_2$, which is not $x_1x_2$ but is $4x_1x_2$; sums and differences are also variant, the transforms of $x_1+x_2$ and $x_1-x_2$ being respectively $2(x_1+x_2)$ and $2(x_1-x_2)$; but ratios are invariant for $\frac{y_1}{y_2}=\frac{2x_1}{2x_2}=\frac{x_1}{x_2}$. Now suppose the law of transformation to be: (2) $y=\frac{1}{2}x$; then neither the property of being an integer, nor value, nor product nor sum nor difference is invariant, but ratio is; so, too, is the property of being a number, as was also the case under (1), though not there mentioned. This latter property is again invariant under the transformation: (3) $y=x+1$; but ratio is not. Is the property of being a number invariant under every transformation of $I$? No, it is not invariant under the transformation converting the integers of $I$ into an endless succession of days, $d_1$, $d_2$ $d_3$, . . . in accord with the law: (4) $n \rightarrow d_n$; but even under (4) we have an interesting invariant,—namely, the property of *nextness*,—for the transform of an integer next after a given one is itself next after the given one's transform. Note that under none of the transformations (1), (2), (3) is the class $I$ invariant as a whole, being converted by (1) and (3) into a part of itself and by (2) into a class including it and another class besides—a class of fractions. But $I$ as a whole is invariant under many transformations—for example, under the reciprocal one-to-one transformation: (5) $1 \longleftrightarrow 2$, $3 \longleftrightarrow 4$, . . . .

For a different sort of example, consider the following. Let $D$ denote something very strong and solid, say a diamond at a certain time and place; suppose it removed gently to another place, and now denote it by $D'$; $D$ and $D'$

are two objects of thought, for they are evidently not the same in all respects. Let us now suppose $D$ transformed into $D'$, $(D \rightarrow D')$, by a transformation $T$—I do not mean transmutation, I mean association of $D$ with $D'$, for that, as we have seen, is at bottom what a mathematical transformation is, that and nothing more. Among the variant properties of $D$ under $T$ are certainly time and place, and possibly weight and distance from the Moon. What, if any, are the invariants? Subject to some correction by the refinements of modern physics, it is yet instructive to answer that among the invariant properties of $D$ under $T$ are shape, size, mass, degree of hardness, capacity for light absorption, and so on. Some of these will, of course, not be invariant under transformation of $D$ into $D'$, where $D'$ denotes $D$ crushed.

For an example drawn from a very different field, let $P$ denote the personality of John Smith at the age of 15, $P'$ his personality at the age of 30, and let someone, say Smith himself at the age of 45, transform $P$ into $P'$, $(P \rightarrow P')$, by a transformation $T$—again I do not mean transmutation, the mysterious process of a boy's becoming a man. The variant properties of $P$ under $T$ are obviously many—years, for example, wisdom, folly, interests, hope, and so on; among invariants are the properties of being a son, of being a man, of being a human, of being what Count Korzybski calls a time-binder, of being a visible object; another one—of extraordinary interest—is the property called *personal identity*. This last property, which runs through a long sequence of personalities, exemplifies an immense *class* of important invariants that no one has been able to formulate *precisely* though their existence is manifest: we may call them unformulated or *qualitative* invariants. These are not indeed strictly mathematical invariants, which are formulated precisely; yet they evidently belong, like the latter, to the type of invariantive matter and will more and more approximate or even attain precision in course of the progress of analysis and definition.

Let us now have another little dip in the boundless sea of strictly mathematical invariants. In Lecture IV we were introduced to the *pole-polar* transformation of a plane with respect to a circle; we saw that it converts a point (as pole) into a line (as polar), and a line (as polar) into a point (as pole); it is, you see, a *one-to-one reciprocal* transformation—a twofold affair composed of a transformation (1) of points into lines and the converse transformation (2) of lines into points. Let us first think of (1) alone; we readily detect certain variants and certain invariants under it; the property of being a point is not preserved, since the transform of a point is a line; a range of points loses the range property, since the transform of a range is a pencil (of lines); *distance* is lost, since the distance between two points has for transform the *angle* between two lines (the transforms of the points); now, as you know, a curve has two aspects (called dual aspects), one as the *locus* of its *points*, the other as the *envelope* of its (tangent) *lines*; the property of being a curve is invariant (preserved), for under (1) the transform of a curve is a curve; but the locus aspect is lost, its transform being the envelope aspect; as we saw in Lecture IV, the relation of order and the relation of congruence are exceedingly important invariants under (1); by (1) the ordinary geometry $D_1$ of the plane was transformed into the geometry $D_2$ of lines and pathopencils; and, as $D_1$ and $D_2$ are the same in respect to *form*, we see that under (1) doctrinal form, or logical structure, is invariant. I leave to you the analogous consideration of transformation (2).

Let me suggest that, for a handsome illustration of invariance, you look again at the *inversion transformation*, presented and employed in Lecture V. You will readily see that the inversion plane, as a whole, and the inversion circle, are invariants; that a point's property of being a point and a curve's property of being a curve, as well as its locus aspect, are invariants; that, on the other hand, neither the *magnitude* of distances nor the *sense* of angles is invariant, but that angular magnitudes are preserved; and,—most beautiful of all,—if, as is customary, we regard a straight line as a circle (of infinite radius), then cocircularity of points is an invariant, for, as we saw, the inverse, or transform, of a circle is a circle.

I hope I shall not be overtaxing your interest if, to the foregoing list of somewhat random illustrations, I add a specially chosen one, lying at the heart of Projective Geometry. Let $x$ denote a real number and suppose it transformed into $x'$ in accordance with the law:

$$(1)\quad x' = \frac{ax+b}{cx+d};$$

note that (1) contains three parameters,—the three independent ratios of the coefficients,—say, $a/d$, $b/d$, $c/d$. To each of these we may give any one of the infinitely many ($\infty$)real values; and thus, as you see, there are $\infty^3$ different transformations of the form (1); the transform $x'$ of a given $x$ will, of course, depend upon the particular one of the transformations we choose to employ. Let us now imagine that (1) is some definite one of the transformations; then a given $x$ has a definite transform $x'$; let $x_1$, $x_2$, $x_3$, $x_4$ be any 4 given real numbers and let

$x_1', x_2', x_3', x_4'$ be their respective transforms. Consider the expression, or function,

$$\frac{(x_1-x_2)(x_3-x_4)}{(x_2-x_3)(x_4-x_1)};$$

being very important, this function has a special name; it is called the *anharmonic ratio* of $x_1, x_2, x_3, x_4$ taken in the order as written, and may be denoted by the symbol $R(x_1x_2x_3x_4)$; so we may write

$$R(x_1x_2x_3x_4)=\frac{(x_1-x_2)(x_3-x_4)}{(x_2-x_3)(x_4-x_1)}$$

Now, the transform of $R(x_1x_2x_3x_4)$ obviously is $R(x_1'x_2'x_3'x_4')$. How are these two anharmonic ratios related? To find the answer it is sufficient to replace $x_1', x_2', x_3', x_4'$ of the transform ratio by their respective values

$$\frac{ax_1+b}{cx_1+d}, \quad \frac{ax_2+b}{cx_2+d}, \quad \frac{ax_3+b}{cx_3+d}, \quad \frac{ax_4+b}{cx_4+d}$$

and to simplify the result; by doing so, which is easy, you will find that $R(x_1x_2x_3x_4)=R(x_1'x_2'x_3'x_4')$. We have here, as you see, an exceedingly beautiful specimen of mathematical invariance: namely, under each and all of the threefold infinity of transformations of form (1), each of them converting the entire class of real numbers into itself, the anharmonic ratio, ($R(x_1x_2x_3x_4)$, of any ordered set of four numbers, remains absolutely unchanged.

The invariant in question, though it here appears as a function of pure numbers, lies, as I have intimated, at the heart of Projective Geometry. We may see the thing in geometric light readily as follows. Suppose Fig. 22 to be in a projective plane; let $x$ be the distance from

$O$ to $P$; a value of $x$ is thus associated with a point $P$ of the range $L$, with a line $p$ of the pencil $V$, and conversely; $R(x_1x_2x_3x_4)$ may be called the *anharmonic ratio* of the 4 corresponding *points*, $P_1$, $P_2$, $P_3$, $P_4$, and be denoted by $R(P_1P_2P_3P_4)$, or of the corresponding lines, $p_1$, $p_2$, $p_3$, $p_4$, and be denoted by $R(p_1p_2p_3p_4)$; then, you see, any tetrad of collinear points or any tetrad of copunctal lines has an anharmonic ratio.

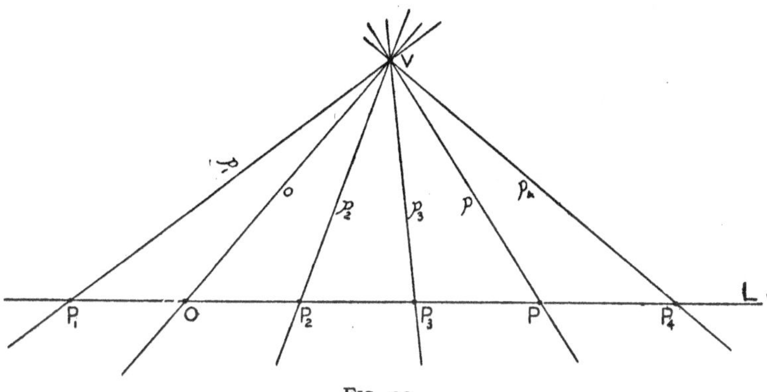


FIG. 22.

Consider a transformation of form (1) in connection with Fig. 23; let us associate the values of $x$ with the points of $L$ and the lines of $V$, and the corresponding values of $x'$ with the points of $L'$ and the lines of $V'$; we have, you see, thus transformed the points of the range $L$ into the points of the range $L'$ and also into the lines of the pencil $V'$; at the same time we have transformed the pencil $V$ into the pencil $V'$ and also into the range $L'$. Now let $P_1$, $P_2$, $P_3$, $P_4$ (or $p_1$, $p_2$, $p_3$, $p_4$) be any 4 points (or lines) of $L$ (or $V$), and let their transforms be $P_1'$, $P_2'$, $P_3'$, $P_4'$ (or $p_1'$, $p_2'$, $p_3'$, $p_4'$) of $L'$ (or $V'$), then, owing to the invariance of anharmonic ratios under our transformation, we have

$$R(P_1P_2P_3P_4) = R(p_1p_2p_3p_4) = R(P_1'P_2'P_3'P_4') = R(p_1'p_2'p_3'p_4')$$

Such a transformation is called a *projective* transformation and, when it has been applied as above, $L$ and $L'$, or

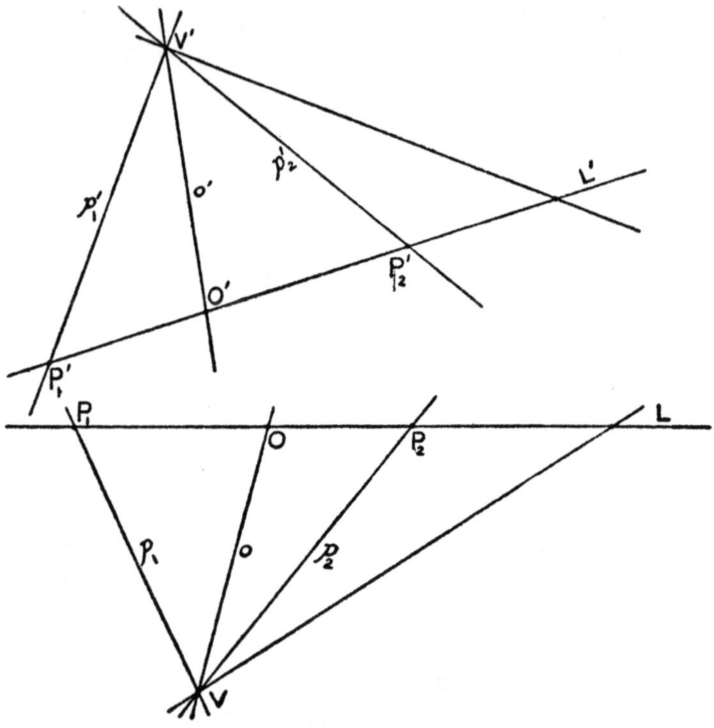


FIG. 23.

$V$ and $V'$, or $L$ and $V'$, or $L'$ and $V$, are said to be projectly related. Why call the transformation projective? Because the correspondence set up by it can be set up by what architects call *projection*—as is shown for the case of $L$ and $L'$ in Fig. 24, where $P_1, P_2, P_3, \ldots$ are projected from $V$ respectively into $P_1', B, A, \ldots,$ and these are then projected from $V'$ respectively into $P_1'$, $P_2'$, $P_3'$, . . ., so that we have finally $P_1$, $P_2$, $P_3$, . . ., corresponding respectively to $P_1'$, $P_2'$, $P_3'$, . . ., as required by the given transformation.

The $\infty^3$ projective transformations of form (1) are included in a yet larger class of projective transformations of the plane, which latter are included in a still larger

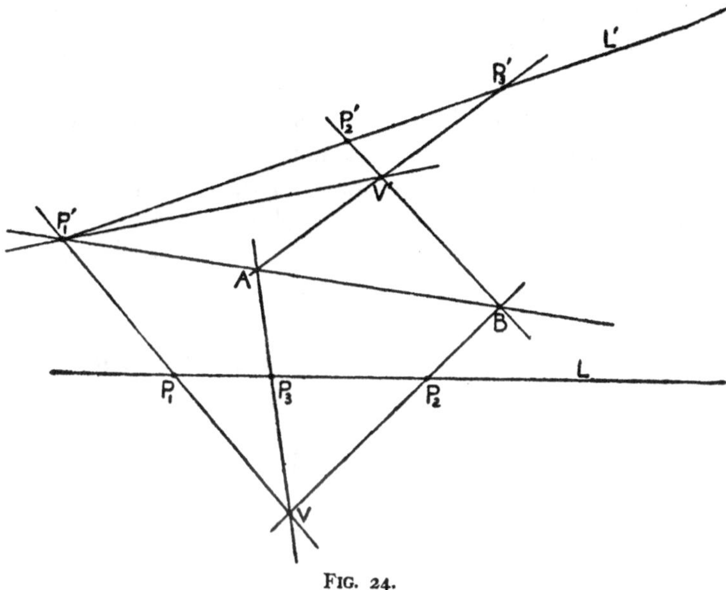


FIG. 24.

class of projective transformations of (ordinary) space, and so on for spaces of higher dimensionality. Imagine two planes $\pi$ and $\pi'$ in ordinary space; let $F$ be some figure in $\pi$; let $O$ be a point in neither plane; the lines through $O$ and the points of $F$ *project* (as we say) the latter points into points of $\pi'$ constituting a figure $F'$. $F$ and $F'$ have certain *properties in common;* that is, certain properties of a figure are *invariant* under projection.

Projective Geometry is the study of such properties; these are all of them expressible in terms of anharmonic ratios and that is why I said that the invariance of the anharmonic ratio is at the heart of projective geometry.

The *idea* of invariance—of permanence in the midst of change,—of abiding realities in a fluctuant world,—is very, very old,—far older than history,—as old probably as the race of man—certainly as old as the dream of eternal things, of everlasting goods. On this account and especially because mathematics has always been peculiarly concerned with eternal things, it seems a bit strange that the *mathematical theory* of invariance—the doctrine, I mean, having invariance consciously for its subject-matter—is a strictly modern theory. Yet such is the case. Why it is so is a question I shall not here attempt to answer. It is but a minor one of a large class of very interesting questions belonging to a great unwritten history—the history of the development of intellectual Curiosity,—a subject requiring for its treatment philosophical genius and learning of the highest order. The mathematical theory of invariance is about as old as American independence. Like most other great doctrines, it began, not in ratiocination, but in an observation, and not in an observation of a great fact by a small mind but in an observation of a small fact by a great mind. I allude to the observation by Lagrange in 1773 of the little fact that the discriminant of the quadratic expression, or form, (1) $ax^2+2bxy+cy^2$, remains unchanged when (1) is transformed by replacing $x$ by $x+\lambda y$. In high school or college, you learned what the discriminant of (1) is and what it signifies. May I remind you? It is $b^2-ac$; and it means that the roots of the equation (2) $ax^2+2bxy+cy^2=0$,—the two values of $x/y$ that satisfy (2)—are equal, or real and unequal, or conjugate imaginary, numbers according as $b^2-ac$ is equal to, or greater than, or less than zero. Transforming (1) by replacing its $x$ by $x+\lambda y$, you will readily find that the transform of (1) is (1′) $ax^2+2(a\lambda+b)xy+(a\lambda^2+2b\lambda+c)y^2$, and that the transform of (2) is the equation (2′) $ax^2+2(a\lambda+b)xy+(a\lambda^2+2b\lambda+c)y=0$. The discriminant of the transform (1′) or (2′) is $(a\lambda+b)^2-a(a\lambda^2+2b\lambda+c)$, and this, you see, reduces to $b^2-ac$ exactly. You will notice that by allowing $\lambda$ to vary in value, we obtain an infinity of transformations—as $x+2y$, $x+\frac{1}{2}y$, $x-\frac{1}{3}y$, $x+\sqrt{14}y$, and so on —all of them similar in type, and a corresponding infinity of transforms (1′) or (2′) of the same expression (1) or equation (2); owing to the invariance of $b^2-ac$ under each and all of these transformations, we know that, if the roots of (2) are equal, or real and unequal, or imaginary, then the same is true of the roots of each one of the infinitely many transform equations (2′). I have explained this simple fact,—small indeed as a mustard seed, —so fully because, as already said, Lagrange's observation of it was the germ of the now vast and still growing mathematical theory of invariance. Its early history owns great names: Gauss who in 1801 showed the discriminant of the ternary quadratic, $ax^2+by^2+cz^2+2dxy+2exz+2fyz$, to be invariant under the transformation replacing $x$ by $Ax+By+Cz$, $y$ by $Dx+Ey+Fz$ and $z$ by $Gx+Hy+Iz$; Boole who in 1841 established, among other interesting results, the invariance of the discriminant of expressions involving an arbitrary number of variables; Cayley who, incited by Boole's beautiful results, assailed the problem of ascertaining all invariant functions of the coefficients of an equation of degree $n$ and produced in rapid succession his great memoirs on

Quantics; Sylvester who soon joined Cayley in the field, brilliantly rivalled his researches there and conceived the subject more poetically as the Calculus of Forms; Eisenstein, Aronhold, Hermite, Clebsch, Gordan and others; names representing, you see, both Great Britain and the continent of Europe. The result is a truly colossal doctrine variously styled the algebra of quantics, the theory of algebraic invariants and the theory or calculus of forms—a doctrine which has not yet ceased to grow and to which American mathematicians have recently made valuable contributions. But this algebraic theory is by no means the whole of the mathematical doctrine of invariance; it is only the oldest and most elaborate part of it; every division of mathematics has its problem of invariants; and vital portions of many subjects,—number theory, for example, differential equations, various function theories, all the multifarious branches of geometry,—belong to the doctrine in question.

It would be a great mistake to imagine that the interest of mathematicians in the matter of invariance is peculiar to them; their method of handling it,—in the abstract by logical means,—is indeed peculiar to them; but the matter itself is not; for a little reflection suffices to show that search for Constance,—constant quantities, constant properties, constant relations,—in our world of ceaseless change, is just as much the concern of religion, of philosophy, of political science, of education, of art, and of natural science as it is of mathematics, though each of the former enterprises has, like mathematics, a method of its own, and differs from its allies in respect to the type of invariants it seeks.

Consider natural science, for example. What is it? For our present purpose, it is sufficiently characterized by its conscious aim, and that aim is discovery of those uniformities in the course of Nature which men of natural science are wont to call natural laws. What, pray, is a natural law? A natural law,—if, strictly speaking, there be such a thing outside the conception thereof,—is fundamentally nothing more nor less than a constant connection among inconstant phenomena: it is, in other words, an invariant relation among variant terms. It is necessary to notice the sense in which the term " relation " is here used. In the preceding lecture, where we spoke of dyadic relations, it was said that such a relation, determined by a propositional function $F(x, y)$, consists of all the couples $(x, y)$ of verifiers of the function. In accordance with that view,—which is the *extensional* view of relations,—the relation determined by the function, $x$ is the father of $y$, is the class of all the couples $(x, y)$ such that the $x$ is some male that begot the $y$, and all such couples are regarded as coexisting and thus constituting the relation once for all even though most fathers and children are either dead or unborn. There is, however, another view of relations,—the *intensional* view,—in which the concept of father is the concept of a constant relation which does indeed subsist between the terms of such a couple if and when the latter exists but which would continue to exist in its full integrity as a relation at an instant or during an interval when there were neither fathers nor children. This view makes it possible to speak, in a " natural " way, of a relation as being itself constant while having in the flux of the world a temporal succession of terms or sets of terms. And this intensional or intrinsic sense of the term relation is the sense in which I employ it when I say that a law of nature is simply an invariant relation among variant terms.

In this conception of natural law and the consequent conception of natural science as having for its aim discovery of invariant relations among the things that appear and disappear in the flowing pageant of the world, there is, I believe, nothing new except its setting and its manner. Rankine, for example, in a paper presented before the Glasgow Philosophical Society in 1867, said: "One of the chief objects of mathematical Physics is to ascertain, by the help of experiment and observation, what physical quantities are *conserved.*" And among invariants thus found he instances mass, resultant momentum, total energy, and other things. More embracing are the words of Major MacMahon in his address to the Mathematical Section of the British Association in 1901. "In any subject of inquiry there are," he says, "certain entities, the mutual relations of which, under various conditions, it is desirable to ascertain. A certain combination of these may be found to have an unalterable value where the entities are submitted to certain processes or are made the subject of certain operations. The theory of invariants in its evident scientific meaning determines these combinations, elucidates their properties, and expresses results when possible in terms of them. Many of the general principles of political science can be expressed by means of invariantive relations connecting the factors which enter as entities into the special problems. The great principle of chemical science which asserts that when elementary or compound bodies combine with one another, the total weight of the material is unchanged, is another case in point. Again, in Physics, a given mass of gas under the operation of varying pressure and temperature has the well-known invariant, pressure multiplied by volume and divided by absolute temperature." You doubtless know that similar examples might be cited at great length; and I need hardly say that, if some of those cited by Rankine and MacMahon may have to be withdrawn in view of recent physical refinements, the weight and justice of their main contentions remain unimpaired.

Interest in things that abide,—interest in stable values, transcending time and change,—is as fundamental and regnant in art as in natural science. It is true that the invariants which art seeks in its own way to find and in its own way to disclose or represent are not sharply defined; like personal identity, they are of the class of those which a little while ago we described as unformulated or qualitative invariants; they are none the less genuine invariants, and no defect of their definition can disguise or dim the fact that, like natural science and like mathematics, art,—art in its great moods and proper character as art,—contemplates the world under the aspect of eternity, aims at what is permanent in the "fleeting show," devotes itself to goods that are everlasting. For the fact is manifest in many ways. A thing of beauty is a joy forever. Who does not know, or at all events feel, the deep and proper meaning of this familiar *mot?* It is not that any phrase or picture or poem or symphony or statue or temple will escape the doom of temporal things; nor that the joy they may give you or me will endure; it is that a certain quality,—the quality in virtue of which a thing of beauty is such a thing,—is timeless, unbegotten and, though its temporal embodiments must perish, is itself imperishable. "The purpose of art," said John LaFarge, "is commemoration." *In æternitatum pingo*, said Zeuxis, the Greek painter. One need not be an artist to understand that, in the words of Joshua Reynolds, "The idea of beauty in each species of being is perfect, *invariable*, divine." One need not be a Plato, to know, as I have elsewhere said, "that by a faculty of imaginative, mystical, idealizing discernment there is revealed to us, amid the fleeting beauties of Time, the immobile presence of Eternal beauties, immutable archetypes and source of the grace and loveliness beheld in the shifting scenes of the flowing world of sense." These archetypes,—perfect, unbegotten, everlasting,—these are the invariants which it is the aim and the function of art to discern and to represent.

It seems unnecessary to argue here that what has been said respecting the motivity of natural science and the like motivity of art is, *mutatis mutandis*, equally valid in education, in jurisprudence, in political science, in economics, in philosophy, and in religion, for the sufficient evidence is not far to seek and you have, I trust, an ample clue.

And so we are led to a grave and impressive thesis—a thesis regarding the principle which unites all the great forms of human endeavor. The thesis is that the unifying principle,—the central binding thread of human history,—the tie of comradeship among the spiritual enterprises of man,—is passion and search for things eternal: the thesis is that quest of invariance,—quest of abiding reality,—is itself the sovereign invariant in the changeful life of reason.

You are students of philosophy—students of the life of reason. To you, therefore, with the utmost confidence I commend the thesis as worthy of your best attention. As you meditate upon it, there will arise within you the bright vision of a great and inspiring task—a task that has not been performed nor even essayed. I mean the writing of A Critical History of Thought Viewed as the Quest of Invariance in a Fluctuant World. Taking Thought in its widest sense, embracing all the cardinal enterprises of the human spirit, such a history, if ever it be written, will have a scope greatly exceeding that of any extant " history of philosophy "; in addition to that and far more important, it will have, unlike such histories of philosophy, a natural unity, for it will have a unity derived from the unity of Thought itself. The history of Thought in our ever-growing, ever-perishing universe is the history of human endeavor to answer the question: What abides? The task of criticism as thus conceived is indeed immense. What abides? To collate and name and locate and order; to understand, describe and explain; to compare, judge and evaluate all of the chief responses that the religions and arts and sciences and philosophies and speculations of the ages have made to the question in divers tongues—these are the things which constitute and define the obligations of the historian of Thought. If you be ambitious—but what is ambition? Some one has conceived it to be a great man's desire to cast his shadow endlessly down the course of history. I prefer to regard it as the urge felt by great men to exercise the power of creation; for this power, the power of creative love—peculiar to man—is the power which, inheriting civilization as fruit of dead men's toil, receives it, not as the beasts receive the natural fruits of earth, but as spiritual capital to be more and more augmented—with ever-increasing speed in the course of successive generations—for the well-being of humankind including posterity. If you be ambitious in that sense, the task I have tried to signalize is worthy of your mettle—worthy of whatever genius you may have, of all the learning you can acquire, of all your talent for devotion and toil. Let me say finally, as I have already intimated, that the bearings thereupon of the mathematical theory of invariance are the bearings of a prototype and guide.

## LECTURE XII

### The Group Concept

THE NOTION SIMPLY EXEMPLIFIED IN MANY FIELDS—IS "MIND" A GROUP?—GROUPS AS INSTRUMENTS FOR DELIMITING DOCTRINES—CONNECTION OF GROUP WITH TRANSFORMATION AND INVARIANCE—THE IDEA FORESHADOWED IN THE AGES OF SPECULATION—THE PHILOSOPHY OF THE COSMIC YEAR—THE IDEA OF PROGRESS.

YOU will recall that near the close of the introductory lecture I gave a partial list of those mathematical terms which may be rightly regarded as denoting the pillar-concepts of the science. Among these are function, relation, transformation, invariance and group. In Lecture X we saw that the first three denote three aspects of one and the same thing seen from different points of view, and this thing,—whether we call it function or relation or transformation,—is sovereign—the central support not only of mathematics but of the entire edifice of science taken in the widest sense. In the closest logical connection therewith stand the two great concepts of invariance and group; so that I can hardly overemphasize the importance of your learning to associate the three notions as intimately in your thought as they are associated in fact: Transformation — Invariance — Group. Of transformation I endeavored to give some account in Lecture X and of invariance in Lecture XI. I invite your attention during the present hour to the notion of group. Even if I were a specialist in group theory,—which I am not,—I could not in one hour give you anything like an extensive knowledge of it, nor facility in its technique, nor a sense of its intricacy and proportions as known to its devotees, the priests of the temple. But the hour should suffice to start you on the way to acquiring at least a minimum of what a respectable philosopher should know of this fundamental subject; and such a minimum will include: a clear conception of what the term "group" means; ability to illustrate it copiously by means of easily understood examples to be found in all the cardinal fields of interest—number, space, time, motion, relation, play, work, the world of sense-data and the world of ideas; a glimpse of its intimate connections with the ideas of transformation and invariance; an inkling of it both as subject-matter and as an instrument for the delimitation and discrimination of doctrines; and discernment of the concept as vaguely prefigured in philosophic speculation from remote antiquity down to the present time.

I believe that the best way to secure a firm hold of the notion of group is to seize upon it first in the abstract and then, by comparing it with concrete examples, gradually to win the sense of holding in your grasp a living thing. In presenting the notion of group in the abstract, it is convenient to use the term *system*. This term has many meanings in mathematics and so at the outset we must clearly understand the sense in which the term is to be employed here. The sense is this: as employed in the definition of group, the term system means some definite class of things together with some definite rule, or way, in accordance with which any member of the class can be combined with any member of it (either with itself or any other member). For a simple example of such a system we may take for the class the class of ordinary whole numbers and for the rule of combination the familiar rule of addition. You should note that there are three and only three respects in which two systems can differ: by having different classes, by having different rules of combination, and by differing in both of these ways.

The definition of the term " group " is as follows.

Let $S$ denote a system consisting of a class $C$ (whose members we will denote by $a$, $b$, $c$ and so on) and of a rule of combination (which rule we will denote by the symbol o, so that by writing, for example, $aob$, we shall mean the result of combining $b$ with $a$). The system $S$ is called a group if and only if it satisfies the following four conditions:

($a$) If $a$ and $b$ are members of $C$, then $aob$ is a member of $C$; that is, $aob = c$, where $c$ is some member of $C$.

($b$) If $a$, $b$, $c$ are members of $C$, then $(aob)oc = ao(boc)$; that is, combining $c$ with the result of combining $b$ with $a$ yields the same as combining with $a$ the result of combining $c$ with $b$; that is, the rule of combination is associative.

($c$) The class $C$ contains a member $i$ (called the identical member or element) such that if $a$ be a member of $C$, then $aoi = ioa = a$; that is, $C$ has a member such that, if it be combined with any given member, or that member with it, the result is the given member.

($d$) If $a$ be a member of $C$, then there is a member $a'$ (called the reciprocal of $a$) such that $aoa' = a'oa = i$; that is, each member of $C$ is matched by a member such that combining the two gives the identical member.

Other definitions of the term " group " have been

proposed and are sometimes used. The definitions are not all of them equivalent but they all agree that to be a group a system must satisfy condition (*a*).

Systems satisfying condition (*a*) are many of them on that account so important that in the older literature of the subject they are called groups, or closed systems, and are now said to have "the group property," even if they do not satisfy conditions (*b*), (*c*) and (*d*). The propriety of the term "closed system" is evident in the fact that a system satisfying (*a*) is such that the result of combining any two of its members is itself a member—a thing *in* the system, not out of it.

*Various Simple Examples of Groups and of Systems that Are Not Groups.*—You observe that by the foregoing definition of group every group is a system; groups, as we shall see, are infinitely numerous; yet it is true that relatively few systems are groups or have even the group property—so few relatively that, if you select a system at random, it is highly probable you will thus hit upon one that is neither a group nor has the group property.

Take, for example, the system $S_1$ whose class $C$ is the class of integers from 1 to 10 inclusive and whose rule of combination is that of ordinary multiplication $\times$; $3 \times 4 = 12$; 12 is not a member of $C$, and so $S_1$ is not closed—it has not the group property.

Let $S_2$ have for its $C$ the class of all the ordinary integers, 1, 2, 3, . . . *ad infinitum*, and let o be $\times$ as before; as the product of any two integers is an integer, (*a*) is satisfied—$S_2$ is closed, has the group property; (*b*), too, is evidently satisfied, and so is (*c*), the identity element being 1 for, if $n$ be any integer, $n \times 1 = 1 \times n = n$; but (*d*) is not satisfied—none of the integers (except 1) composing $C$ has a reciprocal in $C$—there is, for example, no integer $n$ such that $2 \times n = n \times 2 = 1$; and so $S_2$, though it has the group property, is not a group.

Let $S_3$ be the system consisting of the class $C$ of all the positive and negative integers including zero and of addition as the rule of combination; you readily see that $S_3$ is a group, zero being the identical element, and each element having its own negative for reciprocal.

A group is said to be finite or infinite according as its $C$ is a finite or an infinite class and it is said to be Abelian or non-Abelian according as its rule of combination is or is not commutative—according, that is, as we have or do not have $a \circ b = b \circ a$, where $a$ and $b$ are arbitrary members of $C$. You observe that the group $S_3$ is both infinite and Abelian.

For an example of a group that is finite and Abelian it is sufficient to take the system $S_4$ whose $C$ is composed of the four numbers, $1$, $-1$, $i$, $-i$, where $i$ is $\sqrt{-1}$, and whose rule of combination is multiplication; you notice that the identical element is $1$, that $1$ and $-1$ are each its own reciprocal and that $i$ and $-i$ are each the other's reciprocal.

Let $S_5$ have the same $C$ as $S_3$ and suppose $\circ$ to be subtraction instead of addition; show that $S_5$ has the group property but is not a group. Show the like for $S_6$ in which $C$ is the same as before and $\circ$ denotes multiplication. Show that $S_7$ where $C$ is the same as before and $\circ$ means the rule of division, has not even the group property.

Consider $S_8$ where $C$ is the class of all the rational numbers (that is, all the integers and all the fractions whose terms are integers, it being understood that zero can not be a denominator) and where $\circ$ denotes $+$; you will readily find that $S_8$ is a group, infinite and Abelian. Examine the systems obtained by keeping the same $C$

and letting o denote subtraction, then multiplication, then division. Devise a group system where o means division.

If $S$ and $S'$ be two groups having the same rule of combination and if the class $C$ of $S$ be a proper part of the class $C'$ of $S'$ (*i.e.*, if the members of $C$ are members of $C'$ but some members of $C'$ are not in $C$), then $S$ is said to be a sub-group of $S'$. Observe that $S_3$ is a sub-group of $S_8$.

Show that $S_9$ is a group if its $C$ is the class of all real numbers and its o is +; note that $S_8$ is a sub-group of $S_9$ and hence that $S_3$ is a sub-group of a sub-group of a group. Is $S_9$ itself a sub-group? If so, of what group or groups? Examine the systems derived from $S_9$ by altering the rule of combination.

The most difficult thing that teaching has to do is to give a worthy sense of the meaning and scope of a great idea. A great idea is always generic and abstract but it has its living significance in the particular and concrete—in a countless multitude of differing instances or examples of it; each of these sheds only a feeble light upon the idea, leaving the infinite range of its significance in the dark; whence the necessity of examining and comparing a large number of widely differing examples in the hope that many little lights may constitute by union something like a worthy illustration; but to present these numerous examples requires an amount of time and a degree of patience that are seldom at one's disposal, and so it is necessary to be content with a selected few. And now here is the difficulty: if the examples selected be complex and difficult, they repel; if they be simple and easy, they are not impressive; in either case, the significance of the general concept in question remains ungrasped and unappreciated. I am going, however, to take the risk—to the foregoing illustrations of the group concept I am going to add a few further ones,—some of them very simple, some of them more complex,—trusting that the former may not seem to you too trivial nor the latter too hard.

Every one has seen the pretty phenomenon of a grey squirrel rapidly rotating a cylindrical wire cage enclosing it. It may rotate the cage in either of two opposite ways, senses or directions. Let us think of rotation in only one of the ways, and let us call any rotation, whether it be much or little, a *turn*. Each turn carries a point of the cage along a circle-arc of some length, short or long. Denote by $R$ the special turn (through 360°) that brings each point of the cage back to its starting place. Let $S10$ be the system whose $C$ is the class of all possible turns and whose o is addition of turns so that $aob$ shall be the whole turn got by following turn $a$ by turn $b$. You see at once that $S$ has the group property for the sum of any two turns is a turn; it is equally evident that the associative law—condition ($b$)—is satisfied. Note that $R$ is equivalent to no turn,—equivalent to rest,—equivalent to a zero turn, if you please; note that, if $a$ be a turn greater than $R$ and less than $2R$, then $a$ is equivalent to $a$'s excess over $R$; that, if $a$ be greater than $2R$ and less than $3R$, then $a$ is equivalent to $a$'s excess over $2R$; and so on; thus any turn greater than $R$ and not equal to a multiple of $R$ is equivalent to a turn less than $R$; let us regard any turn that is thus greater than $R$ as identical with its equivalent less than $R$; we have, then, to consider no turns except $R$ and those less than $R$—of which there are infinitely many; you see immediately that, if $a$ be any turn, $aoR = Roa = a$, which means that condition ($c$) is satisfied with $R$ for identical element. Next notice that for any turn $a$ there is a turn $a'$ such that $aoa' =$

$a' \circ a = R$. Hence $S_{10}$ is, as you see, a group. Show it to be Abelian. You will find it instructive to examine the system derived from $S_{10}$ by letting $C$ be the class of all turns (forward or backward).

Perhaps, you will consider the system suggested by the familiar spectacle of a ladybug or a measuring-worm moving round the rim or edge of a circular tub; or the system suggested by motions along the thread of an endless screw; or that suggested by the turns of the earth upon its axis; or that suggested by the motions of a planet in its orbit.

Do such examples give the meaning of the group concept? Each one gives it somewhat as a water-drop gives the meaning of ocean, or a burning match the meaning of the sun, or a pebble the meaning of the Rocky Mountains. Are they, therefore, to be despised? Far from it. Taken singly, they tell you little; but taken together, if you allow your imagination to play upon them, noting their differences, their similitudes, and the variety of fields they represent, they tell you much. Let us pursue them further, having a look in other fields.

Consider the field of the data of sense,—a field of universal interest,—and fundamental. We are here in the domain of sights and sounds and motions among other things. Are there any groups to be found here? Who, except the blind-born, are not lovers of color? Do the colors constitute a group? I mean sensations of color,—color sensations,—including all shades thereof and white and black. Denote by $S_{11}$ the system whose $C$ is the class of all such sensations and whose rule of combination is, let us say, the *mixing* of such sensations. But what are we to understand by the mixing of two color sensations? Suppose we have two small boxes of powder,—say of finely pulverized chalk,—a box of, say, red powder and a box of blue; one of the powders gives us the sensation red, the other the sensation blue; let us thoroughly mix the powders; the mixture gives us a color sensation; we agree to say that we have mixed the sensations and that the new sensation is the result of mixing the old ones. As the combination of any two color sensations is a color sensation, $S_{11}$ has, you see, the group property. Is it a group? Evidently condition (*b*) is satisfied. Are conditions (*c*) and (*d*) also satisfied?

Let us pass from colors to figures or shapes,—to figures or shapes, I mean, of physical or material objects,—rocks, chairs, trees, animals and the like,—as known to sense-perception. No doubt what we ordinarily call perception of an object's figure or shape is genetically complex, a result of experience contributed to by two or more senses, as sight, touch, motion; let us not, however, try to analyze it thus; let us take it at its face value—let us regard it as being, what it appears to be before analytic reflection upon it, a sense-given datum; and let us confine ourselves to the sense of sight. Here is a dog; its ears have shape; so, too, its eyes, its nose and the other features of its head; these shapes combine to make the shape or figure of the head; each other one of its visible organs has a shape of its own; these shapes all of them combine to make that thing which we call the shape or figure of the dog. Yonder is a table; it has a shape, and this is due to some sort of combination of the shapes of its parts—legs, top, and so on; upon it are several objects—a picture frame, a candlestick, some vases; each has a shape; the table and the other things together constitute one object—disclosed as such to a single glance of the eye; this object has a figure or shape due to the combined presence of the other shapes. In speaking of the dog and the table, I have been using the word "combination" in a very general sense. Can it, in this connection, be made precise enough for our use? Is it possible to find or frame a rule by which, any two visible shapes being given, these can be combined? If so, is the result of the combination a visible shape? If so, the system consisting of the rule and the total class of shapes has the group property. Does the system satisfy the remaining three conditions for a group?

And what of sounds—sensations of sound? Are sounds combinable? Is the result always a sound or is it sometimes silence? If we agree to regard silence as a species of sound,—as the zero of sound,—has the system of sounds the property of a group? There is the question of thresholds: sound is a vibrational phenomenon; if the rate of vibration be too slow or too great,—say, 100,000 per second,—no sound is heard. If you disregard the thresholds, has the sound system the group property? Is it a group? If so, what is the identical element? And what would you say is the reciprocal of a given sound or tone?

Consider other vibrational phenomena—as those of light or electricity. Can you so conceive them as to get group systems? Sharpen your questions and then carry them to physicists. You need have no hesitance—the service is apt to be mutual.

*The Infinite Abelian Group of Angel Flights.*—We are accustomed to think of ourselves as being in a boundless universe of space filled with what we call points any two of which are joined by what we call a straight line. Imagine one of those curious creatures which are to-day for most of us hardly more than figures of speech but which for many hundreds of years were very real and very lovely or very terrible things for millions of men, women and children and were studied and discoursed about seriously by men of genius: I mean angels. Angels can fly. Let us confine their flights to straight lines but impose no other restrictions. I am going to ask you to understand a flight as having nothing but length, direction and sense; if it is parallel to a given straight line, it has that line's direction; if it goes from $A$ towards $B$, it has that sense; if from $B$ toward $A$, the opposite sense; a flight from $A$ to $B$ and one from $C$ to $D$ are the same if they agree in length, direction and sense. Consider a flight $a$ from point $A$ to point $B$ followed by a flight $b$ from $B$ to $C$; you readily see that $a$ and $b$ are two adjacent sides of a parallelogram, one of whose diagonals is the direct flight $d$ from $A$ to $C$; $d$ is called the resultant or flight-sum of $a$ and $b$ because $d$ tells us how far the angel has finally got from the starting place; and so we write $a \circ b = d$. If flight $b'$ goes from $P$ to $Q$ but agrees with $b$ in length, direction and sense, we write $a \circ b' = d$ as before for, as already said, $b$ and $b'$ are one and the same. Now let $S_{12}$ denote the system whose $C$ is the class of all possible angel flights including rest, or zero flight, and whose rule of combination is flight summation as above explained; you see at once that $S_{12}$ is a closed system, has the group property, for the combination of any two flights is a flight; if $a$, $b$ and $c$ be three flights, we may suppose them to go respectively from $A$ to $B$, from $B$ to $C$, and from $C$ to $D$; consider $(a \circ b) \circ c$; $a \circ b = d$, the flight from $A$ to $C$; $d \circ c = e$, the flight from $A$ to $D$; now consider $a \circ (b \circ c)$; $b \circ c = d'$, flight from $B$ to $D$; $a \circ d' = e'$, flight from $A$ to $D$; so $e = e'$ and $(a \circ b) \circ c = a \circ (b \circ c)$; hence summation of flights is associative—condition ($b$) is satisfied. Condition ($c$) is satisfied with zero (0) flight for $i$;

for, if $a$ be any flight, it is plain that $a \circ 0 = 0 \circ a = a$. And condition ($d$) is satisfied for it is evident that $a \circ a' = a' \circ a = 0$ where $a'$ and $a$ agree in length and direction but are opposite in sense. Hence the system of angel flights is a group. And it is easy to see that it is both infinite and Abelian.

What I have here called an angel flight is known in mathematics and in physics as a *vector;* a vector has no position—it has its essential and complete being in having a length, a direction and a sense. And so, you see, the system composed of the vectors of space and of vector addition as a rule of combination is an infinite Abelian group.

*Connection of Groups with Transformations and Invariants.*—Let us have another look at our angel flights, or vectors. I am going to ask you to view them in another light. Let $V$ be any given vector—that is, a vector of given length, sense and direction; where does it begin and where does it end? A moment's reflection will show you that every point in the universe of space is the beginning of a vector identical with $V$ and the end of a vector identical with $V$. Though these vectors are but one, it is convenient to speak of them as many equal vectors—having the same length, direction and sense. Let the point $P$ be the beginning of a $V$ and let the point $Q$ be its end. Let us now associate every such $P$ with its $Q(P \rightarrow Q)$; we have thus transformed our space of points into itself in such wise that the *end* of each $V$ is the *transform* of its *beginning;* call the transformation $T$; let us follow it with a transformation $T'$ converting the beginnings of all vectors equal to a given vector $V'$ into their corresponding ends. What is the result? Notice that $T$ converted $P$ into $Q$ and that $T'$ then converted $Q$

into $Q'$, the end of the $V'$ beginning at $Q$; now there is a vector beginning at $P$ and running direct to $Q'$; and so there is a transformation $T''$ converting $P$ into $Q'$; it is this $T''$ that we shall mean by $ToT'$. Without further talk, you see that our group of angel flights, or vectors, now appears as an infinite Abelian group of *transformations* (of our space of points into itself). Such transformations do not involve motion in fact; it is customary, however, for mathematicians to call them motions, or *translations*, of space; $T$, for example, being thought of and spoken of as a translation of the whole of space (as a rigid body) in the direction and sense of $V$ and through a distance equal to $V$'s length. In accordance with this stimulating, albeit purely figurative, way of speaking, the group in question is the group of the translations of our space.

We are now in a good position to glimpse the very intimate connection between the idea of group and the idea of invariance. Suppose we are given a group of transformations; one of the big questions to be asked regarding it is this: what things remain unaltered,—remain invariant,—under each and all the transformations of the group? In other words, what property or properties are common to the objects transformed and their transforms? Well, we have now before us a certain group of space transformations—the group of translations. Denote it by $G$. Each translation in $G$ converts (transforms, carries, moves) any point into a point, and so converts any configuration $F$ of points,—any geometric figure,—into some configuration $F'$. What remains unchanged? What are the invariants? It is obvious that one of the invariants,—a very important one,—is distance; that is, if $P$ and $Q$ be any two points and if their transforms under some translation be respectively $P'$ and $Q'$, then the distance between $P'$ and $Q'$ is the same as that between $P$ and $Q$; another is order among points—if $Q$ is between $P$ and $R$, $Q'$ is between $P'$ and $R'$; you see at once that angles, areas, volumes, shapes are all of them unchanged: in a word, congruence is invariant—if a translation convert a figure $F$ into a figure $F'$, $F$ and $F'$ are congruent. Of course congruence is invariant under all the translations having a given direction. Do these constitute a group? Obviously they do, and this group $G'$ is a sub-group of $G$. Congruence is invariant under $G'$; it is also invariant under $G$; $G'$ is included in $G$; it is natural, then, to ask whether $G$ itself may not be included in a still larger group having congruence for an invariant. The question suggests the inverse of the one with which we set out. The former question was: given a group, what are its invariants? The inverse question is: given an invariant, what are its groups and especially its largest group? This question is as big as the other one. Consider the example in hand. The given invariant is congruence. Is $G$,—the group of translations,—the largest group of space transformations leaving congruence unchanged? Evidently not; for think of those space transformations that consist in rotations of space (as a rigid body) about a fixed line (as axis); if such a rotation converts a figure $F$ into $F'$, the two figures are congruent. It is clear that the same is true if a transformation be a *twist*—that is, a simultaneous rotation about, and translation along, a fixed line. All such rotations and twists together with the translations constitute a group called the group of *displacements* of space; it includes all transformations leaving congruence invariant. This group, as a little reflection will show you, has many sub-groups, infinitely many; the set of displacements leaving a specified point invariant is such a sub-group; the set leaving two given points unchanged is another. How is the latter related to the sub-group leaving only one of the two points invariant? Is there a displacement leaving three non-collinear points invariant? Do the displacements leaving a line unchanged constitute a group? Such questions are but samples of many that you will find it profitable to ask and to try to answer.

For the sake of emphasis, permit me to repeat the two big questions: (1) Given a group of transformations, what things are unchanged by them? (2) Given something—an object or property or relation, no matter what —that is to remain invariant, what are the groups of transformations, and especially the largest group, that leave the thing unaltered? You may wish to say: I grant that the questions are interesting, and I do not deny that they are big—big in the sense of giving rise to innumerable problems and big in the sense that many of the problems are difficult; but I do not see that they are big with importance. Why should I bother with them? In reply I shall not undertake to say why you *should* bother with them; it is sufficient to remind you that as human beings you cannot help it and you do not desire to do so. In the preceding lecture, we saw that the sovereign impulse of Man is to find the answer to the question: what abides? We saw that Thought,—taken in the widest sense to embrace art, philosophy, religion, science, taken in *their* widest sense,—is the quest of invariance in a fluctuant world. We saw that the craving and search for things eternal is the central binding thread of human history. We saw that the passion for abiding reality is itself the supreme invariant in the life of reason. And we saw that the bearings of the mathematical theory of invariance upon the universal enterprise of Thought are the bearings of a prototype and guide. It is evident that the same is true of the mathematical theory of groups. Our human question is: what abides? As students of thought and the history of thought, we have learned at length that the question can not be answered fully at once but only step by step in an endless progression. And now what are the steps? You can scarcely fail to see, if you reflect a little, that each of them,—whether taken by art or by science or by philosophy,—consists virtually in ascertaining either the invariants under some group of transformations or else the groups of transformations that leave some thing or things unchanged.

*Groups as Instruments for Defining, Delimiting, Discriminating and Classifying Doctrines.*—The foregoing question (2) has another aspect, which I believe to be of profound interest to all students except those, if there be such, who are insensate to things philosophical. I mean that, if and whenever you ascertain the group of all the transformations that leave invariant some specified object or objects of thought, you thereby define perfectly some actual (or potential) branch of science—some actual (or potential) doctrine. I will endeavor to make this fact evident by a few simple examples, and I will choose them from the general field of geometry, though, as you will perceive, such examples might be taken from other fields.

For a first example, consider the above-mentioned group $D$ of the displacements of our space. I say that this group defines a geometry of space, which may be called the geometry of displacements. It defines it by defining, or delimiting, its subject-matter. What is its subject-matter? What does the geometry study? The two questions are not equivalent. It studies all the figures in space but it does not study all their properties. Its subject-matter consists of those properties which it does study. What are these? They are those and only those properties (of figures) that remain invariant under all displacements but under no other transformations of space. The geometry of displacements might be called congruence geometry. It includes the greater part of the ordinary geometry of high school but not all of it, for the latter deals, for example, with similarity of figures; similarity is indeed invariant under displacements, but it is also invariant under other transformations—the so-called similitude transformations, to be mentioned presently.

For a second example, consider the following. I may wish to confine my study of spatial figures to their *shape*. The doctrine thus arising may be called the geometry of shape, or shape geometry. If I tell you that I am studying shape geometry and you ask me what I mean by the geometry of shape, there are two ways in which I may answer your question. One of the ways requires me to define the term shape—shape of a geometric figure; the other way,—the group way,—does not. Let us examine them a little. I have never seen a mathematical definition of shape, but it may, I believe, be precisely defined as follows. We must distinguish the three things: sameness of shape; shape of a given figure; and shape of a figure. I will define the first; then the second in terms of the first; and, finally, the third in terms of the second. Two figures $F$ and $F'$ will be said to have the same shape if and only if it is possible to set up a one-to-one correspondence between the points of $F$ and those of $F'$, such that, $AB$ and $CD$ being any distances between points of

$F$, and $A'B'$ and $C'D'$ being the distances between the corresponding points of $F'$, $AB/CD = A'B'/C'D'$. Two figures having the same shape will be said to be similar, and conversely. Having defined sameness of shape, or similarity, of figures, I will define the term "shape of a given figure" as follows: if $F$ be a given, or specific, figure, the shape of $F$ is the class $\sigma$ of all figures similar to $F$; it is evident that, if $F$ and $F'$ are not similar, the class $\sigma$ and the class $\sigma'$—the shape of $F'$—have no figures in common; it is evident, moreover, that there are as many $\sigma$'s as there are figure shapes. And now what do we mean by the general term shape, or—what is tantamount—shape of a figure? What the answer must be is pretty obvious: shape is the class $\Sigma$ of all the $\sigma$'s. Note that $\Sigma$ is a class of classes and that any $\sigma$ is a class of (similar) figures. Having defined the general term shape, I have, you see, virtually answered your question: what is the geometry of shape?

Let us now see how the question may be answered by means of the group concept. Two congruent figures are clearly similar, and so similarity is invariant under the group of displacements. But you readily see that there are many other transformations under which similarity is invariant. Let $O$ be a point; consider the bundle of straight lines,—all the lines through $O$,—having $O$ for its vertex; every point of space is on some line of the bundle; let $k$ be any real number (except zero); let $P$ be any point and let $P'$ be such a point on the line $OP$ that the segment $OP' = k \times$ segment $OP$; you see that each point $P$ is transformed into a point $P'$; the transformation is called *homothetic;* its effect, if $k$ be positive and exceed 1, is a uniform expansion of space from $O$ outward in all directions; if $k$ be positive and less than 1, the effect is contraction toward $O$; if $k$ be negative, the effect is such an expansion or contraction, followed or preceded by reflection in $O$ as in a mirror; distances are clearly not preserved, but distance ratios are; that is, if $A$, $B$, $C$ be any three points and if their respective transforms under some homothetic transformation be $A'$, $B'$, $C'$, it is evident that $AB/BC = A'B'/B'C'$; accordingly, if $F'$ be the transform of a figure $F$, the figures are similar,—they have the same shape but not the same size,—they are not congruent: similarity is, then, invariant under all homothetic transformations, and hence under combinations of them with one another and with displacements; the displacements and the homothetic transformations together with all such combinations constitute a group called the group of *similitude transformations;* it contains all and only such space transformations as leave similarity unchanged. Here, then, is our group definition of shape geometry: namely, *the geometry of shape is the study of that property of figures which is common to every figure and its transforms under each and all transformations of the similitude group*. Observe that this definition, unlike the former one, employs neither the notion of shape in general, nor that of the shape of a given figure; it employs only the notion of similarity—sameness of shape.

We ought, I think, to consider one more example of how a group of transformations serves to determine the nature and limits of a doctrine and thereby to discriminate the doctrine from all others. I will again take a geometric example, but for the sake of simplicity I will choose it from the geometry of the plane (instead of space). Before presenting it, let me adduce a yet simpler example of the same kind taken from the geometry of points in a straight line. In Lecture IV, I explained what is meant by a projective line—an ordinary line endowed with an "ideal" point, or point at infinity, where the line meets all lines parallel to it. Let $L$ be a projective line. In the preceding lecture, we gained some acquaintance with the transformations of the form

$$(1)\quad x' = \frac{ax+b}{cx+d}$$

where the coefficients are any real numbers such that $ad-bc \neq 0$; we saw that there are $\infty^3$ such transformations and that each of them converts the points of $L$ into the points of $L$ in such a way that the anharmonic ratio of any four points is equal to the anharmonic ratio of their transforms. Distances are not preserved; neither are the ordinary ratios of distances preserved; hence neither congruence nor similarity is invariant; no relation among points—that is, no property of figures (for here a figure is simply a set of points on $L$)—is invariant except anharmonic ratio and properties expressible in terms of the latter; no other transformations leave these properties invariant. By a little finger work you can prove in a formal way that these transformations constitute a group. I will merely indicate the procedure, leaving it to you to carry it out if you desire to do so. The transformations differ only in their coefficients. Let $(a_1, b_1, c_1, d_1)$, $(a_2, b_2, c_2, d_2)$, $(a_3, b_3, c_3, d_3)$ be any three of the transformations; consider the first and second; the rule o of combination is to be: operate with the first and then on the result with the second. The first converts point $x$ into point $x'$:

$$(2)\quad x' = \frac{a_1x+b_1}{c_1x+d_1};$$

the second converts $x'$ into point $x''$:

$$(3)\ x''=\frac{a_2x'+b_2}{c_2x'+d_2};$$

in (3) replace $x'$ by its value given by (2), simplify and then notice that you have a transformation of form (1) converting $x$ directly into $x''$. This shows that the set of transformations have the group property. To show that they obey the associative law, it is sufficient to perform the operations

$$(4)\ (a_1, b_1, c_1, d_1)\text{o}[(a_2, b_2, c_2, d_2)\text{o}(a_3, b_3, c_3, d_3)],$$

$$(5)\ [(a_1, b_1, c_1, d_1)\text{o}(a_2, b_2, c_2, d_2)]\text{o}(a_3, b_3, c_3, d_3),$$

and then to observe that the results are the same. The identical element $i$ is $(a, 0, 0, a)$—that is, the transformation, $x'=x$. The inverse of any transformation $(a, b, c, d)$ is $(-d, b, c, -a)$ for you can readily show that combination of these gives $(a, 0, 0, a)$.

The fact to be specially noted is that this group of so-called homographic transformations defines a certain kind of geometry in the line $L$—namely, its projective geometry. In a line there are various geometries; among these the projective geometry is characterized by its subject-matter, and its subject-matter consists of such properties of point sets, or figures, as remain invariant under its homographic group.

And now I come to the example alluded to a moment ago—the one to be taken from geometries in (or of) a plane. The foregoing homographic group—in a line, a one-dimensional space—has an analogue in a projective plane, another in ordinary 3-dimensional projective space, another in a projective space of four dimensions, and so on *ad infinitum.* What is the analogous group for a plane? In a chosen plane take a pair of axes as explained in Lecture V, and consider the pair of equations

$$(1)\quad \begin{cases} x' = \dfrac{Ax+By+C}{Gx+Hy+I}, \\ y' = \dfrac{Dx+Ey+F}{Gx+Hy+I}, \end{cases}$$

where the coefficients are any real numbers such that

$$(1')\quad \begin{vmatrix} A & B & C \\ D & E & F \\ G & H & I \end{vmatrix} \neq 0;$$

*i.e.*, such that

$$(1')\quad AEI-CEG+CDH-BDI+BFG-AFH\neq 0.$$

The coefficients furnish eight independent ratios,—called "parameters,"—and so we have $\infty^8$ equation pairs of form (1); choose any one of them and notice that it is a transformation converting the number pair $(x, y)$ into a number pair $(x', y')$, and so converting the point $(x, y)$ into a point $(x', y')$; owing to the inequality (1'), any point $(x, y)$ is transformed into a definite point $(x', y')$. In any line $ax'+by'+c=0$, replace $x'$ and $y'$ by their values given by (1), and simplify; the resulting equation is of first degree in $x$ and $y$ and hence represents a line; hence, you see, points of a straight line are converted into points of a straight line—the relation, collinearity, is preserved; so is copunctality—a pencil of lines has a pencil for its transform; you can readily show that order is not preserved, nor distances nor ordinary distance-ratios, nor angles; hence, if the figure $F'$ be the transform of $F$, the two figures are, in general, neither congruent nor similar;

we say, however, that $F$ and $F'$ are projective because, as can be proved, the anharmonic ratio of any 4 points (or lines) of either is equal to that of the corresponding (transform) points (or lines) of the other. By the method indicated for the homographic transformations of a line, you can prove that the $\infty^8$ transformations of form (1) constitute a group.

Just as a *point* of the plane has two coordinates $(x, y)$, so a *line* depends on two coordinates; there are various ways to see that such is the case; an easy way is this: the line, $ax+by+c=0$, depends solely upon the ratios $(a : b : c)$ of the coefficients; these three ratios are not independent—two of them determine the third one; you thus see that the line depends upon only two independent variables—it has, like the point, two coordinates; let us denote them by $(u, v)$. Now consider the transformations

$$(2) \quad \begin{cases} u' = \dfrac{Ju+Kv+L}{Pu+Qv+R}, \\ v' = \dfrac{Mu+Nv+O}{Pu+Qv+R}, \end{cases}$$

where the coefficients are subject to a relation like (1'). We saw that a transformation (1) converts points into points directly and lines into lines indirectly; just so, a transformation (2) converts lines into lines directly and points into points indirectly; hence the group of line-to-line transformations (2) is essentially the same as the foregoing group of point-to-point transformations (1). This latter group is called the group of *collineations* of (or in) the plane.

I am going now to ask you to notice an ensemble of transformations (of the plane) that are neither point-to-point nor line-to-line transformations but are at once point-to-line and line-to-point transformations. These are represented by the pair of formulas

$$(3)\quad \begin{cases} u=\dfrac{ax+by+c}{gx+hy+i}, \\ v=\dfrac{dx+ey+f}{gx+hy+i}, \end{cases}$$

where the coefficients are again subject to a relation like (1′). Any such transformation converts a point $(x, y)$ into a line $(u, v)$; now operate on the points of this line by the same transformation or another one of form (3); the points are converted into lines constituting a pencil having a vertex, say $(x', y')$; thus the combination converts point $(x, y)$ into point $(x', y')$—it is a point-to-point transformation and hence belongs to the group of collineations; you thus see that the set of transformations (3) is not a group; but this set and the collineations together constitute a group including the collineations as a subgroup. This large group is called the *Group of Projective Transformations of the Plane.* Why? Because every transformation in it and no other transformation leaves all anharmonic ratios unchanged.

What is the projective geometry of the plane? The group now in hand enables us to answer the question perfectly. The answer is: Projective plane geometry is that geometry which studies such and only such properties of plane figures as remain invariant under the group of projective transformations.

In reading the essays of the late Henri Poincaré you have met the statement: "Euclidean space is simply a group." The foregoing examples should enable you to understand its meaning. And they should lead you to surmise—what is true—that answers like the foregoing ones are available for similar questions regarding all geometries of a space of any number of dimensions and —what is more—regarding mathematical doctrines in general. Whatsoever things are invariant under all and only the transformations of some group constitute the peculiar subject-matter of some (actual or potential) branch of knowledge. And you see that every such group-defined science views its subject matter under the aspect of eternity.

*A Question for Psychologists.*—Before closing this lecture, I desire to speak briefly of two additional matters connected with the notion of group: one of the matters is psychological: the other is historical. Being students of philosophy, you are obliged to have at least a good secondary interest in psychology. I wish to propose for your future consideration a psychological question—one which psychologists (I believe) have not considered and which, though it has haunted me a good deal from time to time in recent years, I am not yet prepared to answer confidently. The question is: Is mind a group? Let us restrict the question and ask: Is mind a closed system—that is, has it the group property? Some of the difficulties are immediately obvious. In order that the question shall have definite meaning, it is necessary to think of mind as a system composed of a class of things and a rule, or law, of combination by which each of the things can be combined with itself and with each of the other things. We may make a beginning by saying that the required class is the class of mental phenomena. But what does the class include? What phenomena are members of it? Some phenomena,—feeling, for example, seeing, hearing, tasting, thinking, believing, doubting, craving, hoping, expecting,—are undoubtedly mental; others seem not to be—as what I see, for example, what I taste, what I believe, and the like. Here are difficulties. You will find a fresh and suggestive treatment of them in Bertrand Russell's *The Ultimate Constituents of Matter*, found in the author's *Mysticism and Logic* and especially in his *Analysis of Mind*. Let us suppose that, despite the difficulties in the way, you have decided what you are going to call mental phenomena. You have then to consider the question of their combination. We do habitually speak of combining mental things: hoping, for example, is, in some sense, a union, or combination, of desiring and expecting; the feeling called patriotism is evidently a combination of a pretty large variety of feelings; in the realm of ideas,—which you will probably desire to include among mental phenomena,—we have seen that, for example, the idea named " vector " is a union, or combination, of the ideas of direction, sense and length; and so on —examples abound. But does combination as a process or operation have the same meaning in all such cases? It seems not. What, then, is your rule $o$ to be? Possibly the difficulty could be surmounted as follows: if you discovered that some mental phenomena are combinable by a rule $o_1$, others by a rule $o_2$, still others by a rule $o_3$, and so on, thus getting a finite number of particular rules, you could then take for a more general rule $o$ the disjunction, or so-called logical sum, of the particular ones; that is, you could say that rule $o$ is to be: $o_1$ or $o_2$ or $o_3$ or . . . or $o_n$; so that two phenomena would be combinable by $o$ if they were combinable by one or more of the rules $o_1, o_2, \ldots, o_n$. If you thus found a rule by which every two of the phenomena you had decided to call *mental* admitted of combination, then your final question would be: is the result of every such combination a mental phenomenon? That is not quite the question; for under the rule two phenomena might be combinable in two or more ways, and some of the results might (conceivably) be mental and the others not; so your question would be: can every two mental phenomena be combined under your rule so as to yield a mental phenomenon? If so, then mind, as you had defined it, would have the group property under some rule of combination. If you found mind to have the group property under some rule or rules but not under others, you would be at once confronted with a further problem, which I will not tarry to state.

We have been speaking of mind—of mind in general. Similar questions,—perhaps easier if not more fruitful questions,—can be put respecting particular minds—your mind, mine, John Smith's. Has every individual mind the group property? Has no such mind the property? Have some of them the property and others not?

It seems very probable that the answer to the first of the questions must be negative. There are at all events some minds having (presenting, containing) mental phenomena that are definitely combinable in a way to yield mental phenomena that nevertheless do not belong to those minds. What is meant is this: a given mind may possess certain ideas which are combinable so as to form another idea; it may happen that the mind in question is incapable of grasping the new idea. Such minds have no doubt come under the observation of every experienced teacher. I myself have seen many such cases and remember one of them very vividly: that of a young woman who had made a brilliant record in undergraduate collegiate mathematics including the elements of analytical geometry and calculus; and who, encouraged by this initial success, aspired to the mathematical doctorate and entered seriously upon higher studies essential thereto; it was necessary for her to grasp more and more complicate concepts formed by combining ideas she already possessed; after no long time she reached the limit of her ability in this matter,—a fact first noticed by her instructors and then by herself,—and being a woman of good sense, she abandoned the pursuit of higher mathematics. I may add that subsequently she gained the doctorate in history. It may be that some minds are not thus limited. It may be that a genius of the so-called universal type,—an Aristotle, for example, or a Leibniz or a Leonardo da Vinci,—is one whose mind has the group property. May I leave the questions for your consideration in the days to come?

*The Group Concept Dimly Adumbrated in Early Philosophic Speculation.*—The mathematical theory of groups is immense and manifold; in the main it is a work of the last sixty years; even the germ of it seems not to antedate Ruffini and Lagrange. Why so modern? Why did not the concept of a closed system,—of a system having the group property,—come to birth many centuries earlier? The elemental constituents of the concept,—the idea of a class of things, the idea of anything being or not being a member of a class, the idea of a rule or law of combination,—all these were as familiar thousands of years ago as they are now. The question is one of a host of similar questions whose answers, if ever they be found, will constitute what in a previous lecture I called the yet unwritten history of the development of intellectual curiosity. Who will write that history? And when?

The fact that the precise formation of the mathematical concept of group is of so recent date is all the more curious because an idea closely resembling that of group has haunted the minds of a long line of thinkers and is found stalking like a ghost in the mist of philosophic speculation from remote antiquity down even to Herbert Spencer. I refer to those worldwide, age-long, philosophic speculations which, because of their peculiar views of the universe, may be fitly called the *Philosophy of the Cosmic Cycle or Cosmic Year*. This philosophy, despite the spell of a certain beauty inherent in it, has lost its vogue. To-day we are accustomed to thinking of the universe as undergoing a beginningless and endless evolution in course of which no aspect or event ever was or ever will be exactly repeated. In sharpest contrast with that conception, the philosophy of the cosmic cycle regards all the changes of which the universe is capable as constituting an immense indeed but finite and closed system of transformations, which follow each other in definite succession, like the spokes of a gigantic revolving wheel, until all possible changes have occurred in the lapse of a long but finite period of time—called a cosmic cycle or cosmic year—whereupon everything is repeated precisely, and so on and on without end. This philosophy, I have said, has lost its vogue; but, if the philosophy be true, it will regain it, for, if true, it belongs to the cosmic cycle and hence will recur. The history of the philosophy of the cosmic year is exceedingly interesting and it would, I believe, be an excellent subject for a doctor's dissertation. The literature is wide-ranging in kind, in place and in time. Let me cite a little of it as showing how closely its central idea resembles the mathematical concept of a cyclic group.

In his *Philosophie der Griechen* (Vol. III, 2nd edition) Zeller, speaking of the speculations of the Stoics, says:

> Out of the original substance the separate things are developed according to an inner law. For inasmuch as the first principle, according to its definition, is the creative and formative power, the whole universe must grow out of it with the same necessity as the animal or the plant from the seed. The original fire, according to the Stoics and Heraclitus, first changes to "air" or vapor, then to water; out of this a portion is precipitated as earth, another remains water, a third evaporates as atmospheric air, which again kindles the fire, and out of the changing mixture of these four elements there is formed,—from the earth as center,—the world. . . . Through this separation of the elements there arises the contrast of the active and the passive principle: the soul of the world and its body. . . . But as this contrast came in time, so it is destined to cease; the original substance gradually consumes the matter, which is segregated out of itself as its body, till at the end of this world-period a universal world conflagration brings everything back to the primeval condition. . . . But when everything has thus returned to the original unity, and the great world-year has run out, the formation of a new world begins again, which is so exactly like the former one that in it all things, persons and phenomena, return exactly as before; and in this wise the history of the world and the deity . . . moves in an endless cycle through the same stages.

A similar view of cosmic history is present in the speculations of Empedocles, for whom a cycle consists of four great periods: Predominant Love—a state of complete aggregation; decreasing Love and increasing Hate; predominant Strife—complete separation of the elements; decreasing Strife and increasing Love. At the end of this fourth period, the cycle is complete and is then repeated—

the history of the universe being a continuous and endlessly repeated vaudeville performance of a single play.

Something like the foregoing seems to be implicit in the following statement by Aristotle in the Metaphysics:

> Every art and every kind of philosophy have probably been invented many times up to the limits of what is possible and been again destroyed.

And in Ecclesiastes (III, 15):

> That which hath been is now; and that which is to be hath already been.

Even Herbert Spencer at the close of his *First Principles* speaks as follows:

> Thus we are led to the conclusion that the entire process of things, as displayed in the aggregate of the visible universe, is analogous to the entire process of things as displayed in the smallest aggregates. Motion as well as matter being fixed in quantity, it would seem that the change in the distribution of matter which motion effects, coming to a limit in whatever direction it is carried, the indestructible motion necessitates a reverse redistribution. Apparently the universally coexistent forces of attraction and repulsion, which necessitate rhythm in all minor changes throughout the universe, also necessitates rhythm in the totality of changes—alternate eras of evolution and dissolution. And thus there is suggested the conception of a past during which there have been successive evolutions analogous to that which is now going on; and a future during which successive other evolutions may go on—ever the same in principle but never the same in concrete result.

Spencer was, you know, but poorly informed in the history of thought and he was probably not aware that the main idea in the lines just now quoted was ancient two thousand years before he was born. You should note that the Spencerian universe of transformations narrowly escapes being a closed system—escapes by the last six words of foregoing quotation. The cosmic cycles do indeed follow each other in an endless sequence—" ever the same in principle but never the same in concrete result." The repetitions are such " in principle " only, not in result—there is always something new.

One of the very greatest works of man is the *De Rerum Natura* of Lucretius—immortal exposition of the thought of Epicurus, " who surpassed in intellect the race of man and quenched the light of all as the ethereal sun arisen quenches the stars." Neither a student of philosophy nor a student of natural science can afford to neglect the reading of that book. For, although it contains many,—very, very many,—errors of detail,—some of them astonishing to a modern reader,—yet there are at least four great respects in which it is unsurpassed among the works that have come down from what we humans, in our ignorance of man's real antiquity, have been wont to call the ancient world: it is unsurpassed, I mean, in scientific spirit; in the union of that spirit with literary excellence; in the magnificence of its enterprise; and in its anticipation of concepts among the most fruitful of modern science. For such as can not read it in the original there are, happily, two excellent English translations of it—one by H. A. J. Munro and a later one by Cyril Bailey. Of this work I hope to speak further in a subsequent lecture of this course. My purpose in citing it here is to signalize it as being perhaps the weightiest of all contributions to what I have called the philosophy of the cosmic year. The Lucretian universe though not a finite system, is indeed a closed system, of transformations: any event, whether great or small, that has occurred in course of the beginningless past has occurred infinitely many times and will recur infinitely many times in course of an unending future; and nothing can occur that has not occurred,—there never has been and there never will be aught that is new,—every occurrence is a recurrence. Let me say parenthetically, in passing, that such a concept of the universe is damnably depressing but not more so than the regnant mechanistic hypothesis of modern natural science. In relation to this hypothesis you should by no means fail to read and digest Professor W. B. Smith's great address: "Push or Pull?" (*Monist*, Vol. XXIII, 1913). See also Smith's "Are Motions Emotions?" (*Tulane Graduates' Magazine*, Jan., 1914). And you should read J. S. Haldane's *Life, Mechanism and Personality.*

Should you desire to pursue the matter further either with a view to noting speculative adumbrations of the group concept or, as I hope, with the larger purpose of writing a historical monograph on the philosophy of the Cosmic Cycle, the following references may be of some service as a clue.

"The Dream of Scipio" in Cicero's *Republic* (Hardingham's translation).

Michael Foster's *Physiology.*

Lyell's *Principles of Geology.*

The fourth *Eclogue* of Virgil (verses 31–36).

Rückert's poem *Chidher.*

Moleschott's *Kreislauf des Lebens.*

Clifford's "The First and Last Catastrophe" in his *Lectures and Essays.*

Inge's *The Idea of Progress* (being the Romanes Lecture, 1920).

## LECTURE XIII

### Variables and Limits

A GLANCE AT THE SHADOWY BACKGROUND OF SCIENTIFIC IDEAS—THE MEANINGS OF VARIABLE AND CONSTANT—RANGES OF VARIATION AND THE IDEA OF NEIGHBORHOOD—VARIOUS DEFINITIONS OF LIMIT CLARIFIED BY SIMPLE EXAMPLES—THE SCANDAL OF A STARVING NURSE IN THE RICHEST LAND KNOWN TO HISTORY.

In the introductory lecture, I spoke at length on the mathematical obligations of philosophy. In preparing to discharge them it is imperatively necessary for students of philosophy to gain genuine understanding of those great concepts that are as vital organs in the body of mathematics, giving the science not only its life but its character and its power. It is one of the aims of these lectures to assist students primarily interested in philosophy to gain such an understanding. In pursuance of that aim it is, I believe, essential to devote one or two lectures to the nature and the significance of the mathematical concept denoted by the term "limit." There are in current use several concepts of the term, but for the present we may speak as if there were only one. The importance and the power of the concept have been so long recognized by mathematicians that the notion of limit and what is often called the method of limits have found their way down into text-books of algebra, geometry and trigonometry designed for high schools. In high school you probably learned something of the lingo of limits; if you really there grasped the ideas involved, you were extraordinary pupils or were very fortunate in your teachers or both. I say this because in collegiate freshman classes I have met many students who in their preparation for college had been exposed to the notion and method of limits, and I have the impression that among them there were very few or none whose wisdom in the matter was appreciably more than phraseological; in the case of most it was even less. The explanation is not far to seek: the concept of limit is a subtle concept and the right use of it in mathematical argumentation is a delicate process; these two things can not be caught, so to speak, on the fly; they require to be reflected upon again and again; they are among the things that require to be pondered; but such meditation, such deliberation upon elusive scientific abstractions, is one of the things which boys and girls of the indicated age will not do; it is not their fault; if they did it they would not be boys and girls; they would be seasoned logicians and philosophers. If such understanding of the nature and significance of limits as you may be supposed to have acquired in high school has not been deepened and refined by subsequent study, the intervening years have probably so dimmed your impressions of the matter that you are now fortunately able to approach the subject afresh, bringing to bear upon it critical power of sufficient maturity.

In endeavoring to analyze the concept of limit it becomes immediately evident that it involves the notion of *variable* and the notion of *constant*, together with such other notions as that of variable and that of constant themselves involve. What do mathematicians mean by the terms variable and constant? The great majority of professional mathematicians take the meanings of these terms mainly for granted; they do so because they are so occupied in teaching mathematics or in extending its superstructure as to have but little or no interest in "the nice sharp quillets of the law" as revealed in its logical foundations; in the foregoing lectures I have followed the actual practice of mathematicians in respect to the two terms in question; I have, that is, freely employed the terms, not indeed quite unconsciously, but without apology and without explanation, believing that such use would lead to no appreciable misunderstanding. Now, however, before attempting to give a formal definition of the great term "limit," it will be worth while, I believe, to glance at its shadowy background—to reflect a little on the meanings of the yet greater notions upon which the concept of a limit depends.

*Meaning of the Terms Variable and Constant.*—What is the mathematical meaning of the term variable? It is natural to answer, and to answer very confidently, that a variable is something that varies or changes, like the position of an object in motion, the time of day, the length of a burning cigar. We are going to see that this answer, though perfectly natural, is entirely wrong. In a previous lecture I drew your attention to the fact that mathematicians habitually employ highly figurative speech; especially that they are constantly employing dynamic terms in describing static facts. In particular I pointed out that it is a common practice of mathematicians to use the dynamic term transformation,—suggesting change, variation, transmutation,—to denote what is in fact a static thing, namely, a relation—something that is unchanging, eternal. There is thus a striking incongruity between what is said, or the manner of saying it, and what is meant. We are going to see that there is a like incongruity between the meaning of the mathematical term " variable " and the manner in which mathematicians habitually speak of variables. I do not condemn their manner of speech; I approve of it because it is stimulating and economical and because it does not, except in certain very fundamental questions, lead to error; but such incongruity is a thing which you as philosophers should carefully notice in the interest of clarity and critical understanding. Mathematicians do indeed habitually speak of variables as if the mathematical concept of a variable were the concept of something whose essential nature is to suffer change; that is to say, when they use some symbol, say, $x$, to denote what they call a variable, they familiarly speak of the variable $x$ as altering its value, as increasing or decreasing, as growing large or growing small, as approaching or not approaching this or that, and so on; yet, in spite of such a way of speaking, what they really mean by the term " variable " essentially involves no idea of change whatever as " change " is commonly understood. This fact may, I believe, be made sufficiently evident.

To make it evident let us, in seeking the meaning of the term 'variable," recur to the idea of propositional function; for, although some of the things called variables in the logical theory of propositional functions are not so called in traditional mathematics, yet whatever is called a variable in mathematics appears explicitly or implicitly as a variable in some propositional function; for example, the variable $x$ in such a propositional function as—$x$ is a man,—has not yet gained *full* recognition as a *mathematical* variable; on the other hand, the mathematical variable $x$ in an equation, say, $3x^2+2x-9=0$, appears as a variable in a propositional function, for, as you know, such an equation is such a function; and, for another example, when the mathematician says, "I will let $x$ represent any point in a certain line $L$," thereby indicating that he will use $x$ as a variable, he virtually (implicitly) says, "I will let $x$ represent any one of the verifiers of the propositional function $-x$ is a point in the line $L$."

Now let $\phi(x)$ denote some given propositional function involving one and only one of the things called variables. I am going to speak of $\phi(x)$; while I am doing so, you may find it helpful to attach what is said to some simple specific function such as "$x$ is a man" or "$x^2=4$" or "$x$ is a member of this audience." Our function $\phi(x)$ contains, we say, one variable, namely, $x$; $x$, we say, is a symbol; notice that, when speaking precisely, we do not say that the symbol *denotes* the variable, we say that the symbol *is* the variable. The question is: What is meant by saying that as here used the symbol $x$ is a variable? Before attempting to answer, let us reflect that there are terms such that if any one of them be substituted for $x$ in $\phi(x)$ the resulting statement is nonsensical,—non-significant,—and hence neither true nor false; and that there are other terms which, on being thus substituted, yield significant statements—that is, propositions (true or false). You will recall that terms of the former kind,—nonsense-giving terms,—were described in a previous lecture as inadmissible for $\phi(x)$ and that the latter kind,—sense-giving terms,—were described as admissible terms for $\phi(x)$. Now, it is significant statements, —statements that are true or false,—propositions,—and only such that we are concerned with when using propositional functions as instruments in research or in exposition, and hence the admissible terms and only these are important, for it is in virtue of these and only these that a propositional function is, as we have seen, a matrix, mould, or source, of propositions. If now we observe that the admissible terms for a given function $\phi(x)$ constitute a type, or class, of terms, we shall be prepared to answer our question. The answer is: The symbol $x$ in a given propositional function $\phi(x)$ is called a variable because the symbol represents *any one* of the terms of the class of admissible terms for $\phi(x)$ and represents nothing else. There is nothing subtler in human speech and nothing more important than the phrase " any one " as here used; without it, logic, science, philosophy, even the discourse of the workaday world, would be impossible. What does the phrase mean? It does not admit of precise definition, for it is essentially involved (explicitly or implicitly) in the very act of definition. The only or the best way to sharpen our sense of its meaning is to meditate upon examples of its use. A farmer has in his barn three horses—Black, Sorrel and Gray. He says to his servant: " John, fetch me a horse from the barn." John asks: " Which one?" " Any one," replies the farmer. As here used the phrase " a horse " is a variable because it represents " any one " of a certain class of horses; in representing " any one " of the class, it does not refer to a particular horse, for evidently " any one " is not a description or designation of a particular one of the horses; neither does it refer to all of the horses conjunctively—Black *and* Sorrel *and* Gray—John is not to fetch them all; it does refer to *each* of them *disjunctively*—John is to fetch Black *or* Sorrel *or* Gray—no matter which one. So it is in the foregoing definition: in representing " any one " of the class of admissible terms for $\phi(x)$, $x$ does not refer to a specific one of the terms nor to all of them conjunctively; it refers to each of them disjunctively. It is essential and now easy to see clearly that no idea of variation or change is involved: $\phi(x)$ being given, it is timeless, unchanging; the class of its admissible terms is timeless, unchanging; " any one " applied to the terms of this class is timeless, unchanging; $x$'s representation of this " any one " is timeless, unchanging; you thus see that a given mathematical variable is timeless, unchanging; and so you see that, when mathematicians speak of a variable as undergoing change, they speak metaphorically. Such speech is, I have said, very convenient and stimulating, and, now that we recognize its metaphorical character, I shall feel at liberty to employ it freely in this discussion.

A variable being given, the class of terms " any one " of which is represented by it is commonly called the *range* of the variable; thus in the case of our $\phi(x)$, the range of $x$ is the class of admissible terms for $\phi(x)$. The range of a variable may contain only one term, as, for example, when we say, " Let $x$ represent any point common to the given intersecting lines $L$ and $L'$." Such a variable is called a *constant;* thus you see that a mathematical constant, far from being (as vulgarly supposed) the opposite of a variable, is itself a variable. If a variable's range be a null class (an empty class, a class having no terms) we may describe the variable as a null variable. For example, $x$ is such a variable if it denotes any integer greater than 4 and less than 5.

It will be very helpful to illustrate the notion of a variable by means of examples. Before doing so, however, I wish to handle briefly a puzzling question that may have occurred to you in the course of the foregoing discussion. We saw that the admissible terms for $\phi(x)$,—the sense-giving terms,—constitute a class of terms—the range of $x$. Let us denote the class by $C$. The question is: Do the nonsense-giving terms,—the inadmissible terms,—constitute a class? The answer is No; there are such terms, but they do not together constitute a class. The correctness of the answer may be shown as follows: If the terms in question constitute a class, denote it by $C'$; $C'$, being itself a term, is either in $C$ or in $C'$; to see that $C'$ is not a term in $C$, consider any simple example of $\phi(x)$—say, $x$ is a man; in this case $C$ is composed of all the terms such that it is significant,—true or false,—to say that any one of them is a man; our hypothetical $C'$ consists of all the terms such that it is neither true nor false, but is nonsense to say that any one of them is a man; evidently it is neither true nor false, but is nonsense to say, "The class of all the terms such that it is nonsense to say they are men" is a man; hence $C'$, if there be such a class, is not a term in $C$, but is a term in $C'$; it is, however, foolish to say that $C'$ is a term in $C'$, in itself—as foolish as to say, for example, that a class of apples or of points is an apple or a point, or that the class of featherless bipeds is a two-legged thing without feathers. And thus you see that the inadmissible terms for a given propositional function do not constitute a class. The question I have thus summarily treated is of the kind of questions which have led Messrs. Russell and Whitehead—or rather have driven them—to the theory of Types in the *Principia*. In its present state the theory is far from being entirely satisfactory, but it is exceedingly helpful and it undoubtedly faces in the right direction. I desire to recommend it to you for consideration and for improvement.

*Examples of Variables.*—Let us now turn to the task of illustrating the mathematical notion of a variable by means of various examples. Every example of a great idea gives a little light and casts a big shadow; we must try to see the idea in the mingled lights and not lose it in the composite dark of the shadows. Again consider the propositional function $\phi(x)$. The class $C$—the range of the function's variable $x$—is, as you know, the logical sum of two sub-classes: $C_1$, composed of the verifiers of $\phi(x)$; and $C_2$, composed of the falsifiers of $\phi(x)$. We may chance to be interested only in the true propositions derivable from $\phi(x)$ or only in the false ones; accordingly we then restrict our thought to the verifiers or else to the falsifiers; if to the former, then the variable $x$ represents, not any one of the terms in $C$, but any one of the verifiers —$x$'s range being, not $C$, but $C_1$; if to the latter, then $x$'s range is $C_2$. In these cases what determines the variable's range? The answer is: neither the function alone nor our restrictive decision alone, but the two things taken together. The range of a variable is in every case either the class of admissible terms for some propositional function or a sub-class of such a class, the sub-class being determined by some restriction which the function as such does not impose; observe that, if some symbol $x$ is to be a variable, it is *we* who decide what its range is to be, for it is we who choose the function and, if any restriction is added, it is we who impose it. Is our freedom in the matter absolute? No; there is no such thing as absolute freedom; in the matter in question, as in all other purely intellectual matters, we have all the freedom there is, but it is not absolute; we have just seen that we can not have a variable representing " any one " of the inadmissible terms for a given propositional function, for the supposition that we can leads, as we saw, to contradiction. Freedom of thought,—intellectual freedom,—is conditioned, restricted, limited; but it is fundamentally limited by only one Law—the law which says to Intellect, "Thou shalt not incur a contradiction in terms." This law is the eternal guardian of intellectual *integrity;* reverence for it,—the disposition to keep it,—is the absolute invariant of intellectual life; disregard of the law,—I do not mean inadvertent violation of it,—means intellectual extinction: for intellect, disloyalty is death. Incidentally, we thus glimpse another phase of the truth, mentioned before, that mathematics is the study of Fate and Freedom.

Examples illustrating the mathematical concept of a variable are more numerous than the sands of the seashore or indeed the stars of the heavens, even if the multitude of the stars be infinite. In examining the following more specific and more familiar specimens it should be borne in mind that in mathematical discourse the range of a variable is very frequently indicated without explicitly stating the propositional function or functions necessarily involved in determining the range; such statement is, however, always possible and is often made. And now some familiar specimens.

(1) Consider the finite cardinal numbers: 0, 1, 2, 3, 4, 5, . . .; let $x$ represent any one of them; here the symbol $x$ is a variable; its range is, not the endless *row* as such, but the *class* of terms (numbers) *in* the row; the range is the same as would be indicated if we said, let $x$ represent any one of the verifiers of the propositional function—$n$ is a cardinal number. What is here the range of $n$?

(2) A variable's range may be finite or infinite. In (1) the range, you note, is infinite. If we let $x$ represent any cardinal greater than 1 and less than 10, $x$'s range is the

finite class composed of the terms, 2, 3, 4, 5, 6, 7, 8, 9: we can indicate the range by saying, let $x$ represent any one of the verifiers of the function—$n$ is a cardinal whose value is between 1 and 10. What is $n$'s range? What would the range of $x$ be if we simply said " $x$ is a cardinal between 1 and 10"?

(3) If we say, let $x$ be any verifier of the function —$n$ is a cardinal between 5 and 7—, the range of $x$ is—, not 5—but the *class* whose sole member is 5; in this case the varible is a *constant.*

(4) Consider the infinite (endless) series: $1+2+3+4+5+\ldots$; the sum, $S_n$, of the first $n$ terms is $\frac{1}{2}n(n+1)$, so that $S_n=\frac{1}{2}n(n+1)$. Here the language implies that $n$ is being regarded as a variable whose range is the class of all positive integers; perhaps some one doubts the implication; very well, let us explicitly agree to let $n$ be such a variable; you see at once that we then have another variable on our hands, namely, $S_n$—an ordinary (not a propositional) function of $n$, for to each value, as we say, of $n$ (*i.e.*, to each number in $n$'s range) there corresponds a definite sum, a definite value of $S_n$; plainly these sums are: 1, 3, 6, 10, 15, . . .; the class of these is the range of the variable $S_n$. Adopting the usual figurative speech, we may say that, as the variable $n$ runs along the row 1, 2, 3, . . . , the dependent variable, or function $S_n$, glides along the sequence 1, 3, 6, . . . But we must not let such talk make us forget what the ranges are: these are classes and not rows of their members or terms; the terms of either range appear in many different rows, but the range is one thing and each of the rows is another. If we so desire we can make explicit the propositional function involved in determining the range of $S_n$; we can say, for example, that the symbol $S_n$ represents any one

of the terms in the class of verifiers of the propositional function—$x$ is the first term or the sum of the first two terms or the sum of the first three terms, . . . of the row 1, 2, 3, 4, 5, . . . ; such speech is or may become forbiddingly cumbrous, but it serves to remind us that, though propositional functions be not always mentioned, they are yet omnipresent and may, if we so desire, be made manifest to the eye or the ear.

(5) If we agree to let $x$ represent any one of the numbers in the unending row, 1, $\frac{1}{2}$, $\frac{1}{4}$, $\frac{1}{8}$, . . . , the range of the variable $x$ is the class of all the numbers in the row; it is common and often convenient to indicate this row by writing: $1, \frac{1}{2}, \frac{1}{2^2}, \frac{1}{2^3}, \frac{1}{2^4}, \ldots, \frac{1}{2^{n-1}}, \ldots$; the symbol $\frac{1}{2^{n-1}}$ is called the *general* term or the $n$th term (of the row); it is, you observe, a function of $n$. But what is $n$ here? Plainly it is a variable whose range is the class of positive integers; so the symbol $\frac{1}{2^{n-1}}$ is a variable depending on another variable $n$; using our figurative way of speaking, we may say that, as $n$ varies from term to term of its range, the function $\frac{1}{2^{n-1}}$ runs from term to term of its range, the class of numbers in the foregoing row; we may say that, if $n$ starts at 1 and runs along the row 1, 2, . . . , $\frac{1}{2^{n-1}}$ starts at 1 and runs along the row 1, $\frac{1}{2}$, $\frac{1}{4}$, . . . ; or we may say more precisely that $n$ represents any one of the numbers of its range and that $\frac{1}{2^{n-1}}$ correspondingly represents any one of the numbers of its range; the latter

range is identical with the range of $x$, and so you see that in this example $x$ and $\frac{1}{2^{n-1}}$ are, as variables, identical, two symbols playing the same rôle.

(6) Consider the infinite (unending) series,

$$1+\frac{1}{2}+\frac{1}{2^2}+\frac{1}{2^3}+\ldots+\frac{1}{2^{n-1}};$$

its terms are those of the row in (5); it is, you notice, a *geometric* series or progression whose ratio, as it is called,—the ratio of any term (except the first) to its predecessor,—is $\frac{1}{2}$; let us denote the sum of the first $n$ terms by $S_n$; then, as you learned in elementary algebra, $S_n=2-\frac{1}{2^{n-1}}$. Observe that we are here confronted with three related variables: $n$, $\frac{1}{2^{n-1}}$, and $S_n$, the second being a function of the first, and the third a function of the second (directly) and of the first (indirectly, through the second). What are their respective ranges? That of the first is the class of positive integers; that of the second is the class of numbers in the row $1, \frac{1}{2}, \frac{1}{2^2}, \frac{1}{2^3}, \ldots$; that of the third is the class of numbers in the row $1, \frac{3}{2}, \frac{7}{4}, \frac{15}{16}, \ldots$ I leave it to you to describe the situation in the dynamic, or picturesque, language of change, variation, behavior.

(7) The foregoing six examples are very specific. Let us take one that is somewhat less so. Consider the geometric progression

$$(S)\ a+ar^2+ar^3+\ldots+ar^{n-1}+\ldots;$$

for the sum of the first $n$ terms we have, as you know,

$$S_n = a\left(\frac{1}{1-r} - \frac{r^n}{1-r}\right);$$

as in example (6) we have here three related variables, $n$, $r^n$ and $S_n$, if we suppose (as we commonly do) that $a$ and $r$ have given values; and you may wish to tell, as an exercise, how the three variables are related and what their respective ranges are. I desire to call your attention to the fact that, when we say " Let $(S)$ be any geometric progression," we are implicitly treating both $a$ and $r$ as variables, and not only that, but we are explicitly treating $(S)$ itself as a variable. In such case, what are the ranges of $a$, $r$ and $(S)$? We can not answer definitely until we have told what field or domain of number we are working in, as (say) that of the positive integers or that of the rational numbers or that of all the so-called real numbers or that of the ordinary complex numbers, and so on. If we are working in,—confining ourselves to,—the domain of the real numbers, the answer is that the range of $a$ is the class of all the real numbers, that $r$'s range is the same as $a$'s and that the range of the variable $(S)$ is the class of all unending geometric series or progressions formed or formable of real numbers. It is interesting to observe, in passing, that the range of $(S)$ contains a two-fold infinity ($\infty^2$) of geometric series; for assigning some value to $a$, we can obtain as many series whose $a$ has that particular value as there are possible values for $r$,—as many, that is, as there are real numbers,—an infinite multiplicity often denoted by the symbol $\infty$; it is plain that for *each* of these we get $\infty$ series by letting $a$ assume its possible values; so that in the domain of real numbers there exist $\infty \times \infty$, or $\infty^2$, unending geometric series—as many as there are points in a plane or pairs of real numbers. You will find it very interesting to describe in your own way the functional relations among the six variables $n$, $a$, $r$, $r^n$, $S_n$ and $(S)$, and very instructive to state explicitly the propositional functions implicitly involved in determining the various ranges. Instead of fixing upon a particular domain as above, you may say, if you like, " Let our number domain $D$ be any number domain "; then $D$, too, will be a variable and the complications thicken. Variables, you see, are lurking everywhere and are ever ready to leap into the arena of our attention, thus lifting our thought to higher and higher levels of generality. The scale of levels is summitless. Science, it is said, is the study of functions; functions are variables—science is the study of variables; every variable has its range partly determined by the scope of some propositional function—science is the study of propositional functions; the general concept and the name of propositional function are only beginning to gain a little recognition; the philosophy of science is in its infancy; the infant's nurse is the philosophy of mathematics; in the richest nation known to history,—made such by science,—the nurse can hardly contrive to live—she can hardly even publish her works because they are not profitable commercially; the nation is vain and boastful. May God deliver us.

In the foregoing list of examples the ranges of the variables are composed of numbers, with one notable exception: the range of $(S)$ is composed of series; these are indeed constructed out of numbers but they are not themselves numbers; and, though they are called geometric series, they are not spatial entities and have no *essential* connection with geometry. In citing additional examples, it will be easy to include among them some that are strictly geometric and some that are neither geometric nor numerical.

(8) Consider the class $\Sigma$ of the spheres having a given point $C$ for center and any radius greater than zero. Each of the spheres has many points, many tangent lines and many tangent planes. Each sphere has an interior,—the region or room bounded by the sphere,—and an exterior—the region outside of it. If you let $S$, $R$, $R'$, $r$, $P$, $L$, $\pi$, $l$, $A$, $V$, $P'$, $P''$ respectively represent any one of the spheres, any one of the interiors, any one of the exteriors, any one of the radii (taken as line segments), any one of the points on any one of the spheres, any one of the tangent lines, any one of the tangent planes, any one of the radial lengths, any one of the sphere areas, any one of the sphere volumes, any one of the points common to all the interiors, any one of the points common to all the exteriors, then the symbols $S$, $R$, $R'$, $r$, $P$, $L$, $\pi$, $P'$, $P''$ are geometric variables, and $l$, $A$, $V$ are arithmetic, or numerical, variables. Of the former ones the respective ranges are: the class $\Sigma$, the class of all the interiors of the spheres in $\Sigma$; the class of all the sphere exteriors; the class of all the radii; the class of all the points of our space except the point $C$; the class of all the lines of space except the lines of the line-sheaf having $C$ for vertex; the class of all the planes of space except those of the bundled vertexed at $C$; the class whose sole member is the point $C$ ($P'$ being thus a constant); and the null (empty) class of points (the range of $P''$ containing no terms); the ranges of $l$, $A$ and $V$ are the same, namely, the class of all the positive real numbers (zero not included).

Let me remind you of something and then submit a few questions that should give you a happy hour or so of cloistral meditation. You know, I believe, that when mathematicians say that two variables $V$ and $V'$ (having $K$ and $K'$ for their ranges) are functions of each other, they mean that to any given term in either range there corresponds a definite term (or terms) in the other range; you know that, if to each term in (say) $K'$ there corresponds only one term in $K$, then $V$ is called a *one*-valued function of $V'$, and that, if two or more terms in $K$ correspond to a term in $K'$, $V$ is called a *many*-valued function of $V'$, where the "many" may be finite or infinite. The questions I wish to ask are these: Is each one of the foregoing 12 variables, $S$, $R$, $R'$, $r$, and so on, a function of each of the 11 others? Which ones, if any, of these functions are one-valued? Which, if any, are many-valued? Which, if any, are infinitely many-valued? What sort of a function is $l$ of $r$? $r$ of $l$? Is there in $\Sigma$ a least sphere? A largest one? If we choose to regard the common center $C$ as a sphere of zero radial length, and, in describing $\Sigma$, drop the requirement that every radius must exceed zero, what, if any, changes must be made in the answers to the foregoing questions? If we fancy ourselves working in projective space—which, as you know, has one and but one plane (an infinite sphere) at $\infty$ and if we let $\Sigma$ be the class of spheres having a given common center $C$ and radius equal to or greater than zero, what are the ranges of our 12 variables?

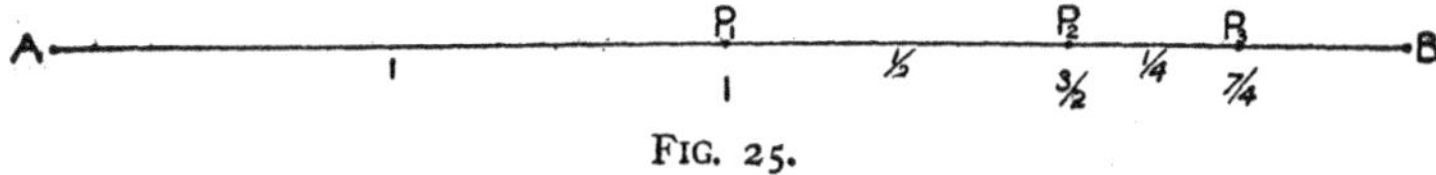


FIG. 25.

(9) For further examples of variables, some of them geometric, some of them numerical, consider the following: Let the line-segment $AB$ have a length of 2 units. Denote

by $P_1$ the mid-point of $AB$, by $P_2$ the mid-point of segment $P_1B$, by $P_3$ the mid-point of $P_2B$, and so on endlessly, so that $P_n$ will denote the mid-point of the segment $P_{n-1}B$. Suppose we let the symbols $P$, $S$, $S'$, $l$, $l'$ represent respectively any one of the indicated mid-points, any one of the segments ($AP_1$, $P_1P_2$, $P_2P_3$, . . . ), any one of the segments beginning at $A$ and ending at a mid-point, any one of the lengths of the $S$-segments, any one of the lengths of the $S'$-segments; the ranges are respectively the class of the mid-points, the class of segments in the row $AP_1$, $P_1P_2$, $P_2P_3$, . . . , the class of segments in the row $AP_1$, $AP_2$, $AP_3$, $AP_4$, . . . , the class of numbers in the row $1, \frac{1}{2}, \frac{1}{2^2}, \frac{1}{2^3}, \ldots$, and the class of numbers in the row $1, \frac{3}{2}, \frac{7}{4}, \ldots$ Observe that $P$, $S$ and $S'$ are geometric variables and that $l$ and $l'$ are numerical. Observe that the rôle of $P$ is the same as that of $P_n$ and so we may write $P=P_n$; similarly, we may write $S=P_{n-1}P_n$ (understanding that $P_0$ is $A$), $S'=AP_n$, $l=\frac{1}{2^{n-1}}$ and $l'=2-\frac{1}{2^{n-1}}$; and so, you see, the variables $P$, $S$, $S'$, $l$, $l'$ are functions of the variable $n$ whose range is the class of positive integers. Are the functions one-valued or are they many-valued? Describe the so-called variation of these functions as $n$ "varies." As $n$ increases more and more will $P_n$ ever reach $B$? Why not? Can you, by increasing $n$, make the length of the segment $P_nB$ smaller than any length you choose if you don't choose zero length?

(10) In Fig. 26 we have a circle $K$ of given center $C$ and radius of given length $R$. By circle $K$, I mean the curve. Denote the length of $K$ by $L$ and the area of $K$ by $A$. Do not fail to distinguish geometric from numerical

things: observe, for example, that $K$ is geometric,—a specific curve,—and that, like $R$, $L$ and $A$ are numerical—specific numbers. Consider the indicated inscribed regular polygons—the triangle (or 3-side), hexagon (or 6-side), do-decagon (or 12-side), 24-side, 48-side, and so on endlessly; note that the $n$th polygon of the unending row has

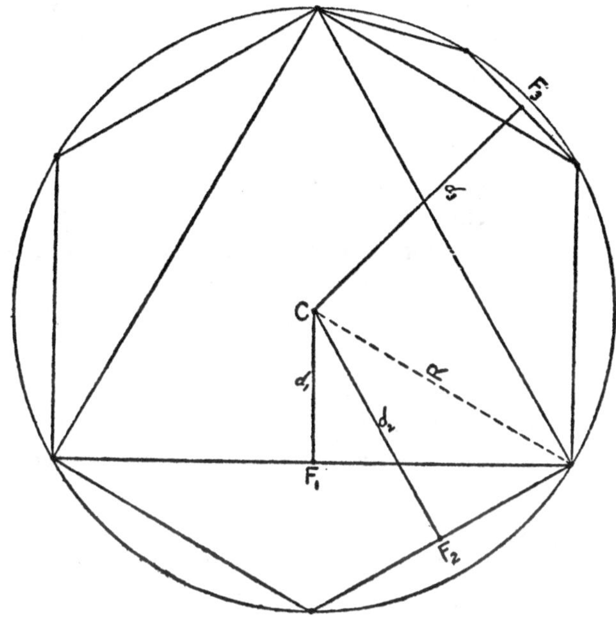


FIG. 26.

$3 \times 2^{n-1}$ sides. You see that $n$ is a variable whose range is the class of positive integers. Denote by $P_n$ the $n$th polygon, by $L_n$ the perimeter (length) of $P_n$, by $A_n$ the area of (size or magnitude of) $P_n$, and by $d_n$ the length of the perpendicular from $C$ to a side of $P_n$. The symbols $P_n$, $L_n$, $A_n$ and $d_n$ are variables. Which ones are geometric? Which numerical? They are functions of $n$.

What kind? What are their ranges? Are these finite or infinite? Do the functions increase or decrease with $n$? All increase with $n$ except $P_n$; how justify the exception? Can $n$ so increase that $P_n$ will coincide with $K$? Or $L_n$ with $L$? Or $A_n$ with $A$? Or $d_n$ with R? Why not? the differences $L-L_n$, $A-A_n$, $R-d_n$ are variables, functions of $n$—are they not? If I name a positive number $\sigma$ as small as I please, can you then choose such a value for $n$,—such a term in $n$'s range,—that the foregoing differences will be less than $\sigma$? Having thus chosen $n$, if you let $n$ take still larger and larger values, will the differences in question keep always less than $\sigma$? Will they ever vanish—that is, be equal to zero? Why or why not?

(11) Let us now consider two variables that are neither geometric nor numerical. Referring to Fig. 26, let us write $K-P_1$, $K-P_2$, . . . ; these seem to be symbols but what, pray, do they symbolize? They are indeed suggestive, but they have at present no definite meanings, for neither $K$ nor any of the $P$'s is a number, $K$ being just a circle—a specific circle—and each $P$ a specific polygon. We may, however, assign meanings to the ostensible symbols, thus making them genuine symbols. What meanings shall we assign? A little reflection upon the make-up of Fig. 26 naturally suggests that we let $K-P_1$, $K-P_2$, $K-P_3$, . . . , denote respectively the endless polygon-*rows* indicated at the right of them below:

$$(S)\left\{\begin{array}{l} K-P_1 : P_2, P_3, P_4, P_5, \ldots ; \\ K-P_2 : P_3, P_4, P_5, \ldots\ldots ; \\ K-P_3 : P_4, P_5, \ldots\ldots\ldots ; \\ \quad . \quad . \quad . \quad . \quad . \quad . \quad . \\ \quad . \quad . \quad . \quad . \quad . \quad . \quad . \\ \quad . \quad . \quad . \quad . \quad . \quad . \quad . \end{array}\right.$$

Note that we have here an endless (downward running) row ($S$), of endless rows of polygons. It is obviously natural to denote by $K-P_n$ any one—the $n$th one—of the rows in the row ($S$). The symbol $K—P_n$ thus becomes a variable whose range is, not ($S$), but the class of rows in ($S$). Our new variable is plainly not numerical; neither is it geometric for, though the rows,—the terms in its range,—are composed of polygons, a row of geometric figures is, in strictness, no more a geometric figure than a row of men is a man.

For an example of a variable more evidently, though not more actually, non-geometric and non-numerical than the preceding one, we may take $p$ where $p$ represents any one of the propositions derivable from some given propositional function $\phi(x)$ by means of its verifiers. The range of $p$ is obviously the class of true propositions having $\phi(x)$ for matrix.

*Conceptions of Limit.*—It is evident that much of our human thinking,—I strongly suspect that all of it,—is concerned with variables. One of the lessons which the history of thought and our personal experience in thinking teach very clearly is that we can not deal with variables logically or with any close approximation to rigor without the help of the notion (or notions) which mathematicians denote by the term " limit." We must, accordingly, try to understand what the term means. I have just now used the plural—conceptions of limit—instead of the singular. I have done so because there are various meanings of the term and I intend to present more than one of them. The definitions I am going to present are closely related, but they are not equivalent: they differ in content and scope, and you will find it very instructive to compare them in these respects. You will observe at once that there are two respects in which they agree—they involve the notion of variable and its range (every limit is a limit of a variable) and all but one of them involve the notion of difference. The notion of variable we have discussed at unusual length and it is, I trust, now fairly clear. It will be useful, I believe, to say a preliminary word regarding the notion of difference. It is not my aim to define the general notion; my aim is merely to enliven a little our consciousness of it. In the background of our human thinking, however refined, however precise the ideas we are explicitly handling, there lurk other ideas—shadowy, nebulous, vague—which we have not defined, which we may not have attempted to define, which we may not even be conscious of; yet these background ideas give our so-called precise ones all the meaning the latter have. In the present discussion, the idea of difference is a background idea.

Any given variable, as we have seen, has a range—a certain class of things, or objects, called the terms of the class and commonly spoken of as the variable's " values." A class being given, each of its terms is comparable with each in one or more respects: that is to say, each of them *differs* from each in one or more respects; the respects, and hence the differences, may be very definite or fairly definite or very vague; the differences may be differences in respect of position or of magnitude (size) or of number or color or shape or weight or of importance or of dignity or of beauty or of sensibility and so on; we may, therefore, speak of *kinds* of difference as distinguished from amounts of a given kind. It is essential to note the obvious fact that, if each term of a class differs from each of its fellow terms in respect to some specific kind $k$ of difference, it may happen that the terms in the class differ from some terms not in it in respect to the same kind *k*. Examples abound, and need not be cited. If, with respect to a given kind of difference, two terms be identical, we shall say that the amount of their difference (of the given kind) is null or naught; if, as often happens, the given kind of difference be numerical, then the hypothetical identity is numerical equality and we shall say, in accordance with usage, that the amount of difference is zero; we shall thus be using zero as a special variety of the foregoing null or naught. Here, as elsewhere, it is essential to use good sense. It would, for example, be nonsense to speak of two numbers as differing in respect to color or patriotism or loyalty. When we speak of two terms as having an amount of difference of a kind *k*, it is to be understood that *k* is a kind with respect to which the terms can be significantly compared. A kind *k* of difference may be called a *k*-difference; and a given amount *d* of it, the *k*-difference *d*.

I am now going to define a very convenient idea to be called a *k-neighborhood* of a term *t*. Let *t* be a term comparable with one or more terms in respect to a kind *k* of difference, and let *d* be a given amount (greater than null) of such difference; then *d* is said to determine a *k*-neighborhood of *t*. If *d* grow larger or smaller, the neighborhood will do likewise. If *d* be specified, we may speak of the *k*-neighborhood *d* of *t*. Observe that *d* determines the same neighborhood for each and every term comparable with one or more terms in respect to the given difference-kind *k*. A term *t′* will be said to be *in* or not in the *k*-neighborhood *d* of *t* according as *t′* differs from *t* by an amount less than or not less than *d*. You see at once that, if *t′* be in the *k*-neighborhood *d* of *t*, *t* is in the *k*-neighborhood *d* of *t′*. Obviously *t* is itself in all the $k$-neighborhoods obtainable by letting $d$ vary, for the amount of $t$'s difference from $t$ is null and null is less than any amount $d$ is allowed to be.

Let us now give the following broad definition $D_1$ of the term limit, and see what happens.

$D_1$: Let $V$ be a variable, $R$ its range, and $k$ a kind of difference with respect to which the terms in $R$ are comparable; if there be a term $t$ (in $R$ or not) such that, however small a $k$-neighborhood of $t$ be chosen, some term of $R$ is in the neighborhood, then $t$ is a $k$-limit of $V$.

When, as is usual, the context indicates what particular $k$ is under consideration, it need not be mentioned explicitly and we may speak of a $t$ (satisfying the foregoing conditions) simply as a limit of $V$.

It is easy to see that, under definition $D_1$, every term in the range $R$ of a variable $V$ is a limit of $V$, for, if $t$ be in $R$, and $k$ be an admissible kind of difference, the amount of $t$'s $k$-difference from $t$ being null, $t$ is in every $k$-neighborhood of $t$, and is, therefore, a $k$-limit of $V$. A null variable has no limit; every other one has. Unless the contrary be clearly indicated, let us understand that the variables under discussion are not null variables. Can a $V$ have a limit that is not in the $R$ of $V$? If $t$ be not in $R$ and if $t$'s $k$-difference from some term in $R$ be null, then clearly $t$ is a $k$-limit of $V$; if an outside $t$ be a $k$-limit of $V$, it may or may not be a $k'$-limit of $V$, if $k$ and $k'$ be not the same. For example, let $V$'s range have but one term—say, a sphere $S$ of given color, mass and volume; let $k$-difference, $k'$-difference and $k''$-difference be respectively difference in color, in mass and in volume; then any object having the same color as $S$ will be a $k$-limit of $V$, though not in general a $k'$-limit or a $k''$-limit; and so on.

Query: Can a $V$ have a $k$-limit $t$ not in the $R$ of $V$

if the amounts of $t$'s $k$-difference from the terms in $R$ be each more than null? If $R$ be a finite class, the answer is evidently no. If $R$ be an infinite class, the answer depends on $k$. Here we must make an important distinction. The difference-kind $k$ may be such that, given any two amounts of it, there are at most only a finite number of intermediate amounts. Such a kind may be called a *discrete* difference-kind. An example is the difference-kind we have or should have in mind when, confining our attention to zero and the positive integers, we talk of the " differences " (strictly, the amounts of difference) found by subtraction; of such amounts, the smallest is zero, the next smallest is 1, the next 2, and so on, and there are no other amounts of the kind of difference we are here dealing with. If we were talking of " differences,"—amounts of difference—of (say) rational fractions, we should have in mind a different kind of difference. As in the foregoing example, so if $k$ be any discrete difference-kind, there is, as you readily see, an amount of the $k$-difference next greater than the null of it. On the other hand, a difference-kind may be such that, given any two amounts of it, there is one amount (and hence infinitely many amounts) intermediate to the given ones. Such a kind may be called a *compact* or *dense* difference-kind. An example is the difference-kind involved when, confining our attention to the rational numbers, we say the amount of difference of this fraction and that is so-and-so. It is perfectly clear that no amount of dense difference-kind is next greater than the null amount of it.

Let us now return to our query. If the $k$-difference be discrete, the answer is negative, even though $R$ be infinite. For let $d$ be the smallest amount of $k$-difference except null; then no term $t'$ of $R$ is in the $k$-neighborhood *d* of *t* for, if *t′* were in the neighborhood, then the amount of *k*-difference of *t* and *t′* would be less than *d*, but there is no such amount except null, and null is excluded by the hypothesis of the query. On the other hand, if the *k*-difference be dense, the answer is affirmative: there are *V*'s having the sort of limit required. For it is sufficient that the range *R* be such that for some *t* not in *R* there is, for any chosen *k*-neighborhood *d* of *t*, however small the neighborhood, an *R* term in the neighborhood—that is, an *R* term differing from *t* by more than null and less than *d*; and the existence of such *V*'s may be shown by letting *V* be the variable whose *R* has for its terms zero and all ordinary fractions less than 1; you note that 1 is a *t* not in *R*; that in comparing the terms in question we employ a dense difference-kind; that, however small a neighborhood of 1 be chosen, *R* has terms in the neighborhood; and that 1 is, therefore, a limit of *V*. For another example consider the variable *x* in (5) of our foregoing list of variables; show that zero, which is not in the range, is a limit of *x*. You will find it interesting and very instructive to examine all the variables of the cited list with a view to ascertaining which of them have limits outside their ranges and what the limits are.

The concept of a limit as defined by the definition $D_1$ has, you see, some striking properties. It is, however, too broad for certain highly useful purposes; for example, it does not sufficiently discriminate variables among themselves; according to it all variables (except null ones) have limits, as we have seen, and every term in a variable's range is a limit of the variable. It is obviously desirable to classify variables with respect to the character or constitution of their ranges. Let us accordingly try a somewhat narrower definition $D_2$ of the term limit.

$D_2$: Let $V$ be a variable, $R$ its range, and $k$ a suitable kind of difference; if there be a term $t$ (in $R$ or not) such that, however small a $k$-neighborhood of $t$ be chosen, some $R$ term differing from $t$ by more than null is in the neighborhood, then $t$ is a $k$-limit of $V$.

Note that the sole distinction between $D_1$ and $D_2$ is due to the presence in $D_2$ of the phrase—" differing from $t$ by more than null." The distinction, seemingly slight, is very grave, as we shall see. In the first place we easily see that no $V$ has a $k$-limit if $k$ be a discrete difference-kind. For suppose $t$ to be such a limit; let $d$ be the smallest amount of the $k$-difference greater than the null amount; then $R$ has a term, say $t'$, in the $k$-neighborhood $d$ of $t$; and $t'$ differs from $t$ by more than null and less than $d$, which is impossible. Again, no $V$ whose $R$ is finite has a limit. For suppose such a $V$ to have a limit, say $t$; let the difference-kind $k$ be dense—the preceding proposition makes it superfluous to consider the case of $k$ discrete; the number of terms in $R$ is finite, say, $n$; there are at most $n$ non-null amounts of $k$-difference between them and $t$; one of these amounts, say $d$, is as small as any of them; by supposition there is an $R$ term, say, $t'$, in the $k$-neighborhood $d$ of $t$, and $t'$ differs from $t$ by more than null and less than $d$, which is impossible.

From the two propositions just now proved it follows that if a $V$ have a limit under definition $D_2$, it is necessary both that $V$'s range be infinite and that the difference-kind concerned be dense. Are these necessary requirements also sufficient? A simple example will suffice to show that they are not. Let $V$'s $R$ be the class whose terms are the *rational* numbers: 0, 1, 2, 3, 4, . . . ; the appropriate difference-kind is dense (since we are dealing with rationals) and $R$ is infinite; if $V$ have a limit $t$, $t$ is a rational number; of all the non-null (non-zero in this example) amounts of difference between $t$ and the $R$ terms, one of them, say $d$, is as small as any of them, obviously; hence no $R$ term differing from $t$ by more than zero is in the neighborhood $d$ of $t$; and so $V$ is limitless.

Query: Are there any variables which, under $D_2$, have limits? The answer is yes; such variables abound in endless number and variety. Is the $x$ in (1) of our list of variables one of them? No; for the difference-kind appropriate to cardinal numbers is discrete. For a like reason neither is the variable $S_n$ in (4) one of them. But, as you easily see, the variable $x$, or $\frac{1}{2^{n-1}}$, in (5) has a limit, namely, zero (0); so has $S_n$ in (6), the limit being 2; so, too, has $S_n$ in (7) if $r$ be numerically less than 1; for the range of $n$ is the class: 1, 2, 3, . . . ; the range of $r^n$ is the class: $r^1, r^2, r^3, \ldots$ ; the range of $\frac{r^n}{1-r}$ is the class: $\frac{r}{1-r}, \frac{r^2}{1-r}, \frac{r^3}{1-r}, \ldots$ ; and the range of $S_n$ is the class

$$a\left(\frac{1}{1-r}-\frac{r}{1-r}\right),\ a\left(\frac{1}{1-r}-\frac{r^2}{1-r}\right),\ a\left(\frac{1}{1-r}-\frac{r^3}{1-r}\right),\ldots$$

Consider these ranges. Let $d$ be any given number greater than 0; as $r$ is less than 1, it is plain that there are terms in the range of $r^n$ exceeding 0 by less than $d$—terms that is, in any prescribed neighborhood of 0, however small; and so, 0 is the limit of $r^n$; evidently it is also the limit of $\frac{r^n}{1-r}$; without further talk you see that in any prescribed neighborhood of $\frac{a}{1-r}$ there are terms of the range of $S_n$; hence $S_n$ has $\frac{a}{1-r}$ as limit. What if $r$ be not numerically less than 1? If $r=1$, the divisor $1-r$ is zero, but the phrase " division by zero " is meaningless, and so, if $r=1$, $S_n$ is limitless. If $r$ be numerically greater than 1, you readily see that the (numerically) smallest difference between $\frac{a}{1-r}$ and the terms of $S_n$'s range is $\frac{ar}{1-r}$; hence, if we choose a neighborhood $d$ of $\frac{a}{1-r}$ where $d$ is smaller than $\frac{ar}{1-r}$ (and we can take it thus smaller, if, as we are assuming, $a$ is not zero), $S_n$'s range has no term in the neighborhood; therefore $\frac{a}{1-r}$ is not a limit of $S_n$, if $r$ be numerically greater than zero. Has $S_n$ any limit at all under the hypothesis? You can readily show that it has not.

I wonder if you are growing weary of this long discussion. I must believe you are, unless your desire to understand the subject be genuine, deep, invincible. If we fail to master some important idea, what is the explanation? It may be stupidity; it is, more probably, unwillingness to pay the price—meditation; but, if we be really interested, meditation is not a price, it is a pleasure—a sustaining joy; the great source of success is abiding interest. I have been counting upon your interest and am going to count upon it yet further.

Consider that somewhat unfamiliar variable ($S$) in (7) of our list. Has ($S$) a limit? Its range, we saw, is the twofold infinitude of geometric series formable from $a$ and $r$ where $a$ and $r$ are variables each representing any real number. As the range of $a$ (or of $r$) is the class of real numbers, it is evident that every real number is a limit of $a$ (and of $r$), for it is evident that in any given neighborhood, however small, of any given real number, there are real numbers whose numerical difference from the given number is more than zero. If we are to talk of ($S$) as having or not having a limit, we must indicate what we are going to mean by " difference " of geometric series as such. This we may do as follows: let $(a_1, r_1)$ be a given pair of real numbers, and let $d$ and $d'$ be given positive numbers; we agree to say that the neighborhood $d$ of $a_1$, and the neighborhood $d'$ of $r_1$, together determine a neighborhood $(d, d')$ of the series $a_1+a_1r_1+a_1r_1^2+\ldots$, and that, if $a_2$ be in the first neighborhood and $r_2$ in the second, then and only then the series $a_2+a_2r_2+a_2r_2^2+\ldots$, is in the third. Now, as we saw a moment ago, $a_1$ is a limit of $a$, and $r_1$ of $r$; hence, the series $a_1+a_1r_1+a_1r_1^2+\ldots$ is a limit of $a+ar+ar^2+\ldots$, that is, of ($S$); hence, every series in the range of ($S$) is a limit of ($S$).

You will recall that in (11) of our little list of variables, $p$ is a variable whose range is the class of all the true propositions having a given propositional function $\phi(x)$ for their common matrix. Can we associate the notion of limit with $p$? We can, as follows: Let $V$ be a variable whose range $R$ is the class of verifiers of $(x)$; denote the range of $p$ by $R'$; if $x_1$ be a given term in $R$, then the proposition $\phi(x_1)$ is a definite term in $R'$—to each $R$ term there thus corresponds an $R'$ term, and conversely; let $k$ be a suitable difference-kind for the $R$ terms; it will evidently be a suitable difference-kind for the $R'$ terms, for, if and only if the amount of $k$-difference between the $R$ terms $x_1$ and $x_2$ be null, the corresponding $R'$ terms $\phi(x_1)$, $\phi(x_2)$ are identical propositions—indistinguishable with reference to $k$; we will regard the amount of $k$-differences between $x_1$ and $x_2$ as the measure of $k$-difference between $\phi(x_1)$ and $\phi(x_2)$; let $t$ be a term (in $R$ or not) comparable with the $R$ terms with respect to $k$; then $\phi(t)$, whether in $R'$ or not, is comparable with the $R'$ terms with respect to $k$; to the $k$-neighborhood $d$ of $t$ corresponds the $k$-neighborhood $d$ of $\phi(t)$; if an $R$ term $t'$ be in the former neighborhood, the corresponding $R'$ term $\phi(t')$ is in the latter neighborhood; and so, you see, if $t$ be a $k$-limit of $V$, the proposition $\phi(t)$ is a $k$-limit of $p$. Observe that, according as a $k$-limit of $V$ is or is not an $R$ term, the corresponding limit of $p$ is a true or a false proposition.

For a very simple example of the foregoing, suppose $\phi(x)$ to be: $x$ is a term in the class $\alpha$ of the rational numbers $\frac{1}{2}$, $\frac{1}{3}$, $\frac{1}{4}$, . . . ; $V$'s range is $\alpha$; $p$'s range is the class $\alpha'$ of propositions: $\frac{1}{2}$ is a term in $\alpha$; $\frac{1}{3}$ is a term in $\alpha$; $\frac{1}{4}$ is a term in $\alpha$; . . . . Zero being a limit of $V$, the false proposition—zero is a term in $\alpha$—is the corresponding limit of $p$. What change of supposition will make the limit of $p$ a true proposition?

Permit me to recommend strongly that, as an exercise, you determine which of the variables in the above-given list of variables have limits under definition $D_2$ and what the limits are; that you similarly examine a goodly variety of variables not in the list; and that you consider the question: if a $V$ has a $k$-limit $t$ under $D_2$, has it the same limit under $D_1$?

There are two reasons why I am inviting you to consider various non-equivalent definitions of the term limit. One of the reasons is that such consideration helps to deepen, refine and clarify our understanding of the great conceptions—variable and range thereof. The other reason is that mathematicians use the term " limit " in a variety of senses differing in scope. In any discussion involving the term " limit," mathematicians, when they speak carefully (which, being human beings, they do not always do), indicate explicitly or contextually the sense in which the term is being employed. I have now presented two widely differing definitions of the term— $D_1$ and $D_2$. I am not aware that the former one has been hitherto given. On the other hand, $D_2$ (or some virtual equivalent otherwise stated) is of very frequent use in the mathematical literature of the last half-century. In the next lecture I shall invite you to consider additional meanings of the term in question.

# LECTURE XIV

## More About Limits

FURTHER DEFINITIONS OF LIMIT—LIMITS AND THE INFINITESIMAL CALCULUS—CONNECTION WITH ORDER, SERIES AND SEQUENCES—LIMITS AND LIMIT PROCESSES OMNIPRESENT AS IDEALS AND IDEALIZATION IN ALL THOUGHT AND HUMAN ASPIRATION—IDEALS THE FLINT OF REALITY—GENIUS AND GENERALIZATION.

THERE are two additional definitions of the term "limit" with which it is, I believe, very important for philosophical students to get well acquainted. Both of them are closely, indeed essentially, connected with what mathematicians variously call a *linear order* or a *serial relation* or a *series* or a *sequence.* Before presenting them we must recall clearly to mind some matters briefly explained in Lecture X and then join therewith certain kindred ideas and distinctions. You will recall that a propositional function, say $\phi(x, y)$, containing two variables, is said to determine a (dyadic) relation; that, if $\phi(x_1, y_2)$ is a true proposition, then and only then the pair or couple $(x_1, y_2)$ is called a constituent or element of the relation; that the class of all such constituents,—the class of all the pairs verifying (satisfying) $\phi(x, y)$, *is* the relation; that, if we denote the relation by $R$, we say "$x$ has the relation $R$ to $y$" by writing $xRy$; that $R$ accordingly has a *sense*—so that, if $(x_1, y_1)$ be a constituent, we have $x_1Ry_1$, but, in general, not $y_1Rx_1$; that the domain of $R$ is the class of all the $x$'s,—the class of all the terms,—such that each of them has the relation to something or other; and that the codomain of $R$ is the class of all the $y$'s,—the class of all the terms,—such that, given any one of them, something or other has the relation to it. I may add that the terms in the domain of $R$ are often called the *referents* of $R$ and that the terms in the codomain are called the *relata* of $R$. Some relations have *fields;* others, not. $R$ has a field if and only if the domain and codomain are of the same type,—that is, are composed of individuals or else of classes of individuals or else of classes of classes of individuals, and so on,—and the field is, if there be one, the logical sum of the domain and codomain,—the class, that is, containing every term in the domain or in the codomain and no other term. Thus, if $\phi(x, y)$ be—$x$ is a husband of $y$—then, if $y_1$ be a wife of $x_1$, the couple $(x_1, y_1)$ is a constituent of the relation; the relation "husband of" is the class of all such couples; the domain is the class of husbands; the codomain is the class of wives; the field is the class of husbands and wives; each husband and nothing else is a referent; each wife and nothing else is a relatum; observe that in this example, the domain and codomain have no common terms. If $\phi(x, y)$ be—$x$ is a positive integer less than a positive integer $y$—then the relation is the class of all couples $(x_1, y_1)$ such that $x_1$ and $y_1$ are positive integers of which the former is the less; every integer is a referent; every integer except 1 is a relatum; and so, you see, the domain includes the codomain, but the converse is not true. If $R$ were identity, for example, or equality or diversity, then, as you easily see, the domain and the codomain would each include the other—they would coincide. Of relations as a subject I have already repeatedly indicated the immensity and the first-rate importance. At present, I am asking you to consider only so much of it as is necessary and sufficient for our present purpose, which is that of preparing us to understand certain highly important meanings of the term "limit."

Relations are endless in number and in variety and they are omnipresent as well in practical life as in abstract thought. There is one variety (including a vast multitude of sub-varieties) to which I am going now to ask your best attention. Before defining it, it will be helpful to consider a simple specific example of it. The example I am going to use is one of the relations instanced a moment ago. I mean the relation determined by the propositional function: $x$ is a positive integer less than a positive integer $y$. Let us denote the relation by $P$. Observe what $P$ is. It is the class of couples: (1, 2), (1, 3), (1, 4), (1, 5), . . . ; (2, 3), (2, 4), (2, 5), . . . ; (3, 4), (3, 5), . . ; . . . ; . . . ; and so on endlessly. Note that $P$ has a field—the class of all the positive integers. The relation $P$ has numerous properties; let me ask you to inspect very carefully just three of them. The three are these: (*a*) if $n$ be in $P$'s field, $(n, n)$ is not a constituent of $P$,—that is, $nPn$ is a false proposition,—that is, $P$ is not a relation which, like identity, holds between a term and that same term; (*b*) if $n$ and $n'$ are in $P$'s field, then either $(n, n')$ or else $(n', n)$ is a constituent of $P$—that is, $nPn'$ or else $n'Pn$; (*c*) if $nPn'$ and $n'Pn''$, then also $nPn''$. Because the relation has these three properties, it is called a serial relation, or a series, or a sequence, or a specimen of linear order. You detect at once how to define these equivalent terms. The definition is as follows: A serial relation (or series or sequence or linear order) is a relation $R$ such that: ($a$) if $xRy$, $x$ and $y$ are not the same; ($b$) if $x$ and $y$ are terms in $R$'s field, then $xRy$ or else $yRx$; and ($c$) if $xRy$ and $yRz$, then $xRz$. In this discussion, let us use the shorter names, sequence and series, for such an $R$, instead of the other names. Evidently the above $P$ is a specific instance of a sequence or series. Consider another instance, say $P'$, where $P'$ is determined by the propositional function: $x$ is a positive integer greater than a positive integer $y$. You see that $P'$ is indeed a sequence. Notice that $P$ and $P'$ are different sequences: for example, the couple (1, 2) is a constituent of $P$ but (2, 1) is not, while (2, 1) is a constituent of $P'$ but (1, 2) is not. Yet the field of $P$ is the same as the field of $P'$—namely, the class of positive integers. You readily see that, if the field $F$ of a given sequence be infinite, there are infinitely many different sequences having $F$ for their field. It is plain that the smallest class that can be the field of a sequence is a class having two and only two members, say, $a$ and $b$; even in this case, there are two sequences having the field in common; one of them consists of the couple ($a$, $b$), the other of the couple ($b$, $a$). Let me, in passing, propose an instructive little exercise. Given a class of three terms, $a$, $b$ and $c$, show that there are six sequences having the class for field, that each sequence is a class of three couples, and write down the couples for each case.

In our introductory study of sequences, or series, it is desirable to learn something more of the subject's language; for as supersimians, we must chatter about the subject, and as supersimian philosophers, we must try to chatter intelligibly. If the relation $R$ be a sequence we say that the referents of $R$ are predecessors—predecessors for $R$, or $R$ predecessors; that the relata of $R$ are successors—successors for $R$, or $R$ successors; that, if $xRy$, $x$ is a predecessor of $y$ and $y$ is a successor of $x$—more precisely, that $x$ is an $R$ predecessor of $y$ and $y$ is an $R$ successor of $x$. You see immediately that every term in $R$'s domain is an $R$ predecessor, that every term in $R$'s codomain is an $R$ successor, and that every term in $R$'s field is either a predecessor or a successor and is generally (not always) both. Thus in the case of our example $P$, 1 is a predecessor but not a successor, while every other integer in the field is both; on the other hand, in the case of $P'$, 1 is a successor but not a predecessor, while every other positive integer is again both. If a term $t$ be an $R$ predecessor but not an $R$ successor, the sequence $R$ is commonly and conveniently said to have a beginning $t$—to begin at $t$; thus $P$ has a beginning, it begins at 1. If $t$ be a successor but not a predecessor, the sequence has an end $t$—it ends at $t$; thus $P'$ has an end, it ends at 1; $P$ is endless, $P'$ is beginningless. A sequence may have both beginning and end or neither. An example of the former is the sequence $P''$ determined by the propositional function: $x$ is a positive integer less than a positive integer $y$ not greater than 10; you see that $P''$ is a sequence and that it has a beginning, 1, and an end, 10. The field of $P''$ is finite. Can a sequence whose field is infinite have both beginning and end? Yes; consider the sequence determined by the propositional function: $x$ is a real number (equal to or greater than 1) less than a real number $y$ (not greater than 2); you see that the relation determined by the function is a sequence, that the sequence begins at 1 and ends at 2, and that the field is infinite—the class whose terms are 1, 2 and all the intervening real numbers. For an example of a sequence having neither beginning nor end, we may take the series $P'''$,—the class of couples $(x, y)$,—determined by the propositional function: $x$ is a real number (greater than 1) less than a real number $y$ (less than 2); it is clear that $P'''$ is beginningless and endless—each term in its field is both a predecessor and a successor. If $t$ be a term in the field of a sequence and be at once a successor of $t_1$ and a predecessor of $t_2$, then $t$ is said to be *between* $t_1$ and $t_2$. If the sequence be $S$, we may say that the terms *between* $t_1$ and $t_2$ are $S$-intermediate to $t_1$ and $t_2$. If between every two terms in the field of a sequence there is a term of the field, the sequence is said to be *dense;* thus, $P'''$, for example, is dense, while $P$, $P'$, $P''$ are not. You will hardly confuse the notion of a dense sequence with that of a dense difference kind $k$. The amounts of such a kind constitute a field of a dense sequence but the difference-kind is not itself a sequence.

It is noteworthy that, in dealing with a sequence, mathematicians do not usually state explicitly a propositional function determining it, though it is always possible and often helpful to do so; neither do they usually indicate explicitly (as above done in the case of $P$) the class of couples constituting the sequence, though this, too, can be done if desired. For example, if a mathematician wishes to invite attention to our sequence $P$, he will ordinarily say: Consider the sequence

$$1, 2, 3, 4, \ldots, n, n+1, \ldots$$

He will probably talk as if the *row* of numbers were the sequence, though it is not—the sequence being, as we have seen, a certain class of couples; the numbers in the row constitute the field of $P$ but the field as such he will probably not mention; he will speak of the numbers as the terms of the sequence, though they are merely the terms of the field, the terms of the sequence (a class of couples) being the couples in the class; and he will ordinarily have you understand that a number on the left of another in the row is a predecessor of the latter and that the latter is a successor of the former. This usual row method of indicating sequences has obvious advantages,—it is mechanical, spatial, visual, diagrammatic,—but it has to be used with care if confusion and error are to be avoided, for, as you already see and will further see, it disguises some of the nicer logicalities involved. For example, the mathemetician may indicate the sequence $P'$ (instead of $P$) by writing the foregoing row of numbers; in this case, a number to be a predecessor of another must be on the right of the latter instead of on the left of it; you see that the notion left-right (or its like) is not that of predecessor-successor; the former is spatial and sensuous, the latter logical and supersensuous; in the case of $P$, 3 is the predecessor of 4, not because 3 is on the left of 4 in the row, but simply because $3P4$; and in the case of $P'$, 4 is the predecessor of 3, not because the former is on the latter's right, but because $4P'3$; again $P$ begins at 1, not because 1 begins the row of numbers, for, you see, $P'$ ends at 1, despite the fact that the row begins at 1. Keeping such precautions in mind, we may often very conveniently employ the row method of indicating or representing sequences.

We are at length almost prepared for a certain new definition of the term " limit "; but there remains to be explained one further preliminary. It is a sfollows: If $xRy$ implies $xR'y$, the relation $R$ is said to be *included* in the relation $R'$; in other words, $R$ is included in $R'$ if every couple $(x, y)$ in the class of couples constituting $R$ is a couple in the class of couples constituting $R'$; if $R$

be included in $R'$ we say naturally that $R$ is a *part* of $R'$. I need hardly point out the fact that a relation includes itself and is thus a part of itself. $R$ and $R'$ are *identical* when and only when each is a part of the other. Now suppose $R$ to be a part of $R'$ and suppose $R'$ to be a sequence (series); then $R$ is also a sequence obviously; and thus, as you see, one sequence may be a part of another. It is plain that, if a sequence $R$ be a part of a sequence $R'$, the field $F$ of $R$ is a part of the field $F'$ of $R'$; and conversely that, if a class $F$ (of two or more terms) be a part of the field $F'$ of a sequence $R'$, then $F$ is the field of a sequence $R$ included in $R'$. Consider, for example, our familiar friend $P$; its $F$ is the class of positive integers; take any part of $F$,—any part containing two or more terms,—say, the class $C$ composed of 2, 3 and 7; $C$ is the field of the sequence composed of the couples (2, 3), (2, 7), (3, 7); this sequence is a part of $P$. It will be enlightening to notice that $P$ is itself a part of another sequence; let $Q$ be the sequence determined by the propositional function, $x$ is a positive real number less than a positive real number $y$; you see that $Q$ is a sequence, that its field is the class of all positive real numbers, that this field includes the field of $P$; that every couple in $P$ is also in $Q$; and that $P$ is a part of $Q$. You readily see that if one sequence be a part of a second, and the second a part of a third, the first is a part of the third.

When we speak of the (amount of) difference between a term $t$ in the field of a sequence $S$ and a term $t'$, let it be always understood that $t'$ is either in the field of $S$ or in the field of a sequence including $S$.

I hope we are now prepared to grasp the following definition of the term " limit."

$D_3$: Let $V$ be a variable whose range $R$ is included in

the field $F$ of a sequence $S$, and (as before) let $k$ be an available difference-kind; if there be in $F$ a term $t$ such that, however small a $k$-neighborhood of $t$ be chosen, there is in the neighborhood an $R$ term $t'$ differing by more than null from $t$ and being such that all $R$ terms between $t$ and $t'$ are in the neighborhood, then $t$ is an $S$ $k$-limit of $V$.

The meaning of $D_3$, which is a bit subtle and sly, may be made evident by means of a few examples. In adducing examples it will be convenient to make some use of the customary row method of representing sequences.

For a simple example, let $S$ be the sequence determined by the propositional function: $x$ is a fraction (having 1 for numerator and a positive integer for denominator) greater than a fraction $y$ (having 1 or zero for numerator and a positive integer for denominator). $F$ is composed of the numbers in the row

$$\frac{1}{1}, \frac{1}{2}, \frac{1}{3}, \frac{1}{4}, \ldots \text{ (ad infinitum)}, 0.$$

Let $V$ be the variable whose range $R$ is the class of all the $F$ terms except 0; let $k$ be the kind of difference in respect of which we compare the values or magnitudes of rational fractions (as when we say $\frac{1}{2}-\frac{1}{3}=\frac{1}{6}$). The question is: Has $V$ an $S$ $k$-limit $t$? The answer is yes: $t$ is such a limit if $t$ be zero (0). To prove it, suppose chosen a $k$-neighborhood $d$ of 0, however small; there is no restriction upon the choice of $d$ save that $d$ must be a positive rational number; it is plain that there is in $R$ a number $t$ differing from 0 by more than null (zero) and by less than $d$; it is evident that such a $t$ and all the $R$ terms between t and 0 are in the chosen neighborhood; and hence 0 is, as said, an $S$ $k$-limit of $V$.

It should be said in passing that a $V$ having, under

$D_3$, zero (or null) for limit is called an *infinitesimal*—a term of great importance in most branches of mathematics. We will return to it if time permits.

Let us choose another difference-kind and see what happens when $V$, $F$ and $S$ have the same meanings as above. Observe that the denominators of the $F$ terms are positive integers, for 0 may be written $0/n$ where $n$ is a positive integer. We may compare the $F$ terms with sole reference to the values of their denominators—with reference, that is, to the difference-kind $k'$ in respect of which we compare the magnitudes of positive integers *as such*. Of $k'$ the only amounts $d$ are: *null*, 1, 2, 3, 4, . . . ; hence the smallest $k'$-neighborhood $d$ of 0 is that for which $d=1$; as no $R$ term differs from 0 $(0/n)$ by an amount of difference-kind $k'$ more than null and less than 1, it is seen that no $R$ term $t'$ differing by more than null of the difference-kind $k'$ from 0 is in the neighborhood of 0 for which $d=1$; therefore, 0 is not an $S$ $k'$-limit of V. And so is justified the mention of $k$ in $D_3$.

We have just seen that, though a $t$ be an $S$ $k$-limit of $V$, it may not be an $S$ $k'$-limit of $V$ if $k$ and $k'$ be not the same. We may now show that, though a $t$ be an $S$ $k$-limit of $V$, it may not be an $S'$ $k$-limit of $V$, if $S$ and $S'$ be different sequences. Consider the numbers in the row

$$\tfrac{1}{2}, \tfrac{1}{3}, \tfrac{1}{4}, \ldots \text{ (\textit{ad infinitum})}, \tfrac{1}{1}, 0.$$

Let $S'$ be a sequence having the class of these numbers for its field and let $S'$ be such that, if $a$ is an $S'$ predecessor of $b$, then $a$ and $b$ are in the row, $a$ on the left of $b$, and that any number in the row is an $S'$ predecessor of all the numbers on its right and an $S'$ successor of all the numbers on its left. $S'$, as you see, is now completely determined—

we know all the couples constituting it; and you note that its field $F$ is the same as that of $S$ above. Let $V$ be the same as before and let $k$ be the same as in the above paragraph, in which 0 was found to be an $S$ $k$-limit of $V$. The question is: is 0 an $S'$ $k$-limit of $V$? The answer is *no*, as you readily see; for choose a $k$-neighborhood $d$ of 0, where $d$ is, say, $\frac{1}{2}$; any $F$ term $t'$ differing by more than null (zero) from 0, if it be in the chosen neighborhood, is, as you see, a predecessor of $\frac{1}{1}$; and so $\frac{1}{1}$, though it is between such $t'$ and 0, is not in the chosen neighborhood; accordingly, as said, 0 is not an $S'$ $k$-limit of $V$.

Comparing $D_2$ with $D_3$ you observe that $D_3$ contemplates $V$'s range as a part of the field of a sequence and that $D_2$ does not; you notice, too, that $D_3$ contains the same conditions as $D_2$ contains and one other—the " between " condition (which would indeed be meaningless in $D_2$ inasmuch as $D_2$ does not regard $V$'s range as included in the field of a sequence). It follows that if a $V$ have a limit $t$ under $D_3$, the same $V$ has $t$ for limit under $D_2$. Is the converse true? It is easy and instructive to show by an example that it is not. Consider the numbers in the row

$$(\tfrac{1}{2}+1),\ (\tfrac{1}{4}-1),\ (\tfrac{1}{8}+1),\ (\tfrac{1}{16}-1),\ (\tfrac{1}{32}+1),\ (\tfrac{1}{64}-1),\ \ldots;$$

which are the same as the numbers in the row

$$\tfrac{3}{2},\ -\tfrac{3}{4},\ \tfrac{9}{8},\ -\tfrac{15}{16},\ \tfrac{33}{32},\ -\tfrac{63}{64},\ \ldots;$$

let $S$ be a sequence such that, if $aSb$, $a$ and $b$ are in the row, $a$ on the left of $b$, and that each number in the row is an $S$ predecessor of every number on its right. $S$'s field $F$ is the class of the numbers in the row. Let $k$ be the difference-kind appropriate for comparing (as by subtraction) the values of real numbers. Let $V$ be the variable whose range is $F$. It is easy to see that 1 and $-1$ are both of them limits of $V$ under $D_2$ and that neither of them is an $S$ limit of $V$ under $D_3$. For choose a neighborhood $d$ of 1, no matter how small; plainly there is an $F$ term (a positive number in the row, but not a negative one) differing from 1 numerically by less than $d$, and such a term is in the chosen neighborhood; accordingly 1 is, by $D_2$, a limit of $V$; but 1 is not an $S$ limit of $V$ for 1 is not even a term of $S$'s field. Like reasoning would show that $-1$ is a limit of $V$ under $D_2$ but is not, under $D_3$, an $S$ limit of $V$, where $S$ is the sequence above indicated.

We have seen that, if two sequences $S$ and $S'$ have a common field $F$ and if $V$ be a variable whose range $R$ is a part of $F$, a term $t$ may be an $S$ limit of $V$ without being an $S'$ limit of $V$. This fact is so important that it seems advisable to give it further exemplification. Let $F$ be the class of all the positive rational numbers and zero. Consider the following sequences $S_1$, $S_2$, $S_3$, having $F$ for their common field.

$S_1$ is to be such that, if $xS_1y$, $x$ and $y$ are in $F$ and $x$ is less than $y$; and such that, if $x$ and $y$ be in $F$, then $xS_1y$ if and only if $x$ is less than $y$. We commonly say that $S_1$ as defined arranges the terms of $F$ in the order of increasing magnitude.

To define $S_2$ consider the row

$$(a)\ \tfrac{1}{1}, \tfrac{1}{2}, \tfrac{1}{3}, \ldots;\ 0, \tfrac{2}{1}, \tfrac{2}{3}, \tfrac{2}{5}, \ldots;\ \tfrac{3}{1}, \tfrac{3}{2}, \tfrac{3}{4}, \ldots;\ \ldots;\ \ldots$$

You observe that the row contains all and only the terms of $F$; $S_2$ is to be such that, if $xS_2y$, $x$ is on the left of $y$ in the row, and that any term in the row is an $S_2$ predecessor of every number on its right.

To define $S_3$ consider the array.

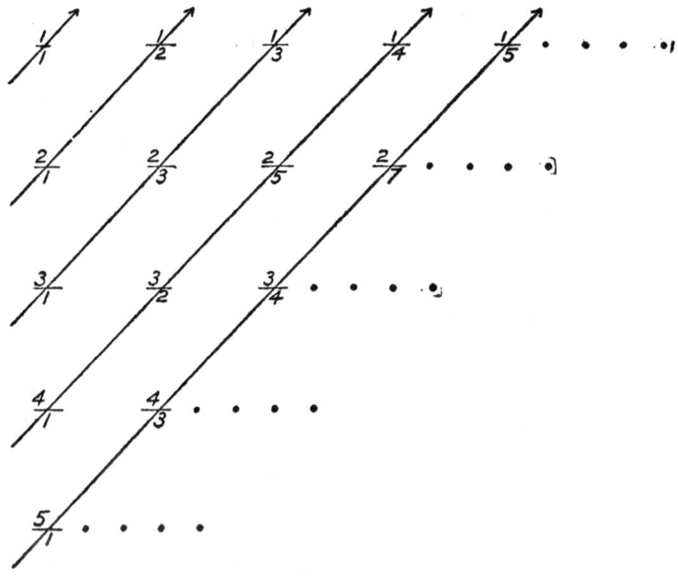


FIG. 27.

A little inspection shows that the array contains all and only the terms of $F$, zero excepted. The arrows indicate how the terms of $F$ may be arranged as in the row

$$(b)\ \tfrac{1}{1}, \tfrac{2}{1}, \tfrac{1}{2}, \tfrac{3}{1}, \tfrac{2}{3}, \tfrac{1}{3}, \tfrac{4}{1}, \tfrac{3}{2}, \tfrac{2}{5}, \tfrac{1}{4}, \ldots ;\ 0.$$

Notice that the scheme gives each $F$ term a definite place in the row. $S_3$ is to be such that, if $xS_3y$, $x$ is on the left of $y$ in $(b)$ and that every term in $(b)$ is an $S_3$ predecessor of every term on its right.

Now let $V$ be the variable having for its range the class of all $F$ terms *except* zero, and let $k$ be the familiar difference-kind we have in mind when we say the (amount of) difference between this real number and that is such-and-such. Applying $D_3$ to $V$ you will readily find that zero and all positive rational numbers are $S_1$ limits of $V$; that zero and nothing else is an $S_2$ limit of $V$; and that $V$ has no $S_3$ limit whatever. Two additional facts are worth noting here: one of them is that, under $D_2$ zero and every positive real number is a limit of $V$; the other is that, under $D_3$, zero and every positive real number will be an $S$ limit of $V$, if $S$'s field be the class of positive reals and zero, and if $S$ arrange the terms of $F$ in the order of increasing magnitude.

I shall leave it to you to practise to your heart's content in applying $D_3$ to such various variables and sequences as you can readily find or devise.

Presently, I shall ask you to consider a fourth conception of limit. Before doing so, I wish to call your attention to a curious nice little dispute that now and then arises respecting the notion of limit as defined by $D_3$ or by a virtual equivalent of $D_3$. The dispute arises out of confusion due partly to the row method of indicating sequences and partly to the custom of speaking figuratively of a variable as if it actually changed, varied, increased, decreased, and so on, instead of merely representing " any one " of the terms of some specified class. I can best present the matter by means of an example or two. Consider the three sequences indicated by the three number rows.

$$(r_1) : \tfrac{1}{2}, \tfrac{3}{4}, \tfrac{7}{8}, \tfrac{15}{16}, \ldots ; 1$$
$$(r_2) : \tfrac{3}{2}, \tfrac{5}{4}, \tfrac{9}{8}, \tfrac{17}{16}, \ldots ; 1$$
$$(r_3) : \tfrac{1}{2}, \tfrac{3}{2}, \tfrac{3}{4}, \tfrac{5}{4}, \ \ldots ; 1$$

You note that the three sequences are distinct and that their fields are distinct. If the ranges of the variables $V_1$, $V_2$, $V_3$ contain respectively the same terms as the fields except 1, you easily see that, under $D_3$, 1 is a limit of $V_1$, of $V_2$ and of $V_3$. It is customary to say, "As $V_1$ runs along the sequence $(r_1)$ from left to right it approaches 1 as its limit" or to use some equivalent equally figurative speech; and similarly for $V_2$ and $V_3$. It is noticed that the $V$'s in running along their so-called sequences get nearer and nearer to their limits but never reach them. The question arises: Is it possible for a sequence having a limit to be such that the variable, in the course of its approaching the limit, reaches it one or more times? Some say *no;* others say *yes.* The latter attempt to justify their answer substantially as follows: Consider, they will say, the sequence

$$(r)\ \tfrac{1}{2},\ \tfrac{3}{2},\ 1,\ \tfrac{3}{4},\ \tfrac{5}{4},\ 1,\ \tfrac{7}{8},\ \tfrac{9}{8},\ 1,\ \ldots\ (\textit{ad infinitum});\ 1,$$

got from $(r_3)$ by inserting 1 after each of the successive pairs of numbers in $(r_3)$; observe, they will say, that if a $V$ runs along $(r)$, skipping the third term, the sixth and so on, it will approach the same limit (namely, 1) as if it ran along $(r_3)$, and that, if it runs along $(r)$ without skipping, it will again evidently approach the same limit, 1, but in this case will actually reach 1 infinitely often in endlessly approaching it; and so you are expected to see not only that $(r)$ is a sequence having a limit but that, while endlessly approaching it, it actually reaches it again and again and again. You instinctively feel that you are being hocus-pocused by such argument, and your instinct is sound. What is the trick? It is easy to detect. The juggler (we may call him a juggler, though he does not intend to deceive) asks us to regard $(r)$ as a sequence or at all events as indicating a sequence. Let us try to do so in good faith. If $(r)$ be or indicate a sequence $S$, what is the field $F$? The answer is obvious: the terms of $F$ are the numbers in $(r_3)$. Among these is 1, the final number of ($r_3$) and of ($r$). But ($r$) indicates that $F$ contains a host of 1's—the "inserted" 1's; but if these are in $F$, they are $S$ predecessors and successors, and we have $1S1$ contrary to the definition of a sequence. You see that ($r$) neither is, nor property indicates, a sequence. (It is of course possible to define the terms "sequence" and "limit" so that a sequence having a limit may be such that the variable in running towards the limit reaches it one or more times.) Here is a good place to emphasize the fact that the field of a sequence never contains two identical terms. Why not? Because a field is a class, and a class contains all and only the verifiers of some propositional function, say, $\phi(x)$; if $x_1$ be a verifier of $\phi(x)$, then $x_1$ is a term or member of the class; it is evident that as such a member, it occurs but once. We do indeed often speak (unprecisely) as if such were not the case; but when we speak of $a$ and $a'$ as being identical members of a class, we mean that $a$ and $a'$ are two different symbols for one and the same member of the class and we do not mean that the two symbols are themselves members of the class.

*Serial (Ordinal) Definition of the Term "Limit."*—We have now before us three definitions—$D_1$, $D_2$, $D_3$—of the term. It is important to observe that each of them essentially involves the notion of *quantity;* they involve it, for they involve the notion of the neighborhood of a term, and this notion is quantitative; a given neighborhood has a size; another one is larger or smaller; neighborhoods are among the things differing from one another in respect of magnitude—quantity is of their essence. We should not fail to observe, too, that, while the three definitions thus agree in involving the notion of quantity, $D_3$ involves also the notion of a sequence or series,—

a non-quantitative, purely ordinal notion,—and that $D_1$ and $D_2$ do not. I mean that $D_3$ contemplates the variable's range as being the field (or a part of the field) of a series, or sequence, and that $D_1$ and $D_2$ do not. As ontologists you may no doubt contend that the terms of any given class and hence the terms of any given variable's range are, quite independently of our intention or will, arranged once for all and eternally in every variety of sequence of which they are capable. I am not disputing the justice of that contention; conceding it to be just, granting the eternal existence of all the sequences possible for a given range, I am merely signalizing the fact that $D_1$ and $D_2$ disregard them each and all, and that $D_3$ does not; $D_3$ regards the variable's range as an ordered class of terms; $D_1$ and $D_2$, disregarding order, regard the variable's range as an *orderless* collection. We may say, then, that $D_1$ and $D_2$ are quantitative definitions and that $D_3$ is mixed—both quantitative and serial.

It is natural to ask whether the term " limit " sometimes denotes a purely serial conception. The answer is affirmative. The following definition presents such a definition of the term.

$D_4$: Let $V$ be a variable whose range $R$ is included in the field $F$ of a sequence (series) $S$; if an $F$ term $t$ be such that, given any $S$ predecessor $t'$ of $t$ among the $R$ terms, there is an $R$ term between $t'$ and $t$, or such that, given any $S$ successor $t'$ of $t$ among the $R$ terms, there is an $R$ term between $t$ and $t'$, then $t$ is an $S$ limit of $V$.

Let us at once cite some simple examples. Consider the sequence.

$$S_1 : 1, 2, 3, 4, \ldots \textit{(ad infinitum)}, \tfrac{1}{2}, \tfrac{1}{3}.$$

Let predecessor-successor mean left-right; let the terms

of $F$ be the numbers in the row; let the terms of $V$'s range $R$ be the row's integers. You see that $\frac{1}{2}$ is an $S_1$ limit of $V$ and that $\frac{1}{3}$ is not. Why not? Notice that $\frac{1}{2}$ is not an $S_1$ limit of $V$ under $D_3$ nor a limit of $V$ under $D_2$. If a term be a limit of a $V$ under $D_3$ or $D_2$, must it be a limit under $D_4$? If, in the foregoing example, we suppose the $R$ to include $\frac{1}{2}$ (besides the integers), will the new $V$ have an $S$ limit? Why not? If we strike out $\frac{1}{2}$ and $\frac{1}{3}$ and suppose $R$ to coincide with the new $F$ (of the new $S_1$), will $V$ have an $S_1$ limit? Why not?

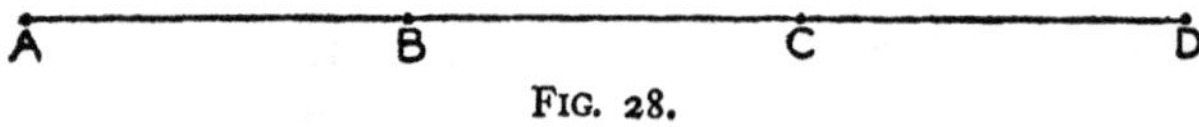


FIG. 28.

For other examples consider the following. Let $S_2$ be a sequence having for its $F$ the points of the line segment $AD$, including $A$ and $D$. Let us take predecessor-successor to mean, as before, left-right; let the $R$ of $V$ be composed of $A$ and the other points preceding $B$; you see that all the points in $R$ and the point $B$ but no other $F$ points are $S_2$ limits of $V$. Notice that the same would be true if we supposed $R$ to include $B$. Suppose the $F$ of $S_2$ to be composed of $B$, $B$'s predecessors, $C$ and $C$'s successors; and, as before, let the terms of $R$ be $B$'s predecessors; you readily see that the $S_2$ limits of $V$ are $B$, $B$'s predecessors and $C$; that all of these except $C$ are $S_2$ limits of $V$ under $D_3$ and limits of $V$ under $D_2$; and that, under $D_4$, $C$ is not an $S_2$ limit if $R$ include $B$ (as well as its predecessors). Why not? Next suppose the $F$ of $S_2$ to consist of all the points of $AD$ except $B$ and $C$, and let the terms of $R$ be the midpoint $A'$ of $AB$, the midpoint of $A'B$, and so on, the midpoint $B'$ of $BC$, the midpoint of $B'C$, and so on, the midpoint $C'$ of $CD$, the midpoint of $C'D$, and so so; then show that $D$ is the only $S_2$ limit of

$V$ under either $D_4$ or $D_3$ but that, under $D_2$, $B$ and $C$ as well as $D$ are limits of $V$.

With the foregoing non-quantitative—purely serial—conception of limit, you can make yourselves familiar by applying the definition to numerous examples which you can readily construct or easily find, for they abound on every hand.

I have now spoken of limits and limit conceptions at far greater length than I had originally intended to do. If I have thus exhausted your interest and patience, I assure you that I have by no means exhausted the subject. There are in use yet other conceptions of the term limit and connected therewith many interesting and important refinements,—refinements of refinements,—with which, however, I do not intend to trouble you. There remain two questions which must have occurred to you and which I am sure you will desire to consider before we take final leave of the subject. One of them is easy and admits of a brief answer. The question is: In view of the variety of senses in which mathematicians employ the term "limit," how do they manage, if they do manage, to avoid confusion—confusion of themselves and others? The answer is: They do not always avoid it, but in general they do, and they do so, as I have already intimated, by indicating either explicitly or contextually, when speaking of a limit, the sense in which the term is to be understood. The second question relates to the scientific and philosophic importance of the term. Both by dwelling on it so long and by explicit statement, I have said that its importance is very great. I wish now to show that the estimate is just and how it is so.

*Scientific and Philosophic Importance of the Term Limit.*—As to its scientific importance, the task of showing it is very easy if we take "scientific" in its stricter and narrower sense. We may go at once to the heart of the matter by reflecting a little upon the most rigorously scientific of scientific subjects and procedures—the Differential and Integral Calculus—and upon its ramifications and its applications. Some of you have had a beginner's course in the calculus; others of you, not; I am not going to offer here an introduction to it but will merely state succinctly, by way of reminder or of information, a few such facts respecting it as will make indubitable the great scientific importance of the term in question. One of the facts is that the Calculus is primarily and mainly concerned with what mathematicians call *continuous* functions (or variables), and that both functional continuity and functional discontinuity, with which latter the calculus is also concerned, are not only defined by means of limits, but are indeed not otherwise definable. Another of the salient facts is that among the host of ideas met with in the Calculus three ideas are supreme—namely, those denoted by the terms Derivative, Antiderivative (or Indefinite Integral) and Definite Integral—and the three essentially involve a limit conception, the first and third of them directly, the second one indirectly. It follows, as you see, that in all the multifarious ramifications and applications of the Calculus, whether in differential equations or function theories or geometry or mechanics or astronomy or physics or chemistry or other fields into which the calculus has found or is inevitably finding its way, some variety of limit conception is continually playing an indispensable scientific rôle. Indeed it is only by prolonged meditation upon the matter that one can even fairly begin to realize how very deeply the progress of science and therewith of civilization depends upon ideas denoted by the modest little five-lettered word—*limit.*

What, broadly speaking, we may call its philosophic significance is less well understood for the reason that it has been neglected. It has been neglected because but few mathematicians have been interested in it and but few philosophers have been mathematically qualified to treat it. If only the concept of limit and the rôle thereof had been familiar in the days of Plato! How it would have enriched and fortified his dialectic. In his hands the concept would have been a new spiritual instrument of immeasurable power; in his thought it would have opened new ways to the inner vision of supernal light; in his brightest pages it would have been the secret and source of a yet stranger and brighter glory. His shining Absolutes,—absolute justice, absolute beauty, absolute truth, absolute good,—whose " perception by pure intelligence " brings us, said he, " to the *end* of the intellectual world," would not have appeared as ends, or final terms, of any sequences or progressions in the intellectual world nor even as limits of such progressions but, as I intimated in the initial lecture of the course, the absolutes would have appeared as supernal ideals, over and above every type of excellence in which intellectual progress is possible. And thus the Platonic philosophy would have advanced, in even greater measure than it did advance, the science of Idealization—the science, I mean, which has for its appropriate subject-matter those spiritual phenomena of life which the terms, ideal and idealization, rightly understood, denote. In saying this, I have in some measure anticipated the outcome of considerations not yet adduced, and so I must ask you to reserve your judgment for a little time.

In order to arrive at a fair estimate of the philosophic significance of limit concepts and limit processes,—in order, that is, to win a fair sense of their function and service in the life of Thought taken in all its varieties and scope,—it is necessary as a preparation to examine the matter a little further in mathematical light for it is here and not elsewhere that concepts of limit and limit processes are seen, and seen at work, in their nakedness and purity. As beheld in that light, conceptions of limit, apart from any question regarding their instrumental value, are objects of no little interest—a fact well worthy of passing mention, though I do not insist upon it in this connection. Regarding instrumental value, we have seen that limit concepts enable us to discriminate and classify variables with reference to the constitution of their ranges and to the connections of these with series; we have seen that limit concepts are essential to the formation and so to the meanings of innumerable other concepts, many of them of great import, as that of functional continuity or that of derivative, instanced a moment since; you know, or (if not) you can quickly learn by glancing at mathematical literature that limit concepts play an indispensable, perhaps the chief, rôle in the conduct of proofs, or demonstrations, in all branches of Analysis and its applications. I wish now to invite your best attention to the fact that, over and above the foregoing types of service, limit concepts render an invaluable service of a radically distinct kind in connection with that very familiar yet always strange thing which we are wont to call "generalization." I mean the kind of generalization which consists in our somehow contriving so to extend the meaning of an established concept as to bring within its enlarged scope,—as under the unity and order of a new empire,—what had been seemingly unconnected, reciprocally alien provinces of thought.

The meaning and justice of what I have just now said may be made evident by means of simple examples. Three or four little ones will suffice, and we can both shorten the task and enliven it by speaking of our variables in the customary dynamic fashion.

In the first example, I am going to ask you to imagine that we have arrived at a stage of mathematical evolution where we are familiar with the ordinary fractions, or ratios, including such as $\frac{1}{1}$, $\frac{2}{1}$, . . . , which for convenience we will write 1, 2, . . . ; and that we know nothing of so-called irrational numbers. Let $S$ be the sequence of the ratios arranged in the natural order of increasing magnitude. Let $V_1$ represent any ratio less than 2 (*i.e.*, $\frac{2}{1}$) and let $V_2$ represent any one greater than 2. You immediately see that, under either $D_3$ or $D_4$, 2 is a limit of $V_1$ and also of $V_2$, $V_1$ approaching it from below and $V_2$ from above. Observe that neither of the $V$'s can reach the limit; one of them is always less, the other always greater, than 2; they can, however, so close in upon 2 as to make the difference between them less than any preassigned positive ratio, however small,—we can make the $V$'s as near together as we please if only we do not please to make them meet—between them stands their common limit, 2, fringed on both sides with a row of ratios which the $V$'s in their race towards 2 can never run through. Now consider very carefully two other variables, $V$ and $V'$, the former representing any one of the ratios whose *square* is *less* than 2, and the latter any one whose *square* is *greater* than 2. Note that the new $V$'s, like the old ones, can come indefinitely near together; observe that as they approach each other, one of them growing continually larger, the other continually smaller, though they can never meet, yet they, like the old $V$'s come to differ by less than any preassigned positive amount, however small. Undoubtedly the new $V$'s, like the old ones, *seem* thus to close in upon a common limit. Do they do so in fact? If they do, what is the limit? If there be one, it must, we are sure, be something having a square and having 2 for the square; but this something, if it exist, can not be one of the things which we and our race have hitherto meant by number, for, by hypothesis, the only numbers we know in our present stage of evolution are cardinals, integers and ordinary fractions, and none of these has 2 for its square. You sense vividly, I trust, the painful situation into which our limit idea has brought us. Do you know how we will behave under the circumstances? How we will try to escape? By what means we will endeavor to reach a reconciliation? We are to suppose ourselves to be dealing with the difficulty as the mathematicians have dealt with it. Accordingly, we will not all of us behave in the same way—some of us will resort to one means of extrication and some to another. (*A*) Some of us will say: $V$ and $V'$ have not a common $S$ limit, but they have a common $S'$ limit where $S'$ is a sequence of things we have not yet learned, but must learn, to recognize and handle; this common limit, though not a number in the accepted sense of the term, is something we must regard as having a square and as having 2 for its square; we will denote the thing by the symbol, $\sqrt{2}$, and call it a number of a new kind—an irrational number to distinguish it from the old familiar ratios to be henceforth called rational. (*B*) Others of us will say: $V$ and $V'$ have no common limit of any kind familiar or unfamiliar; it is, however, manifest that they *ought* to have one for the sake of our *convenience*, and, as none exists, we will create one; we will call the creature "the square root of 2," denote it by the symbol, $\sqrt{2}$, regard it as a number, and describe it as irrational to distinguish it from the old sort of numbers,—the ratios,—to be henceforth described as rational. ($C$) Yet others of us, not so numerous but harder-headed and more critical, will say: it is evident that to escape decently from our predicament we must somehow enlarge our conception of number; not, however, by asserting that $V$ and $V'$ have a mysterious sort of common limit, for they evidently have no common limit; nor by pretending to "create" one for them, which we can not do; but by discovering that certain existing things, not hitherto regarded as numbers, ought to be so regarded—a discovery that, briefly sketched, runs as follows: we reflect that the ranges of $V_1$ and $V$ (we could equally well use $V_2$ and $V'$) are classes of ratios ordered by $S$; we observe that neither of the ranges contains a maximum term, a largest ratio, though one of them (or its variable $V_1$) has an upper limit, 2, and the other has no upper limit; giving the name *segment* to such ratio ranges, that is, to such of them as have no maximum, we see that a segment may or may not have an upper limit; we readily see that segments have certain properties (summability, and so on) very like the properties of what we have been calling numbers; we accordingly and naturally agree to call the segments themselves numbers; they are a new kind of numbers——not ratios, but certain classes thereof; we call the new numbers *rational* if the segments have upper limits and *irrational* if they have not; thus the segment represented by $V_1$ is a rational number while that represented by $V$ is irrational; we denote the former by 2 because the upper limit of $V_1$ is the ratio $\frac{2}{1}$, but we must not confound this rational 2 (which is a segment of ratios) with the cardinal 2 nor with the integer 2, nor with the ratio $\frac{2}{1}$ (commonly denoted by 2) nor with any other number that blundering custom may yet denote by the same symbol; the irrational number or segment represented by $V$ is denoted by $\sqrt{2}$; and so on analogously for analogous cases; to each ratio $\frac{a}{b}$ will correspond a rational number $\frac{a}{b}$, so that, for example, to the ratio $\frac{2}{3}$ will correspond the conceptually distinct rational number $\frac{2}{3}$; to the irrational numbers, however, no ratios will thus correspond; the rationals and the irrationals taken together we will call *real* numbers; these may be arranged in the order of increasing magnitude by a sequence $S'$, and, if we then let $V$'s range be, not the class of ratios less than the ratio 2, but the class of rationals less than the rational 2, $V$ will indeed have 2—rational 2—for upper limit; and so at length the mystery is dispelled—what fooled us before was our confounding the familiar class of ratios with the then unknown, yet vaguely felt, class of rationals, corresponding to but logically distinct from the ratios.

Well, what is it that has happened here in our racial history? I hope you see that what has happened is this: we have made a leap, an immense forward leap, in the course of mathematical evolution; we have made a great number-generalization; we have, that is, extended our old familiar well-established concept and name of numbers so as to make it include and cover two immense new varieties, namely, the rationals,—which are as multitudinous as the infinite host of our old traditional ratios,— and the irrationals,—which may be shown to be infinitely more in multitude than all the old numbers taken together.

Do not fail to observe how the tremendous generalization, so copiously enriching our human world of mathematical ideas, was brought about: we were operating in a certain domain,—the domain of ratios; we were there employing the notion of limit; using the notion, we found ourselves looking for a limit where we were suddenly and painfully astonished to find there was none; we were baffled, we wondered, felt a need,—the need of a deeper view, of a larger vision, of a more embracing conception to extricate us; and we found it—how? By means of the limit idea; that which got us into the difficulty got us out of it and, in doing so, gave us a larger world.

Did the limit concept *compel* the generalization? No; such generalization is never compelled,—it is suggested, recommended, stimulated, even urged,—but not compelled as a conclusion from premises,—generalization always involves an act of *will*,—a choice between a smaller, meaner view and a larger, nobler one; and in the present instance it was, you see, the notion of limit that gave man's will the necessary suggestion, incitement and guidance.

There is another aspect of the matter which you as philosophers must on no account fail to notice very carefully, for it is a phenomenon of all genuine generalization. It is this: the world of the real numbers, though itself a strictly actual world once it is found, yet is, for any possible point of view *in* the domain of ratios, a strictly ideal world—ideal in the just sense that, though it is suggested by phenomena in the domain of ratios, it is itself wholly outside thereof and can in no wise be attained by pursuing sequences, however endless, within the domain; a generally neglected fact of the utmost importance, not only in discussing the spiritual bearings of mathematics, but also and especially in understanding the ways of spiritual life—the ways of truth to men.

Not to wish to dwell in this insight long enough to make it our own would show us unworthy—stupid or perverse. We might indeed illustrate it in many ways and in many connections. We might show in detail how limit concepts at work in the domain of real numbers, especially in connection with equations failing to have roots in that domain when the variable coefficients are allowed to approach certain limits, make us keenly aware that the domain of reals, vast as it is, is yet too meagre for our purposes, and how we are thus led to effect another immense number-generalization—that one, I mean, which gave us what we call the complex numbers ($x+iy$, $x$ and $y$ being reals, and $i$ being the so-called imaginary unit, $\sqrt{-1}$), now the subject of a stately theory having wide application in physics and even in engineering; we might show in detail, little step by step, how limit concepts at work in geometry have availed so to extend or generalize such fundamental notions as length, area and volume—formerly clear and well defined only in connection with broken lines, or polygons, and solids bounded by planes—that we can now confidently and understandingly use the notions in connection with all manner of curves and curved surfaces. But such details would require certainly more time and perhaps more patience than we now have at our disposal.

I must, however, once more insist upon the matter which I mentioned a moment ago and which I have emphasized elsewhere. The matter is this: a limit-begotten generalization always originates in the work of some limit concept operating in some established domain (such as that of our ratios, for example) wherein the concept leads us into the presence of baffling phenomena, waking our wonder, giving us a painful sense of failing to see something we ought to see, a sense of logical suffocation, of being hampered, hemmed in; we seek emancipation and at length achieve it, not solely by purely logical means, but partly by observation (as in the case of the segments), partly by reasoning and partly by an act of will—in short, by generalization; this deed gives us a new domain of thought—a new field of ideas (as, for example, the domain of real numbers); the new domain, once thus established, is as actual for us as the old one; with reference, however, to any viewpoint *within* the old one, the new domain is and forever remains a sheer ideal, not to be attained by any process or operation—however oft repeated, swift or prolonged—within the old domain; and finally, a new domain (as that of the real numbers, for example) may in its turn become, in the manner indicated, an old one in relation to another domain (as, for example, the domain of the complex numbers) which, though itself actual, is, with respect to the former, an eternal ideal.

*Mathematical Limit Processes Viewed as Species of Idealization.*—In nearing the close of this second long lecture on variables and limits we come now to what I most desire to signalize as being for students of philosophy the most significant aspect of the whole matter. It is this: In mathematics the great rôle of what we there call limits and limit processes is in *kind* identical with the momentous rôle of that which in other fields of interest we call *ideals* and *idealization.* In the light of the foregoing discussion the fact is evident, and it shows us again very clearly what we have repeatedly seen in other connections—that —far from being detached from common life or alien thereto,—mathematics is a refined model or prototype of that which in life is most precious and, strange to say, most omnipresent, too, though the presence be often disguised. With sequences,—many of them finite, many of them potentially infinite,—our concrete life is indeed replete: sequences of potential degrees of knowledge, of potential degrees of wisdom, of potential degrees of skill, of justice, of beauty, of righteousness, of authority, of power, of freedom, of potential degrees of innumerable forms of excellence in the achievements and dreams and aspirations of mankind. I hope you will not fail to see clearly that just as in the mathematical prototype of idealization,—in the theory, that is, of limits and limit processes,—so here, in our concrete and passionate life, ideals are of two kinds: namely, ideals which we pursue endlessly from degree to degree of excellence of a given type, as a variable having a limit endlessly pursues it without attaining it; and those higher ideals which are indeed not as limits of endless sequences of degrees of excellence of a given type (that of jazz music, for example, or that of a Beethoven sonata), but which require us to rise from given types to higher types by a species of idealization corresponding to that which, in the model, we have called limit-begotten generalization. It is thus evident that ideals are not things to gush over or to sigh and sentimentalize about; they are not what would be left if that which is hard in reality were taken away; ideals are themselves the very flint of reality, beautiful, no doubt, and precious, without which there would be neither dignity nor hope nor light; but their aspect is not sentimental and soft; it is hard, cold, intellectual, logical, austere. Idealization consists in the conception or the intuition of ideals and in the pursuit of them. And ideals, I have said, are of two kinds. Let us make the distinction clearer.

Every sort of human activity,—shoeing horses, abdominal surgery or painting profiles,—admits of a peculiar type of excellence. No sort of activity can escape from its own type, but within its type it admits of indefinite improvement. For each type there is an ideal,—a dream of perfection,—an unattainable limit of an endless sequence of potential ameliorations within the type and on its level. The dreams of such unattainable perfections are as countless as the types of excellence to which they respectively belong and they together constitute the familiar world of our human ideals. To share in it,—to feel the lure of perfection in one or more types of excellence, however lowly,—is to be human; not to feel it is to be sub-human. But this common kind of idealization, though it is very important and very precious, does not produce the *great* events in the life of mankind. These are produced by the kind of idealization that corresponds to what we have called, in the mathematical prototype, limit-begotten generalization,—a kind of idealization that is peculiar to *creative* genius and that, not content to pursue ideals within established types of excellence, creates new types thereof in science, in art, in philosophy, in letters, in ethics, in education, in social order, in all the fields and forms of the spiritual life of man.

We have here, you see, a new way—I think it a most fruitful way—to study the phenomena of spiritual life, whether our own or that of mankind in general. I leave it to you to pursue it if you will, for that is what philosophy is,—the study of the phenomena of the spiritual life man,—and if it is not that, it is nothing. In relation thereto I will merely say, in closing, that of the two kinds of ideals and idealization, it is by meditating on the higher kind, in the light of the mathematical theory of limits, that I have been led, as already indicated, to regard the great Platonic Absolutes as supernal ideals, indicated indeed in the world of logic but there indicated as having their locus above it so that they appear like downward-looking aspects of an over-world.

# LECTURE XV

## Infinity

MATHEMATICAL INFINITY—ITS DYNAMIC AND STATIC ASPECTS—NEED OF HISTORY OF THE IMPERIOUS CONCEPT—THE RÔLE OF INFINITY IN A MIGHTY POEM—NO INFINITY, NO SCIENCE.

IT is inconceivable that a course of lectures having the aim of the present course should be altogether silent respecting the mathematical concept of infinity. For among the great mathematical concepts that are accessible to laymen there is none which surpasses this one in importance or in power; there is none that appeals more strongly to the imagination of such as are qualified to receive it; and the nearer mile-posts of its endless avenue of increasing wonders are not difficult to reach. On these accounts the temptation to devote at least one lecture to an elementary exposition of the idea is strong. I have decided, however, not to yield to it for, if I did so, I should be wasting your time; I should be only adding another one to already numerous expositions of the kind, some of them so simple and clear as to leave no excuse for not acquiring a fair knowledge of the matter except the melancholy excuse of spiritual inaptitude therefor. In the *Revue de Métaphysique et de Morale*, for example, the elements of the subject are handled, often admirably, in a variety of papers, some of them by notable philosophers and some of them by distinguished mathematicians. One of the best expositions I have seen is that by Professor E. V. Huntington in his *The Continuum and Other Types of Serial Order* where, moreover, you can learn what mathematicians mean by the highly important term "continuum,"—the Grand Continuum as Sylvester called it,—an idea with which for lack of time I was unable to deal in the two preceding lectures—and where, too, you will find an introduction to the transfinite numbers of Georg Cantor, masterful primate among all who have contributed to our understanding of mathematical infinity. For another excellent account I may refer you to the already mentioned *Fundamental Concepts of Algebra and Geometry* by Professor J. W. Young. Clear indication of the philosophic significance of the idea in question is found in an article on the "Concept of the Infinite" by the late Professor Royce (*Hibbert Journal,* Vol. I) and in the Appendix to *The World and the Individual,* by the same author. Perhaps no one else has treated the matter with so much deserved emphasis and with so much freshness and facility as Bertrand Russell in his more popular works. For some indication of my own views respecting the bearings of the concept upon certain fundamental questions of philosophy, theology and religion, I may be permitted to refer you to *Science and Religion;* to *The New Infinite and the Old Theology;* and to the articles—"The Walls of the World," "The Axiom of Infinity," and "Mathematical Emancipations"—contained in *The Human Worth of Rigorous Thinking.* In the foregoing works, you will find an ample clue to the extensive literature of the subject, both that which is more popular and that which is, I will not say more scientific, but more technical, if indeed you should, fortunately, desire to pursue the doctrine in its elaborate and recondite developments.

To an audience of philosophical students it need not be said that some notion of infinity has figured conspicuously, often fundamentally and dominantly, throughout the whole historic period of philosophy, and speculation East and West. It may, however, be said to such an audience, and I think it should be said, that a critical history of the concept of infinity—or rather of the concepts thereof, for there have been many of them, for the most part but ill defined—would be an invaluable contribution to the history of Thought—an incomparably more important contribution than the philosophical doctor dissertations commonly accepted. There can hardly be a doubt, I believe, that the mentioned task of historical criticism will sometime be performed. Why should it not be done by one of you? You are, of course, aware that the doing of it calls for an extraordinary kind of composite scholarly preparation—linguistic, historical, philosophical, scientific, and especially mathematical. Our American universities have long been amply equipped with adequate machinery for the giving of such preparation. Perhaps one of you will demonstrate that they have at length acquired the necessary spirit and purpose and atmosphere and temper.

In any adequate historico-critical survey of the rôle which the notion of infinity has played in our human thinking, the thought of many thinkers, widely distributed in time and in space, would have to be passed in review—analyzed, understood, and appraised. Among the questions which the critic would have to ask and try to answer respecting each thinker are such as these: What did he mean by infinite? Did he employ the term to denote a definite concept or at best a vague and emotional intuition? Was his thought and use of it mystical, or logical and analytical, or both? Did he regard his infinite as a fact or as an hypothesis, and why? Was it time? An extension in time? Space? An extension in space? Was it matter or mind or both? Was it physical or spiritual? Concrete or abstract? Did he define it and, if so, in what terms? Or did he take it as a primitive, and, if so, did he do it consciously? Did he think of it as magnitude or as multitude or as both? Had he but one infinite or many of them? If many, were they coordinate or hierarchical? If the latter, was the hierarchy crowned or summitless? Was his infinite subordinate in his thought or central and dominant? Did he employ it consistently or confusedly? Was its function poetic or scientific or both? What was its relation to the modern concept of mathematical infinity?

It has seemed to me that I could best serve you in this hour by sketching what I conceive should be an important chapter in such a critical work. The sketch, which will be very imperfect, is offered, not as a model, but only as a concrete suggestion. I have selected for the purpose the philosophy of Lucretius, in which, as you are no doubt aware, the notion—or some notion—of infinity is very conspicuous. The question is: *what* notion and what is its significance there?

It will facilitate the discussion if we first remind ourselves of the meaning of mathematical infinity and, in connection therewith, note one or two distinctions and make the acquaintance of two important technical terms —*equivalence of classes,* and *denumerability.* In the lecture on the nature of mathematical transformation, we met the notion of an infinite class of terms, or objects of thought; we there saw, for example,—what any one but a fool *can* see,—that we can set up a one-to-one correspondence between the integers of the entire class of integers and the integers (say the even ones) composing a part, or sub-class, of the entire class, by the simple law or device of making 1 correspond to 2, 2 to 4, 3 to 6, 4 to 8, . . . , $n$ to $2n$, and so on endlessly. Two classes between which it is possible to set up, by some law of transformation, a one-to-one correspondence are said to be *equivalent* classes; and a class that is equivalent to a part, or sub-class, of itself is called an infinite class. Infinite classes,—of numbers, of points, of lines, of curves, of surfaces, of propositions, of relations, of functions, and so on,—abound on every hand; theoretically, and therefore practically, infinite classes are more important than finite ones, even though the spiritually blind are unable to see the fact; without infinite classes, as the late Henri Poincaré repeatedly said, there could be, strictly speaking, no such thing as science. Science is indeed the study of infinity.

If an infinite class be equivalent,—in the sense defined,—to the class of positive integers, it is said to be *denumerable*—a denumerably infinite class. Many infinite classes are denumerable which have not the appearance of being so. A striking example is the infinite class of our ordinary ratios, or fractions. Between any two integers,—nay, between any two fractions, however near to each other in value,—there are infinitely many fractions—and yet the entire class of fractions is precisely equivalent to the class of integers: an astonishing fact readily shown as follows. In the preceding lecture we saw that the fractions can be arranged in a row by means of a certain rectangular array with arrows. To see that the equivalence in question actually exists, you have now merely to observe that we can associate the first fraction of the row with the integer 1, the second with 2, and so on, thus using each fraction and each integer once and but once. Such an astonishing result makes one wonder whether *every* infinite class is denumerable. The answer is, *No*. It is well known that the class of real numbers,—even the class of the irrational numbers, even the class of points in a microscopically short line-segment,—is non-denumerable. Such classes are infinite but they are infinite of higher order. It is known that infinities rise above infinities in a summitless hierarchy. At present, the denumerable grade and that of the real numbers are the most important. In time to come such may not be the case; no one knows enough to say.

Here we must make a distinction. A class is a multitude—not a magnitude such as length, for example, or weight or area or volume or distance or the like; an infinite class is thus an infinite multitude; it has its root in the question—how many? A class is not a variable in the ordinary sense of this term,—it is a fixed thing,—a constant,—a datum given once for all in the world of thought—in logic the members of a class do not succeed each other in time—they coexist; and so you see that an infinite class is a *static* infinity. Long before *this* conception of infinity established itself in mathematics there was, as there is now, another conception of infinity,—of a sort of dynamic infinity,—namely, the conception of a changing magnitude or function capable of growing to exceed any given amount denotable by any integer however large. The idea may be conveyed as follows: let $n$ denote any given positive integer, no matter how large; in the fraction $\frac{n}{x}$, let $x$ be a variable representing a real number; let us treat $x$ as an infinitesimal—a variable having zero for its limit; as $x$ grows smaller and smaller, the function $\frac{n}{x}$ grows larger and larger; if we prescribe any *finite* amount—an amount, that is, such that we can denote it or a larger one by a positive integer, then, as $x$ decreases towards zero, $\frac{n}{x}$ will come to exceed the prescribed amount, however large; we express this obvious fact by saying that, for $x$ approaching zero, $\frac{n}{x}$ approaches infinity or becomes infinite or is an infinite variable or function; by such speech mathematicians do not mean that there is a definite quantity called infinity ($\infty$) and that the ratio becomes equal to it when $x$ takes the value zero; for when $x$ takes this value, the indicated division becomes meaningless and the ratio ceases to exist; what the speech means—and it means nothing else,—is, as said, that for $x$ decreasing as indicated, the ratio becomes larger than any prescribed finite amount. Such is the conception of an infinite function or variable,—a *dynamic* infinity, as we may call it to distinguish it from the other,—the static infinity. The dynamic type has its root in the question—how *much?* It is obvious that the two conceptions, though radically distinct, are intimately related. I shall leave it to you to compare them deeply if you will.

The "new infinity," as it is sometimes called, means the static infinity. It was introduced into mathematics something more than a half-century ago by Bernhard Bolzano, Richard Dedekind and Georg Cantor. Long before that, however, it was grappled and wrestled with by two geniuses of the first rank,—Galileo (in *The Two New Sciences*) and Pascal (in the *Pensées*, Havet's edition).

Is the concept, as some non-mathematicians have contended, a mere curiosity? The contention springs out of the unpardonable academic sin of stupidity.

With the preliminaries in mind, let us turn to the *De Rerum Natura* of Lucretius.[1] This work of an Italian poet I have already mentioned, in the lecture on the notion of group, in connection with what I have there called the philosophy of the cosmic cycle, or cosmic year. And I have mentioned it as one of the greatest works, not of a Roman as such, but of Man. Memorable on numerous accounts the Romans were. For the construction of palaces, temples, roads, aqueducts and other public works—with a measureless appalling waste of material and human energy, owing to pathetic ignorance of science; for inventions in the art of war, conquest and public murder; for elaborate, sometimes clever, often crude and vulgar imitation of Greek letters, eloquence, and art; for a manifold development of an imperious jurisprudence; for the theory and practice of empire over subjugated peoples; for the unintentional dissemination of Hellenic culture, which most of them despised, throughout vast portions of the world; for the establishment,—by conquest, exploitation and robbery,—of an unrivalled luxury and sensual magnificence rotting the moral fiber of both rulers and ruled: on these and similar accounts, the Romans are indeed memorable forever. But they

[1] The following discussion is partly embodied in my article "The Rôle of the Concept of Infinity in the Work of Lucretius" in the *Bulletin of the American Mathematical Society*, April, 1918. The article was reprinted in *The Classical Weekly*, January 27, 1919.

are equally memorable on other accounts—for their lack of any genuine spirit of philosophic enquiry, for their lack of reverence for human beings as human, for their stupid belief that the things of wisdom could be *purchased,* and especially for their brutal lack of scientific curiosity, scientific imagination and scientific achievement. Even the one really great exception—*De Rerum Natura*—is, in respect of its content, Greek in origin—it is, as you know, Epicurean; it is, nevertheless, the "one really great exception," for the thought of the Greek thinker stirred the great genius of the Italian poet to its depths; Lucretius understood it and he cast it in immortal form for the edification of all posterity. So far, however, as Romans were concerned, the Lucretian work "was stillborn, into a suffocating atmosphere of vile wealth and military oppression. The true figure to represent the classical Roman attitude to science is not Lucretius, but that Roman soldier who hacked Archimedes to death at the storming of Syracuse."

Most of the many great merits of the work of Lucretius have been long, though not generally nor even widely, recognized. One of its recognized merits is, as I have already said, its superb daring,—the unsurpassed magnificence of its enterprise, which was nothing less than to show forth a method for explaining all phenomena (whether mental or not) without having to resort to any hypothesis of divine intervention; another of its merits, —a very striking one,—is its probably unequalled union of literary excellence with scientific spirit and aim; still another—which includes many, being a highly composite merit—is its confident and often acutely argued presentation, sometimes in detail and sometimes in clear outline only, of ideas and doctrines whose just recognition had to await the coming of modern science. I refer to such scientific concepts and dogmas as: natural law—the atomic constitution of matter—conservation of mass—conservation of energy—organic evolution—spontaneous or chance origination and variation of organic life forms—struggle for existence in partly friendly and partly hostile environment—survival of the fit (the well adapted) and destruction of the ill adapted—and sensation as the ultimate basis of knowledge and as the ultimate test of reality —not to mention other equally brilliant anticipations of "modern" scientific thought.

In extant appreciations of the work of Lucretius his employment of the notion of infinity is indeed commonly indicated but it is indicated only more or less incidentally, without due signal of that notion's rôle in the poet's thought. For example, in Masson's large and, in many ways, excellent volume,—*Lucretius, Epicurean and Poet,* —the term infinite has only a subordinate place in the index of important terms; in the very extensive Notes to Munro's famous translation the term receives but scant attention; and it receives even less in the *Notes* found in Cyril Bailey's recent and deservedly much praised English translation of the poem. What is missed in such appreciations and commentaries and what I wish especially to signalize here is the fact that the concept of infinity,—of infinite multitude and infinite magnitude,—is not merely one among the many ideas employed by Lucretius but is indeed *the* dominant idea in his system of thought. A critical examination of the work as a scientific structure can hardly fail to discover that in the author's judgment the concept of infinity was not only the most powerful of his logical instruments but also—which is quite another matter—the one most obviously indispensable to the prosperity of his great undertaking.

This is not the place to give a detailed account of the Lucretian principles and procedure. For our present purpose it is sufficient to point out that among the propositions of *De Rerum Natura* there are three major ones and that these owe their efficacy and indeed their control of the entire discourse to the fact of their postulating the existence of infinite multitude and infinite magnitude. "Postulating" I have said, although Lucretius regarded the propositions, not as mere hypotheses or assumptions, but as indubitable certitudes. What are these three basic propositions? They are: that the universe of space is a region or room of infinite capacity—infinite extent; that time is an infinite duration composed of a beginningless infinite past and an endless infinite future; and that the universe's matter consists of an infinite multitude of absolutely solid (non-porous) and non-decomposable atoms —"seeds of things"—always moving hither and thither in an infinite variety of ways and ever so distributed throughout infinite space that of all spheres none but such as are microscopically minute could at any instant fail to enclose one or more of the "seeds." Without these infinitudes, explanation of the phenomena of the world was, in the poet's belief, impossible; with them, supplemented by certain other principles, such explanation was possible. In the view of Lucretius cosmic history was an eternal (infinite) drama enacted by an infinitude of unoriginated and indestructible "seeds," atoms, or elements operating upon an infinite stage. The drama was not to be understood except by help of the concept of infinity; and so *De Rerum Natura* may be not unjustly said to be a kind of poetic celebration of what the poet deemed to be the scientific efficacy of that concept.

What did Lucretius mean by infinity? What did he mean by an infinite multitude and by an infinite magnitude? No formal definition of any of these terms is to be found in his work. It is perfectly clear, however, that, if one had asked him whether an infinite multitude of elements was such that it could not be exhausted by removing from it one element at a time, he would have answered in the affirmative; and that, if one had asked him whether the elements of an infinite multitude could be thought of as arranged, like beads on a string, in an endless succession of elements, he would have again answered affirmatively. In short, an infinite multitude signified for Lucretius what mathematicians now describe as a denumerably infinite multitude or class. In his work there is no hint or suggestion that he had any conception or any inkling of any higher order of infinity. It is highly probable or indeed quite certain that, owing to his lack of mathematical discipline, such a conception, had it been suggested, would have seemed to him unintelligible or absurd.

It is in itself noteworthy, and if one is really to understand Lucretius it is essential to note, that, with the possible exception of time, the fundamental Lucretian infinities were not mere variables capable of increase beyond any prescribed finite amount—they were not, that is, what we have called dynamic infinities; on the contrary, they were, like the infinites of Cantor, constant or static affairs; but, unlike the Cantor infinites, those of Lucretius were composed of actual concrete things and not of abstract ones like points, for example, or pure numbers; thus the Lucretian infinitude of atoms, for example, was an infinitude of material particles (taking up room) and they all existed at once.

Let us recall the current definition of an infinite class: An infinite class is a class having a sub-class, or part, equivalent to the whole—equivalent, that is, in the sense that a one-to-one correspondence can be set up between the elements of the part and those of the whole. This definition of infinity was not given by the poet for, as we have seen, he gave no formal definition of it at all. We may ask, however, whether Lucretius was aware of the fact that an infinite multitude, as conceived by him, contained parts, or sub-multitudes, equivalent, as we now say, to the whole. The answer is, yes: not only was he aware of it but he repeatedy employed this characteristic property of infinite multitudes correctly and effectively. This rather astonishing fact is sufficiently interesting to justify citation of one or two passages supporting my assertation of it. If we bear in mind that one of the fundamental Lucretian infinites was the succession of time units (days, say, or generations or other finite stretches) beginning at any given instant and together composing what is called the future, the following famous passage makes it perfectly clear that, according to its author, the removal of any finite multitude of elements from an infinite multitude of them leaves a remainder—a part—exactly equal (or equivalent, as we say) to the whole:

> Nor by prolonging life do we take one tittle from the time past in death nor can we fret anything away, whereby we may haply be a less long time in the condition of the dead. Therefore, you may complete as many generations as you please; none the less, however, will that everlasting death await you; and for no less long a time will he be no more in being, who beginning with today has ended his life, than the man who has died many months and years ago.[1]

[1] Munro's translation, 4th ed., p. 83.

Lucretius, as already said, postulated the existence of an infinitude of atoms. These "seeds of things"—by whose clashings together and interlockings with one another all things (including souls) were produced, to be sooner or later again resolved into their elements by ceaseless hammering of atomic storms—these "seeds," the ultimate constituents of all the world (including minds), were not all of them identical in shape nor in size, though all of them were too minute to be seen singly or to be thus apprehended by any other sense; in respect to shape and size the atoms presented a number of varieties but only a finite number. The atoms of each variety, it was held, constituted an infinite multitude; and so there was some finite number of infinite classes of atoms. The physical functions of the atoms of one class were, in virtue of their size and shape, different from the functions of the atoms of any other class. In respect, however, of multiplicity, these infinite classes were equivalent—they were each of them denumerable—and each of the classes was equivalent to the class which we today should call their logical sum,—to the class, that is, of all the atoms in the universe.

It is sufficiently evident that the poet's conception of infinite multitude was identical with that now employed by mathematicians. If you will carefully scrutinize the poem, you will discover that the same may be said of the author's conception of infinite magnitude. Formal definition of the notion is not present. We are told, however, that all the atoms are, in respect of size, between a finite upper bound and a finite lower bound, and this notion of lower bound is of critical importance—what the lower bound is we are not told but we are told that there is such a bound and that it is finite (not zero),—in other words, atomic size is not infinitesimal,—it is a variable but not one having zero, or null size, for limit; we are told, rightly, that the sum of any finite number of atoms is finite; we are told that the sum of all the atoms of a given atomic form is infinite and that, therefore, their number must be infinite. It is thus evident that the Lucretian conception of an infinite magnitude was that of a magnitude exceeding the sum of any finite number of finite quantities none of which surpasses, in respect of parvitude, a finite size.

It is important to bear in mind that formation of ideas or possession of them is one thing, and that logically correct handling of them in argumentation is quite another. The difference is that between conception and ratiocination. In his use of the ideas in question Lucretius was frequently right and frequently wrong. You would find it a very edifying discipline to determine all the instances of both kinds. Of right use some examples have already been given and it would be easy to cite others. Let us now consider an instance of erroneous use. A remarkable example is found in the following passage (as correctly translated by Munro, page 15)—a passage of exceeding interest apart from the error in question:

> Again unless there shall be a least, the very smallest bodies will consist of infinite parts, inasmuch as half of a half will always have a half and nothing will set bounds to the division. Therefore between the sum of things and the least of things what difference will there be? There will be no distinction at all; for however absolutely infinite soever the whole sum is, yet the things which are smallest will equally consist of infinite parts.

The significance of the passage and the erroneous use it makes of the concept of infinity will be clearer to us if we observe that the passage is a portion of an argument by which Lucretius endeavors to prove that a finite portion of matter is not indefinitely or limitlessly divisible. He assumed, as we have seen, that matter is composed of invisibly small, absolutely solid particles called atoms, or "seeds of things," these atoms being, by hypothesis, the smallest particles capable of existing *spatially* separate from one another. He conceived an atom, however, to be composed of parts, which were, of course, not separable spatially from the atom. His contention was that among the parts of an atom there was a least part—a part, that is, such that none of the parts was smaller. The foregoing quotation is, as I have said, a part of the poet's argument in behalf of this contention. Paraphrased in modern terms this portion of the argument would run about as follows: "If among the parts composing an atom and being such that no two of them have points in common (save points of a common surface) there be no least part, then the atom consists of an infinite number of non-interpenetrating parts; the infinite multitude of atoms in the universe and the infinite multitude of parts of one atom are, as multitudes, equivalent (in the sense of one-to-one correspondence between the atoms in the former multitude and the atom-parts in the latter); the sum of the elements (atoms) of the multitude of the atoms is an infinite magnitude, the total quantity of the universe's matter; so, too, *the sum of the elements (atom-parts) of the infinite multitude of parts of one atom is an infinite magnitude;* but this latter sum is the atom itself; hence, if there be no least part among the parts of an atom, an atom is an infinite magnitude, and as such is no less than the sum of all matter." The error to which I desire to invite your attention,—an error in the poet's use of the concept of infinity,—is his assertion italicised in the foregoing paraphrase. The error is not due to wrong conception of infinity, whether of multitude or of magnitude; it is due solely to the tacit assumption that the sum of the elements of any infinite multitude of elements is infinite,—an assumption which, as you are aware, is false, for, for example, the sum of the elements of the infinite multitude of elements ($\frac{1}{2}$, $\frac{1}{4}$, $\frac{1}{8}$, . . .) is, as you learned in high school, not infinite but is 1—in other words, the limit of the sum, $1-\frac{1}{2^n}$, of the first $n$ terms of the series, $\frac{1}{2}+\frac{1}{2^2}+\frac{1}{2^3}+\ldots, \frac{1}{2^n}+\ldots$ is 1. Such a series,—any series such that the sum of the first $n$ terms has a finite limit for $n$ increasing limitlessly,—is said to be *convergent*. An obvious moral is that a little knowledge of the convergence of series would greatly improve the philosophy of poets and the science of philosophers.

It is astonishing that the mentioned fallacy occurs, as it does, in immediate conscious connection with a line seeming (to us) to refute it: *the half of the half will always have a half and nothing will set bounds to the division*. What is the explanation? It is not to be found in any supposition of stupidity or of momentary nodding. It is doubtless to be found in the author's purpose and point of view. He was here exclusively concerned with natural phenomena, with what he deemed to be existing entities—with bodies (and parts thereof) occupying space, actually filling what would else have been absolute emptiness or void. And so, if you had tried to refute him by means of such a series as $\frac{1}{2}$, $\frac{1}{4}$, $\frac{1}{8}$, . . . , which his own words indeed suggest, he would probably have said in effect: "Composed of man-made symbols like words, your series is not and never can be endless; to speak of the sum of a non-existing endless series is meaningless; moreover, even if we supposed the series to be endless, to be summable and to have 1 for its sum, this would be neither finite nor infinite, for it would not be a magnitude, inasmuch as the summands are themselves not magnitudes but are merely empty abstract symbols; if 1 be said to be a magnitude, in the sense of representing a magnitude, then indeed, if magnitude 1 be composed of two equal magnitudes, I grant that $\frac{1}{2}$ will be a magnitude in the same sense (of representing one); if all the symbols be magnitudes in that same sense, the summation of the series of abstract symbols may be said to be the summation of an endless (infinite) series of magnitudes; but *otherwise, not;* and now what I have contended in my poem is that, if your magnitude 1 be finite, not more than a finite number of the symbols in the series can be magnitudes, and this contention, denying the endless divisibility of finite magnitude,—especially denying that an atom has an infinitude of parts,—is based on physical considerations—on grounds other than that advanced in the passage you have quoted from my argument." If Lucretius thus replied to you, what suitable rejoinder, if any, could you make?

I shall not attempt to recount here, much less to estimate, those other grounds. It must, however, be said, in passing, that one of them is, in point of kind, almost perfectly represented by the following words of Clerk Maxwell (*Theory of Heat*, p. 285):

> What we assert is that after we have divided a body into a certain finite number of constituent parts called molecules, then any further division of these molecules will deprive them of the properties which give rise to the phenomena observed in the substance.

The traditional form of the thesis tacitly invoked by Lucretius to fortify his "other" grounds for holding that among the parts of an atom there is a least part, is exceedingly vague: *all infinites are equal.* Its vagueness helps to account for its ages-long and world-wide vogue. Thus Kanadi, an old Hindu author, employs the thesis to prove that, if every body be infinitely divisible, there can be "no difference of magnitude between a mustard seed and a mountain" (Daubeny's *Introduction to Atomic Theory,* p. 5). In this connection, anyone, philosopher or mathematician, if he be at all interested in the history of the idea of infinity, will be glad to have his attention called to a little-known letter of Newton dealing with the idea. The letter, which is addressed to Richard Bentley (*Works of,* Vol. III, p. 207), is interesting on several accounts: it points out the vagueness and falseness of the above-mentioned thesis, which Dr. Bentley had assumed to be true; it itself repeatedly employs the term "infinite" in a sense not less vague and indeterminate; and it virtually asserts that, if two infinite magnitudes be equal, the addition of any finite magnitude to either of them will destroy the equality—a proposition which we now know to be false.

I have said that a thorough-going Critical History of the Concept of Infinity would be a highly valuable contribution to our knowledge and understanding of humankind. The account I have now given of the rôle of infinity in the work of Lucretius is submitted, let me say again,—not as a model, for it is too imperfect for that, —but as only a concrete suggestion of what one chapter of such a history might contain. In closing the lecture, I desire to guard against the danger of leaving a false impression. The mere *correctness* of the Lucretian *concept* of infinity does not of itself account for the significance and power of the author's work. The secret lies in the fact that the imagination of a great thinker and poet was so stimulated by the concept as to cause him to express in immortal form a body of ideas which he had acquired from an elder and alien world and which after the long lapse of centuries are found to be among the most fruitful scientific ideas of our time.

## LECTURE XVI

### Hyperspaces

MEANING OF DIMENSIONALITY—SPACES OF FOUR OR MORE DIMENSIONS—THE MODE OF THEIR EXISTENCE —DISTINCTION OF IMAGINATION AND CONCEPTION —LOGICAL EXISTENCE AND SENSUOUS EXISTENCE— OPEN AVENUES TO UNIMAGINABLE WORLDS.

IT is the aim of this lecture to explain in simple ways what mathematicians mean by Hyperspace and to convey some sense of the scientific and philosophic importance of the concept which the term denotes. Let us understand in the first place that the terms,—"hyperspace," "multi-dimensional space," "space of *n* dimensions," "*n*-dimensional space," and some other readily recognized variants upon them,—are but different names for one and the same idea; they are employed interchangeably as equivalents; and the like may be said of the terms,—"geometry of hyperspace," "multi-dimensional geometry," "geometry of *n* dimensions," and their variants,—each of which simply denotes the geometry, or the science, of a space having more than three dimensions.

The concept of hyperspace, though it is a modern notion, is not strictly new,—it goes back three or four generations and is now, among enlightened mathematicians, as classic and orthodox as the ordinary multiplication table. Though only a short while ago it was regarded by mathematicians of the conservative and reactionary type with a good deal of suspicion as being, if not crazy, at least a bit queer, over-romantic, and unsound, it is now constantly employed as a great convenience by mathematicians everywhere and even by physicists (say in the kinetic theory of gases) quite without apology. The literature of the subject is large and growing. In Sommerville's *Bibliography of Non-Euclidian Geometry, Including the Theory of Parallels, the Foundations of Geometry, and Space of n Dimensions* (1911) there are listed 1832 references on $n$ dimensions.

The concept has not, indeed, so great scientific and spiritual dignity as some others,—as that of function, for example, or relation or transformation or group or invariance or infinity or limit,—yet it is a very grave notion, and it has, moreover, a certain double distinction: it is, I mean, one of the few among the important concepts in modern mathematics that philosophers have seriously grappled with and one of the still fewer that have piqued the curiosity of the educated public. The results of such popular curiosity are themselves a little curious. Not long ago, for example, I heard and read an address on hyperspace which a professional astronomer had ventured to make before an audience of university students. It was not a happy performance; not only did the speaker confound the idea of $n$-dimensional space with that of non-Euclidean space, but he made it pathetically evident that he had grasped neither the one idea nor the other, nor did anyone in the evidently interested audience appear to observe the fact. I have cited this instance because, if a reputable astronomer can err so egregiously in a matter that is not remote from the field of his special studies, we ought not perhaps to be astonished at the meagreness of the educated layman's understanding of it. And yet the fact *is* astonishing. For interest in the concept of hyperspace and especially in what is naïvely called the idea of "the fourth dimension" is, as you know, widespread among educated laymen; the concept itself, as we are going to see, is not a very difficult one; and fair accounts of it have been given from time to time in popular and semi-popular magazines and books. Nevertheless, understanding of the matter, outside the circle of professional mathematicians, is exceedingly rare. What is the explanation? What has been the trouble? No doubt part of it is that competent mathematicians have, in general, been unwilling, sometimes haughtily unwilling, to explain their ideas in popular terms lest they should seem to be thus seeking the applause of the gallery,—not aware of the fact that such haughtiness is itself one of the most effective means of impressing the gallery without enlightening it, winning its applause of what it is permitted to believe is a kind of mysterious intelligence so high and mighty as to be inaccessible to all mortals save the few who are endowed with mathematical genius; no doubt another part of the trouble has been that, though the concept of hyperspace has indeed aroused wide curiosity, it has not been pursued diligently in our industrial generation as it would have been had it seemed to have practical or bread-winning value,—if, in other words, instead of being only a form of spiritual wealth, it had carried the promise of material wealth. These considerations, however, do not, I believe, explain the matter fully. The main trouble has been that, though the idea in question is not very difficult to acquire, yet the acquisition of it does demand some patient meditation, some precision of thought, and the exercise of a little genuine wit, and this is a price that the vast majority of "educated" laymen are unwilling to pay; their interest in scientific ideas is neither steady nor deep; the ideas they acquire are such as can be taken, so to speak, on the fly, not such as require to be pursued and pondered; amusement is preferred to instruction; it is easier to read newspapers or novels or history of the romantic type or even philosophy of the verbalistic variety than to acquire solid knowledge; it is easier to feel the galvanic effect of a poem than to discern the beauty and feel the inspiration of a scientific work; and far easier to acquire the lighter lingo of knowledge sufficient for the dabbling conversation of a "smoker" or an afternoon tea than it is to think and to know. What I have just now said requires an important qualification,—the public's interest in science can be greatly improved if those who are expert in a branch of science will teach those who are not,—but such teaching has been very slight.

I have just now alluded to "precision of thought" and "the exercise of wit"—"genuine" wit. Perhaps you will allow me to digress a little in this connection. A short while ago I read a review, by a distinguished man of letters, of Professor George Santayana's *Character and Opinion in the United States.* The reviewer tells us that the work has, besides other excellences, the qualities of precision, wit, and beauty. I have read much of Santayana's writing, including the poems and the five volumes of *The Life of Reason.* Undoubtedly, his writing is beautiful—that is why I have read it—and it is bright, too, sparkling, and full of surprises; perhaps we may say that it has, in one sense of the term, wit also; to me its wit appears to be scintillation rather than genuine wit for in this latter there is an element of gravity which Santayana lacks; at all events if he have wit, it is not wit in the sense in which that quality is found in Kant, for example, or Hume or Spinoza or Descartes or Pascal or Aristotle. As for the quality of precision, it is certain that neither Santayana nor his reviewer has it or indeed knows the meaning of it as it is rightly understood by logicians or mathematicians, for the writers in question are not logicians. They do indeed produce literature—beautiful literature—but it does not belong to the literature of knowledge, which is also beautiful; the literature it belongs to is the literature of opinion, some of which is not even beautiful, for though it includes such beautiful writing as that of Benedetto Croce, for example, yet it embraces work like Professor Bliss Perry's *The Present Conflict of Ideals,* which we must allow is a kind of literature even though it remind one of a traveling salesman displaying his wares or, less dimly, of an indiscriminate "feeder" who loves to talk of the things he has tasted and who sometimes ascribes to bad food or bad *cuisine* distress that is due to enfeebled or feeble digestion. Indeed Professor Santayana, like his reviewer, is, primarily and essentially, a poet; but there are three kinds of poetry: there is the poetry of pure thought,—the poetry of Logic,—and there are two other kinds—the hypological and the hyperlogical. Each of the kinds has a muse of its own; that of the first kind, as I said in a previous lecture, is called Logical Rigor, an austere goddess, guardian of precision, mistress of the silent harmonies of perfect thought. By that muse poets like Santayana have not been inspired.

Returning from the digression, let us now endeavor to answer the main question of this lecture: what is the meaning of the term "hyperspace" or "*n*-dimensional space," where $n$ is greater than three? I have said "*the* meaning" as if there were only one. As a matter of fact, the term has three meanings, which, though they are closely related, it is essential to distinguish if we are to avoid confusion. I shall try to explain them clearly, reserving the most interesting one for the last. Certain refinements of refinements I shall avoid as likely to obscure and hinder rather than to clarify and help in a first presentation. It will be very advantageous as a preliminary to speak of some simple matters connected with the system of real numbers.

There is, as you doubtless know, an extensive and very refined theory of the logical genesis and properties of these numbers. The theory is to be found in the numerous works dealing with functions of a real variable, the profoundest treatment of the subject being that in the *Principia Mathematica.* I am not going to assume that you are familiar with that theory. I take it for granted, however, that you are sufficiently acquainted with the system of real numbers to understand fairly well what I purpose to say about it here. You are aware that the system is composed of the infinitude of positive and negative integers, the infinitude of rational fractions, and the infinitude of irrational numbers like $\sqrt{2}$, for example, and including the so-called transcendental numbers such, for example, as the familiar specimens, $\pi$ and $e$. Historically these numbers were called " real " to distinguish them from the so-called " imaginary " numbers, like $\sqrt{-2}$, for example, which latter were long regarded, quite unjustly, as an ungenuine kind of number. Today the old adjectives, real and imaginary, are still regularly employed, but they no longer signify that the numbers thus designated are either more or less genuine than other numbers.

The system of real numbers is a vast system or class—the multitude of the numbers in it is indeed very great. The extent of the multiplicity,—the number, if you please of all the real numbers,—is often conveniently denoted by the familiar symbol for infinity, $\infty$. The same symbol is used to indicate how many things, elements or members there are in any other equally numerous system or class—say, that of the points in a straight line. If $x$ be a variable representing " any one " of the real numbers, we say that $x$ has *one* degree of freedom; we say also that the system of real numbers is a *one*-dimensional system or class; in like manner we say that a point $P$, if free to move along a straight line, has, in virtue of that fact, *one* degree of freedom, and that a line, regarded as a system of points, has *one* dimension or is *one*-dimensional.

Now, a real number is one thing and a *pair* of them is another. Such pairs constitute a system (of pairs). How many pairs are in the system? It is easy to tell. Let the symbol $(x, y)$ be a variable representing any one of the pairs; give $y$ some definite value, say, $y_1$, and let $x$ vary—$x$ can take $\infty$ values; each of these taken with $y_1$ gives a pair, and so we get $\infty$ pairs; in *each* of these we may replace $y_1$ by any of the $\infty$ of values that $y$ may take; we so obtain in all, as you see, $\infty$ times $\infty$, or $\infty^2$, pairs. It is plain that the variable $(x, y)$, an arbitrary or undetermined pair of the system, has *two* degrees of freedom, owing to the variability of the two parameters, or coordinates, $x$ and $y$. Thus the system of all the pairs of real numbers is a *two*-dimensional system (of pairs). You see at once that the system of *triads* or triplets of real numbers has three dimensions, or is a tri-dimensional system (of triads); that the system contains $\infty^3$ triads; and that an undetermined triad $(x, y, z)$ enjoys three degrees of freedom, one for each of the mutually independent coordinates, or parameters, $x$, $y$ and $z$. The generalization is obvious and easy; if we think of the system of all the sets $(x_1, x_2, \ldots, x_n)$ of real numbers, each set composed of $n$ numbers, it is readily evident that the system is an $n$-dimensional system (of sets); that it contains $\infty^n$ sets; and that a free, or undetermined, set of the system possesses $n$ degrees of freedom.

I have been speaking mainly of numerical things. I am now going to speak of things geometrical or spatial, and will first make a little more precise what I said a moment ago regarding a line. Let $L$ be a straight line; choose a point of $L$ for *origin* of distances and mark it $O$; let $P$ denote any point of $L$; denote by $x$ the distance (in terms of some chosen unit) from $O$ to $P$, it being understood that $x$ is positive or negative according as $P$ is on the one side of $O$ or on the other—of course $x$ will be zero if $P$ coincide with $O$. Thus a reciprocal one-to-one correspondence is set up between the real numbers of the system thereof and the points of $L$; $x$ represents $P$ numerically, $P$ represents $x$ geometrically; and $x$ is the coordinate or parameter of $P$. Notice that $L$ is here the field or the space of operation and that we are regarding it as a field or a space of *points*. In the light of the preliminary discussion respecting numbers, you see immediately that a straight line, regarded as a space of points, is *one*-dimensional, since it perfectly matches, as indicated, the one-dimensional system of the real numbers. In other words a line is a point-space of one dimension. You catch the idea: if a point of a line depended upon two or more coordinates instead of only one, we should say that a line is a point-space of two or more dimensions. All this is obvious. I have stressed it because it is essential to a good understanding of any one of the meanings of hyperspace. Let us turn to analogous considerations in the case of a *plane*.

In Lecture V we saw that, by means of a pair of rectangular axes and a unit of length, a one-to-one correspondence can be established between the points $P$ of a plane and the pairs $(x, y)$ of real numbers. The pair represents the point, and the point the pair; the $x$ and $y$ of a pair are at once the coordinates of the pair and of the corresponding point. There are as many points in the plane as there are pairs of real numbers in the system of such pairs. Hence a plane contains $\infty^2$ points; a point that is free to move in a plane and is confined thereto has two and but two degrees of freedom; a plane, regarded as a space of points, is a *two*-dimensional space; and you see why it is so called,—it is because the points of a plane match, in one-to-one fashion as we have seen, the pairs of real numbers in the two-dimensional system of such pairs.

And now what shall we say of ordinary space? What, I mean, shall we say of that immense region or room in which we are immersed and, with us, our floating world and the stars? Let us think of it as a field, or a plenum, of points. What, then, is its dimensionality? It is easy to ascertain.

Choose three mutually perpendicular planes; they have a common point $O$, called the origin; they determine three lines, $OX$, $OY$, $OZ$, called axes; agree that a distance measured parallel to an axis shall be positive or negative according as it is reckoned in the sense of the arrow or in the opposite sense; note that the three planes divide the whole of space into eight compartments; choose a unit of length; let $P$ be any point and, as in the figure, denote its distances from the coordinate planes by $x$, $y$ and $z$, called the coordinates of $P$; they are also the coordinates of the triad $(x, y, z)$. It is plain that a $P$ determines a

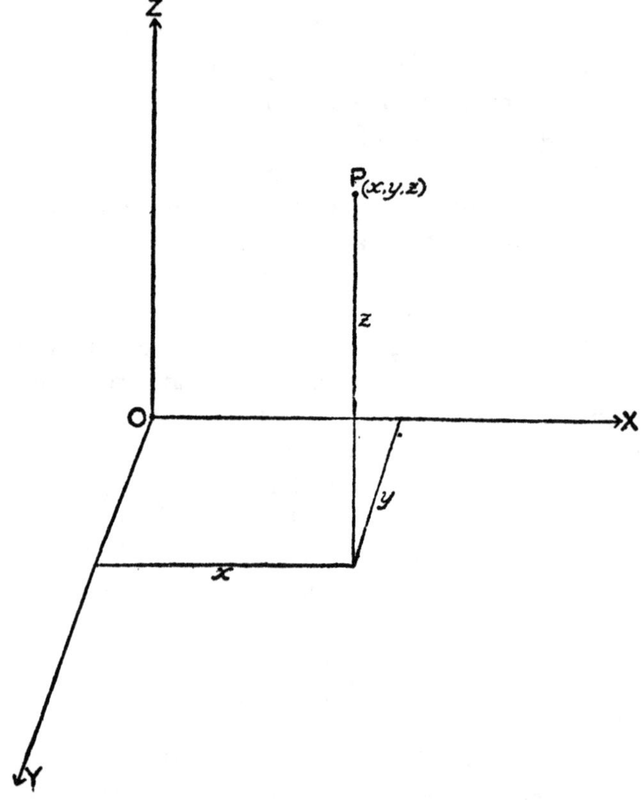


FIG. 29.

triad, and a triad a $P$. You see at once that ordinary space, if regarded as a plenum of points, has *three* dimensions, since the points match the triads in the *three*-dimensional system of triads of real numbers. In such space there are, you see, $\infty^3$ points, and a point has therein three degrees of freedom and only three.

We are going very soon to see very clearly one of the three meanings of the term $n$-dimensional space, $n$ greater than three. Do not fail to note that thus far we have regarded the line, the plane, and ordinary space as fields, or plena, or spaces, of *points;* the point, that is, has been taken for *element;* but nothing constrains us to elect the point to that position; we can geometrize just as well, sometimes better, with some other entity taken as element; we may choose for element the point *pair* or point *triad,* and so on; in the case of the plane, we may take the line or the circle or something else for element; in the case of ordinary space we may take for element any of the foregoing entities or a plane, for example, or a sphere, and so on. It is true that, from time immemorial until a little less than a century ago, the point was exclusively employed as geometric or spatial element, but there is nothing in the ten commandments nor even in the Volstead Act to prevent the use of something else. The point's ages-old monopoly was broken up mainly by Julius Plücker (1801–1868)—one of the greatest of geometricians and a distinguished physicist besides,—who geometrized the plane in terms of its lines and geometrized ordinary space in terms of its planes and its lines—thus emancipating geometry forever from its old bondage to points. A geometry in which the point-*pair* is taken for element will deal with the properties of configurations composed, not of points, but of point-*pairs.* Now, in a line a point-*pair* has two coordinates, two degrees of freedom; in a plane it has four; in ordinary space it has six; a line has $\infty^2$ point-pairs; a plane, $\infty^4$ of them; and ordinary space, $\infty^6$; and so you see that, regarded as a plenum or space of point-pairs, a line has *two*, a plane *four*, and ordinary space *six*, dimensions. You see that the dimensionality of a space depends, not only upon the space itself, but also upon the entity employed as element. You see easily that, in respect to point-*triads*, the dimensionality of a line is *three*, that of a plane is *six*, and that of ordinary space is *nine;* and you see that, if we take for element the point-*set* containing $n$ points, then a line is an $n$-dimensional space, a plane has $2n$ dimensions, and the dimensionality of ordinary space is $3n$. There is no sense in simply saying that a plane, for example, or that ordinary space has such-and-such a dimensionality (or number of dimensions); what we have to say is that it has such-and-such a dimensionality when it is conceived as a space or plenum of elements of such-and-such a kind. When people say simply, as they often do, that space (meaning ordinary space) has three dimensions, they mean—though they do not know well what they mean—that it has three dimensions as a space of *points*. If you think they know what they mean, ask them what they mean. For additional examples showing that space dimensionality depends upon space element, consider the following. In a previous lecture we saw that in a plane a *line* has two coordinates, two degrees of freedom—a plane being precisely as rich in lines as in points; and so a plane is a two-dimensional space of lines, as it is, we have seen, of points. What is the *line* dimensionality of ordinary space? It is easily seen to be *four*. To see it, reflect that a line is determined by two points, say a point in the plane of the floor of this room and a point in the plane of the ceiling; each of the points (kept in its plane) has two coordinates, two degrees of freedom, and so, you see, the line has four.

To distinguish a line of ordinary space from all its other lines, it is necessary and sufficient to tell four independent facts about it; ordinary space contains $\infty^4$ lines, and you see that Plücker's famous line geometry (of ordinary space), which studies configurations composed of lines (and not of points), is a four-dimensional geometry. Let us return for a moment to the plane; think of it as a plenum of *circles*. Each of its points is the center of an $\infty$ of circles, and it has $\infty^2$ points; and so, you see, a plane has $\infty^3$ circles; in a plane the circle has three degrees of freedom,—three coordinates or parameters; a plane of circles is a *three*-dimensional space—as rich in circles as in point-*triads*—as rich in circles as ordinary space in points. You can readily show that ordinary space is four-dimensional in *spheres*, as we have seen it to be in *lines*, five-dimensional in flat line-*pencils* (explained before), *six*-dimensional in *circles*, and so on and on to your heart's content.

I venture to believe that the foregoing illustrations have sufficiently disclosed one of the meanings of the term "hyperspace": that meaning, namely, according to which the term signifies an ensemble of geometric, or spatial, entities, or elements, of such a kind that an undetermined (or arbitrary) one of them has, in the ensemble, four or more degrees of freedom. This statement is not designed to be a definition of the meaning, but only a good-enough description of it. In *this* meaning of hyperspace there certainly is nothing to mystify; for, in order to find examples of such hyperspaces, one is not obliged to perform the familiar mathematical feat,—which many good people seem to find difficult or even impossible, —of going beyond the great domain of Imagination into the infinitely vaster domain of pure Conception. I have sometimes felt that no student is intellectually fit to be graduated from college who does not easily and habitually recognize the immense and fundamental difference between those domains. Such a student is as meagrely disciplined as one who believes that, if two things or persons be each of them indispensable, they are therefore of equal importance—a rank fallacy vitiating 90 per cent of current social philosophy throughout the world. I once heard a railway section-hand argue that, because his work was indispensable, he was just as important as the railway president.

Have you observed that among the hyperspaces which we have so far taken occasion to notice there is no hyperspace of *points?* The examples have been hyperspaces of point-*pairs* or of point-*triads*, . . . or of *n-sets* (of points), . . . or of *lines* or of *circles* or of line-*pencils* or of *spheres*, and you are now doubtless prepared to extend the list of such examples indefinitely, for "the clue, grown familiar to the hand, lengthens as we go and never breaks." But it is not this kind of hyperspace that mystifies the layman. What he desires to have you make clear to him, —though he may not be able to say so very clearly,—is the conception of a hyperspace of *points*. When he asks you to "explain the fourth dimension," he is really asking you to explain the idea of a space that is 4-dimensional in *points* in the sense in which ordinary space is 3-dimensional in points, a plane 2-dimensional and a line 1-dimensional. And so we see that our further task is thus defined. I have said that hyperspace has three intimately related meanings. One of them has been explained. The other two attach to the term, hyperspace of points, or—what is tantamount—point-space of four or more dimensions. And we are now to see what the meanings are. They are not hard to see if we but look attentively.

Let us begin very simply by recalling the fact that in a line a point is *represented* by one coordinate $x$, in a plane by a pair $(x, y)$ of them, and in ordinary space by a triad $(x, y, z)$ of them. Now, instead of always saying that the point is thus " represented," it is very common, because very convenient, to say that the $x$ or the $(x, y)$ or the $(x, y, z)$ *is* the point, and this is done, explicitly or implicitly, in very many ways; thus we say, for example, " consider the point $(x, y)$ " or " consider the line $ax+by+c=0$ " instead of saying, " consider the system of those pairs of values of $x$ and $y$ which satisfy the equation $ax+by+c=0$." This familiar way of speaking *as if* real numbers, pairs thereof and triads thereof were indeed points and *as if* equations were indeed *loci*, is very brief, very neat and very stimulating, too, on account of its keeping the mind continually delighted with the presence of geometric or spatial imagery. You see that, in order to be thoroughly consistent in this manner of speaking, we should have to say that the *system* of real numbers $x$ *is* the *line*, that the *system* of pairs $(x, y)$ *is* the *plane*, and that the *system* of triads $(x, y, z)$ *is* (ordinary) *space*.

And now what I wish to point out is that just such a thoroughgoing geometric way of speaking is often employed by mathematicians when dealing with four or more real variables, $x, y, z, w$, etc. That is to say, if they be handling four variables, they call a tetrad $(x, y, z, w)$ of numbers a *point* and the totality of such points they quite consistently call a point-*space* of four dimensions. And in like manner for any yet larger number of variables.

Query: when mathematicians thus speak, do they suppose that there exists a 4-dimensional space containing points for the number tetrads $(x, y, z, w)$ to represent as there exists a plane (say) containing points for the dyads ($x$, $y$) to represent? The answer is that some of them suppose it and some of them do not; and in this fact is the key to the two meanings of the term, "hyperspace of points." According to one of the meanings, a point-space of $n$-dimensions is, strictly speaking, not a space at all, but is simply and purely an $n$-dimensional system of number sets (each having $n$ numbers); and the theory or science of such a system, verbally geometrized as I have indicated, is not genuine geometry, but is simply a species of $n$-dimensional Algebra or Analysis conducted and couched in geometric speech. Such,—to take an example as early as 1847,—is Cauchy's *Mémoir sur les lieux analytiques* where he says, "We shall call a set of $n$ variables an analytic point, an equation or system of equations an analytical locus," and so on. According to the other meaning a hyperspace of points is held to be a genuine space; the points constituting it, though representable by number sets of $n$ numbers each, are distinct from, and independent of, such sets as the points of ordinary space are distinct from, and independent of, their representative number triads ($x$, $y$, $z$); and the theory of such a space, whether the theory be built up synthetically or analytically, is genuine $n$-dimensional geometry—geniune geometry of a hyperspace of points.

I am sure that in this connection you are impatient to raise the question whether hyperspaces of points may be said to exist, and, if we allow that they may, in what sense of the term "exist." The question evidently involves nice matters both of psychology and of metaphysics. Many mathematicians have not carefully considered those "nice matters" and are quite content (because of convenience as already explained) to speak *as if* the hyperspaces in question exist, without thereby intending either to affirm or to deny that they do exist in fact. After much reflection I have myself no longer any doubt in the premises, and in my *Human Worth of Rigorous Thinking*, p. 256, I have stated my conviction in the words: *hyperspaces have every kind of existence that may be warrantably attributed to the space of ordinary geometry.* The considerations that have led me to that conclusion are set forth at sufficient length in the work cited and need not be restated here.

In relation to the matter in hand, note carefully the sharp difference of temper, attitude and interest between the following two classes of mathematicians: those of the one class, primarily interested in geometry, affirm the existence of a point-space of $n$ dimensions and then investigate its properties—build up its geometry—by the algebraic or analytic method—by applying, that is, the theory of $n$ independent numerical variables (*say*, $x_1$, $x_2$, $x_3$, . . . , $x_n$) to the postulated space; those of the other class, primarily interested in algebra or analysis, employ, in their discourse about the system of $n$ variables, the *nomenclature* of the geometry of a point-space of $n$ dimensions *as if* there were such a space, but do not affirm its existence. (Of course the former class are not *obliged* to employ the analytic method nor are the latter class obliged to employ geometric speech.) Note the " as if " in the following extract from J. J. Sylvester's " A Plea for the Mathematician " (*Mathematical Papers*, Vol. II):

> Dr. Salmon in his extension of Chasles' theory of characteristics to surfaces, Mr. Clifford in a question of probability, and myself in my theory of partitions, and also in my paper on barycentric projection, have all felt and given evidence of the practical utility of

> handling space of four dimensions as if it were conceivable space.

By " conceivable " he here means actual.

It is noteworthy that in the difference between affirming the existence of hyperspaces of points and merely speaking *as if* such spaces existed we have a striking illustration of the Kantian distinction between *postulating* and *feigning*—between *hypothesis* and *fiction*—a much-neglected distinction justly and stoutly insisted upon by Vaihinger in his great *Philosophie des Als Ob* as being of fundamental importance in the philosophy of science and the philosophical history of thought in general. The distinction is indeed very important and very wide in its application. One must be pretty dull not to perceive that the difference is radical between saying, for example, " there is an infinite and all-wise God and hence we ought to live so-and-so " and saying " we ought to live *as if* there were such a God "; or between saying " there is a universal ether having such-and-such properties and that is why light behaves in such-and-such a way " and saying " light behaves as if there were an ether having such-and-such properties "; or between saying " the human soul is immortal and hence we ought to live so-and so " and saying " we ought to live as if the human soul were immortal "; and so on throughout the whole range of thought. A postulate or hypothesis, as here understood, is a proposition and is true or false; but a fiction is not a proposition and is neither true nor false. It would be very enlightening to make a survey of scientific " hypotheses " with a view to ascertaining which of them are genuine hypotheses and which ones are only fictions—only *as ifs*. There can be no doubt that many a scientific worker would be astonished at the results of such a critical survey. An excellent clue to Vaihinger's work is found in Dr. W. B. Smith's penetrating review of it in *The Literary Review* (N. Y. *Evening Post*), July 9, 1921.

The hyperspaces of points are *unimaginable* worlds—unimaginable for us humans, I mean, in our present stage of development—but they are thoroughly *conceivable* worlds; and for mathematical purposes nothing is demanded but thorough conceivability. The importance of that fact is fundamental. Experience has taught me that it is hard to drive the fact home to the average understanding. Wherever the distinction involved in the fact is not understood by "critics," whether scientific or literary or philosophic, criticism is blind and worse than futile, being at once misled and misleading, confused and confusing. It is important to observe and to bear in mind that, with respect to the great powers, or types, of mental activity,—Sensibility (or Sense-perception), Imagination, Conception,—we humans fall into three classes: there are those who have the first power but little of the second; there are those who have the first and the second powers but little of the third; and there are those who have in good measure the three powers. The second class is related to the third very much as the first is related to the second. Beware of the first two classes,—they can give you neither science nor genuine philosophy nor,—properly speaking,—criticism. Are they aware of their limitations? No; at all events not keenly. How could they be?

There are various avenues by which beginners may approach those unimaginable worlds and enter them; and that is a blessing, for the worlds are replete with wonders. One of the ways is that followed by Professor H. P. Manning, for example, in his *Geometry of Four Dimensions*, the reading of which requires no more preparatory knowledge of geometry and geometric method than can be acquired in a good high school.

Another of the ways is the deliberate and patient way of postulate procedure. I am not going to take the time that would be necessary to spread before you here a system of postulates for the geometry of a point-space of $n$ dimensions. To do so would be wasteful, for I may assume that you are now pretty familiar with one or more postulate systems for a space of three dimensions,—as that of Hilbert or that of Veblen or that of Veblen and Young (for 3-dimensional projective geometry),—and only slight alteration of any such system is needed to convert it into a system available for 4, 5 or $n$ dimensions. I will content myself with referring you to page 24 of Veblen and Young's *Projective Geometry*, for example, for a clear indication, if you require to be shown, of the simple sort of alteration that will suffice.

The two foregoing ways of working *into* and working *in* the worlds of hyperspace are, as you observe, the ways of pure, or synthetic, geometry as distinguished from analytic, or algebraic, geometry (which latter, let me remind you, is only a geometric *method*). The " pure " ways are followed with especial frequency by the Italian geometricians, though, of course, the latter often employ the analytic method also.

Of the latter method, I have already said enough for the stated purpose of this lecture. An excellent way for a beginner is found in a somewhat rough *mixture* of the " pure " and the algebraic ways, guided by the chief of all intellectual guides— analogy—as follows:

Let $L$ be a line (say a projective line) of points; it is a space of one dimension, $S_1$; if you like, you may say that a point is itself a space of *zero* dimensions, and denote it by $S_0$; conceive a point $P$ not in $L$ and consider the set of lines joining $P$ to the points of $L$; in this set there are $\infty$ lines; think of the ensemble of all the points of these lines; there are $\infty^2$ of them; they evidently constitute a plane—a 2-dimensional point-space, $S_2$. Now conceive a point $P$ not in $S_2$ and think of the set of all lines joining $P$ to the points of $S_2$; of such lines there are $\infty^2$; plainly the points of these lines together constitute a 3-dimensional space (like the familiar space of ordinary solid geometry); it has $\infty^3$ points; denote it by $S_3$. In the next step *imagination* ceases to accompany our *thought;* so much the worse for imagination, for conception goes on rejoicing, quite as before; if it could not go on endlessly, there could be, strictly speaking, no science. You see, of course, what the next step is; take it boldly: *conceive* a point $P$ not in $S_3$; conceive $P$ joined by lines to all the points of $S_3$; of such lines there are, as you see, $\infty^3$ in all; each of them has an $\infty$ of points and all of them together give you $\infty^4$ points constituting a 4-dimensional space, $S_4$. Repetition of the process yields $S_5$, then $S_6$, and so on till you have the conception of a point-space of any required dimensionality, however high.

Having once formed the concept of hyperspace, what then? What is to be done with it? The answer depends upon you—upon your interest and your ability. Those higher worlds, I have said, are replete with wonders. These are not (yet) shown in the " movies." Neither can I exhibit them here. If you wish to see them you must pay a certain price—that of seriously studying hyperspace *geometry*. Of this geometry the literature is large, is growing, and will continue to grow. An excellent introduction to it is Schoute's *Mehrdimensionale Geometrie* where the matter is handled systematically, elementally and many-sidedly. Among the aims of the course in *Modern Theories of Geometry*, which I have given at Columbia University for many years, is that of helping students to acquire a *working* knowledge of $n$-dimensional geometry. The importance of such a knowledge is by no means restricted to students of so-called pure mathematics. Indeed, a few years ago, I had the honor to give a series of lectures on $n$-dimensional geometry to a group of physicists who had found that without some knowledge of hyperspace methods, they could not read the literature of their own subject, especially that of the kinetic theory of gases. That was before the present relativity rage, which, as we saw in a previous lecture, avails itself of the idea of four-dimensional space. It will take but a minute and it will be instructive to show why students of gas theory are now obliged to acquire some knowledge of $n$-dimensional geometry. It is because some of the foremost writers on the theory,—J. H. Jeans, for example,—have adopted and elaborated the following considerations. Suppose we have a closed vessel, say a sphere, filled with gas. Let us suppose the gas is composed of $N$ molecules. These are flying about hither and thither, all of them in motion. Think of one of them; at a given instant it is at a point $(x, y, z)$; at the same time it is moving so that the components of its velocity along the axes of reference are (say) $u$, $v$ and $w$; if and only if we know the six coordinates of the molecule at an instant, we know where it is and the direction and rate of its going. The $N$ molecules constituting the gas thus depend, you see, upon $6N$ coordinates. At any instant these have definite values. Together these values define the "state" of the gas at that instant. Now, say the writers in question, the

$6N$ values determine a *point* in a space of $6N$ dimensions. Thus there subsists a correspondence between such points and the varying gas states. As the state of the gas changes (owing to the motions of the molecules) the corresponding point generates a path, or locus, in the space of $6N$ dimensions; and so the behavior or history of the gas (as a whole) gets geometrically represented by loci in the mentioned hyperspace. That will suffice as a hint at what has become a recondite mathematical theory —the kinetic theory of gases.

I have said that I cannot here exhibit the wonders to be found in the worlds of hyperspace. To do so in any fair measure would require many lectures as long as this one. I can not refrain, however, from leading you, if you be willing, to see one of the minor wonders met with on the very threshold of 4-dimensional space. We can find it in the " mixed way " we were following a little while ago, guiding ourselves by analogy, and at the same time you will see how you can yourselves discover further wonders. Note the facts carefully and note their analogies as we start at the bottom and ascend the scale. Observe, to begin, that in a *line* ($S_1$) an equation $ax+b=0$ of first degree in one variable ($x$) represents a *point* ($S_0$); in a *plane* ($S_2$) an equation $ax+by+c=0$ of first degree in two variables ($x$, $y$) represents a *line* ($S_1$), and that two such equations taken as simultaneous represent a point ($S_0$)—the common point of the two lines; in an ordinary space ($S_3$) an equation $ax+by+cz+d=0$ of first degree in three variables ($x$, $y$, $z$) represents a *plane* ($S_2$), two such equations taken as simultaneous represent a *line* ($S_1$),—the line common to the two planes,—and that three such equations (if independent) together represent a *point* ($S_0$)—the common point of the three planes. You now have the analogical clue. Following it you see immediately that in a 4-dimensional space ($S_4$) an equation $ax+by+cz+dw+e=0$ of first degree in four variables $(x, y, z, w)$ represents an ordinary *space* ($S_3$)—named a *lineoid* by my colleague, Professor F. N. Cole; that two such equations together represent a *plane* ($S_2$)—the plane common to the two lineoids; that three such equations (if independent) represent a *line* ($S_1$)—the line common to the three lineoids; and, finally, that four such equations (if independent) represent a *point* ($S_0$)—the common point of the four lineoids. You are already, you see, in the midst of astonishing things: you see that an $S_4$—a hyperspace of the lowest dimensionality—contains a fourfold infinity ($\infty^4$) of lineoids (spaces like our own); you see that any two of these have a plane for their intersection, that any three independent lineoids (in $S_4$) have a line in common, and that four of them have one point in common and only one. I spoke of showing you a "minor wonder." It is that in $S_4$ *two planes* (unless they happen to be in a same lineoid) have one and *only* one point in common. To see that this statement is true, consider four independent equations like the last of the foregoing; two of them, as we have seen, represent a plane; the other two represent another plane. What points have the planes in common? The answer is: those points whose coordinates $(x, y, z, w)$ satisfy the four equations. But, as you know, such a system of equations is satisfied by only one set of values. Hence the proposition. There are many other near-lying marvels in $S_4$. One of them is that you can pass from the inside to the outside of an ordinary sphere without going through its surface. Another one is this: if in ordinary space you wish to make a prison bounded by planes, you have to use at least four planes; while in $S_4$ the analogous prison is bounded by *five* ordinary $S_3$'s. I will mention but one more. In ordinary space ($S_3$) two planes have but one angle; in $S_4$ two planes make two angles with each other, so that, if you would bring the planes into coincidence, you must rotate one of them about their common point in two ways. These specimens are *mild* marvels; in $S_4$ their like is inexhaustible; astonishment increases as one ascends the summitless scale of dimensionality, and, with astonishment, also light and edification. Indeed we may say that the science of geometry is, properly speaking, $n$-dimensional geometry.

In closing this long lecture, I need add but little respecting the human significance of the momentous conception with which it has dealt. We have seen that, in the matter of scientific speech, the geometry of hyperspace has clothed pure Analysis with the beauty and strength of a tongue that is at once delightful, stimulating and economical; we have seen that the language, the ideas and the methods of $n$-dimensional geometry are becoming rapidly more and more powerful agencies in the great outlying domain of Physics; we have glimpsed the fact not only that $n$-dimensional geometry is in itself of exceeding great interest, but that the geometry of the higher worlds illuminates that of the lower as the geometry of ordinary space illuminates that of the plane or the line. I have now, finally, to mention what is, in my belief, the chief consideration. Human progress is progress in *emancipation;* and in the Concept of a summitless hierarchy of Hyperspaces is attained, I do not say the most precious, but the amplest, Freedom yet won by the human spirit,—room, I mean, for exterior representation,—for the architecture, if you please,—of every analytic doctrine or theory, even though there be involved in its structure an infinite number of variables.

## LECTURE XVII

## Non-Euclidean Geometries

THEIR BIRTH AND VARIETIES—THEIR LOGICAL PERFECTION — THEIR PSYCHOLOGICAL DIFFERENCES — THEIR SCIENTIFIC AND PHILOSOPHIC SIGNIFICANCE —ALL OF THEM PRAGMATICALLY TRUE—SCIENCE AND TRAGEDY—A PRELUDE ON THE POPULARIZATION OF SCIENCE—SCIENCE AND DEMOCRACY.

> "IN the early part of the last century a philosophic French mathematician, addressing himself to the question of the perfectibility of scientific doctrines, expressed the opinion that one may not imagine the last word has been said of a given theory so long as it can not by a brief explanation be made clear to the man of the street." [1]

The mathematician referred to is Gergonne, one of those who assisted in the second discovery of projective geometry (long after the work of its first discoverer, Desargues, as I said in a previous lecture, had been lost and utterly forgotten). It is to Gergonne that we owe the first enunciation of the great law of Duality—one of the most beautiful and fertile principles of modern geometry. His noble dream respecting the perfectibility

[1] Quoted from address on *Mathematics* written in 1907 and published as final chapter in *The Human Worth of Rigorous Thinking* (Columbia University Press, 1916).

of scientific theories ought to be given in his own words. They are these:

> *On ne peut se flatter d'avoir le dernier mot d'une théorie, tant qu'on ne peut pas l'expliquer en peu de paroles à un passant dans la rue.*

Can the dream come literally true? We are certain that it cannot, for it is an ideal,—a genuine ideal,—and genuine ideals can never be realized fully. Therein is their precious value as lights and lures of the spirit—they are "ever flying perfects," not to be overtaken but to be pursued by us, as they rise and soar and lead, forever.

The ideal of Gergonne is a democratic ideal. To pursue it is, therefore, not merely our privilege; it is a great and solemn duty. Democracy is on trial,—it is an experiment,—the greatest experiment ever undertaken by our humankind. Unless the community be pervaded with ever-increasing scientific intelligence, that supreme experiment,—the sovereign hope of the world,—is doomed to failure. Than that, nothing can be more evident to such as reflect. The affairs of state must be rescued from the hands of ignorant politicans and be committed to scientific management—to the guidance, that is, of *honest* men who *know*. That, too, is as evident as anything can become. How *can* the destiny of the state be committed to the guidance of science if the men and women who constitute the electorate know nothing of science, nothing of its methods, nothing of its content, nothing of its achievements, nothing of its spirit, nothing of its infinite potency for human service? Election is selection. How can the ignorant select the wise?

In view of such considerations, so obvious and so

important, it is indeed strange that scientific men have been so little actuated by Gergonne's beautiful dream. What has been the trouble? What is the secret? Is it that scientific specialists find in the educated public a lack of scientific interest? Tokens of scientific curiosity abound on every hand,—witness, for example, the recent world-wide curiosity manifested by non-specialists in the theories,—most recondite theories,—of Professor Einstein. No doubt such curiosity is often shallow and transitory, but it can be nourished and be thereby made deeper and more enduring. Do scientific specialists really believe that, in general, educated non-specialists have not enough mind to understand scientific ideas, even when these are presented in non-technical speech? If they do, I am convinced that they are mistaken; and if they do, *they* must be convinced, if they have considered the matter, that Democracy is a futile enterprise. The same conclusion would evidently follow if they held that, for the most part, scientific ideas do not admit of intelligible expression in non-technical terms. But they cannot rationally hold that such expression is in fact impossible; they may rightly regard it as difficult, as demanding the patience and skill and humane motivity of a special art, but they must know that it is not impossible; for they know that scientific ideas, however high they be above the level of common experience and common sense, yet have their roots in its homely soil. Was it Lord Kelvin or another sage who said of mathematics that it is just "common sense etherealized"? The statement is as true of science in general as it is of mathematics; it is not indeed a *complete* characterization of either of them, for the process by which they rise out of common sense involves something more than etherealization, something more than purification, more than elimination of dross; it involves, besides, a constructive process, a process of creation. But, though the statement is not a complete characterization of science in general nor of any branch thereof, yet, regarded as a partial characterization, it is fundamentally true of every branch: of all science common experience,—common sense,—is the basic soil. And not only do scientific specialists know that all scientific ideas have their roots in the soil of common sense but they know, too, that every single term in the vast jargon of science ultimately derives its meaning, in one way or another, from generic ideas which, though ill defined in the common consciousness, are present there and are constantly employed by your man-in-the-street. The process of such derivation is a perfectly *natural* one; natural processes are, for the most part, not reversible; but this one is; there is, I mean, no scientific idea whatever, however complicated and refined, and there never will be one, that does not admit of being analyzed and ultimately expressed in the language appropriate to the vulgar elements whence the idea was originally derived. In the case of many ideas, such elemental analysis requires great patience and skill, and their expression in common speech can not be made perfectly clear; it will sometimes, of necessity, be so cumbrous and prolix as to be unprofitable except as an exercise; but the thing can be done, and the point is that, in an immense multitude of cases, it can be done in a way to edify not only the general public, but also the experts who render the service.

The *radical* explanation of the scientific ignorance of the educated public is to be found in the fact that, with rare exceptions, *those who understand do not teach*—do not teach, I mean, save in a manner suitable for the training of specialists like themselves, in terms, that is, that are highly technical and jargonistic. In the course of a good many years of university experience, I have had occasion to attend many public examinations of candidates for the degree of doctor of philosophy, not only in mathematics but in other branches of science. There is one question which I have been accustomed to ask the candidates. The question is: *Can you state intelligibly, in the language current among educated men and women the nature of your research,—the problem you have solved, the methods you have employed, and the results you have obtained?* And in every instance the response has amounted to this: "I have never attempted to do it; I have not thought of it; but I believe it would be very difficult or quite impossible." What is to be said of their estate? I think we may say this: Their estate is pitiable; they have devoted long laborious years to qualifying themselves for a certain ordeal,—the ordeal of demonstrating that they have acquired a certain competence of highly technical scholarship in some field of study and that they have the ability to do independent research in the field; they have, let us say, gone through the ordeal successfully; that, in itself, is well, but the price they have paid is terrible; they have submitted to the painful process by which *men* are converted into *mere technicians;* the education they have acquired in the best years of growth is lacking in the quality of amplitude; they have become narrowly technologized; long confined within the prison walls of a Specialty, sinking deeper and deeper in its profound indeed but narrow shaft, they have become more and more detached from the thronging life of the world, and lost alike the power of sympathy and the power of communion with their fellow men and women;

they have indeed qualified for membership in a small and insulated class of technicians, composed, in the main, of spiritually meagre men; and the worst of it is that, having lost perspective, they are often vain of the distinction; they are apt to fancy themselves investigators, and some of them will be but most of them will not; teaching,—teaching, I mean, in the collegiate sense, they are prone to regard as drudgery with which a cruel fate hampers their genius, while teaching in the larger sense of interpreting science in popular terms for the public enlightenment,—*that* they have been taught to scorn as beneath the dignity of a doctor (which means a teacher) of philosophy. Their estate, I have said, is pitiable; it is pitiable; it is pitiable that men who hold themselves specially trained in the arts of scientific "discovery" should not be able to *discover* the glaringly patent fact that research is often far easier than competent exposition; that every normal child, for example, discovers a world of facts which it seldom has the power to express fittingly; and that little doctors of philosophy are far more numerous, because they are easier to produce, than great expounders and interpreters of scientific truth.

When will scientific specialists, especially those who cherish the hope that the world's human affairs may at length be scientifically controlled by an enlightened Democracy, when, I ask, will such men keenly feel their great obligation to enlighten the public and learn to discipline themselves and their pupils to keep the obligation? In the address alluded to a moment ago, I expressed the hope

> that here at Columbia or other competent center there may one day be established a magazine that shall have for its aim to mediate . . . between the focal concepts

and the larger aspects of the technical doctrines of the specialist, on the one hand, and the teeming curiosity, the great listening, waiting, eager, hungering consciousness of the educated thinking public, on the other.

That hope has not yet been fully realized. But on every hand there are indicia of amelioration: magazines and magazine departments, aiming at scientific enlightenment of the public, are growing in number; scientific expositions in the newspaper press, though often amazingly ignorant and misleading, are becoming less so; books of popular science not only continue to multiply, but they are better than formerly, at all events not quite so bad; one hears more and more frequently of universities providing "omnibus" courses in science—general courses, that is, designed for the scientific enlightenment of students not intending to specialize in science, as one may become intelligent about history, for example, or music or architecture, without becoming a professional historian or an expert musician or architect; it occasionally happens that a university professor, representing some highly specialized subject, undertakes to give instruction in available parts of it in a manner suited to the needs of a general audience; an example of this is the course of mathematical lectures recently given in the University of Illinois by Professor J. B. Shaw and subsequently published in a volume entitled *The Philosophy of Mathematics*. The signs are encouraging; but the best of them remains to be mentioned. I refer to the recent establishment, at Washington, of a new institution, called *Science Service*. The founding and maintenance of this institution was made possible by the liberality of Mr. E. W. Scripps, of California. Its aim is that which I have been here trying to emphasize the importance of—scientific enlightenment of the public. The editor of *Science Service* is Dr. E. E. Slosson. A good account by him of the new institution's charter, scope, purpose, organization and present policy is found in *Science,* April 8, 1921. The work it is attempting to do is of the highest importance but it is exceeding hard. It is hard because there are but few competent scientists who believe in the possibility of popularizing science; because, among those who believe in the possibility, there are but few who are willing to engage in the enterprise; and because, among the willing, there are but few who have acquired the requisite art. What is needed is a more numerous breeding of men like Galileo Galilei and Auguste Comte and W. K. Clifford and Thomas Huxley and John Tyndall and Ernst Haeckel and Joseph Le Conte and Camille Flammarion and Louis Couturat and Ernst Mach and Josiah Royce. Let the multiplication of mute specialists proceed—their service is mighty in its way; but its way is not enough. Democracy demands that the discovery of truth be attended and be followed by exposition, by interpretation, by evaluation, in terms that educated laymen can understand.

Doubtless, you have been wondering why these meditations upon the popularization of science have been inserted as a prelude to a lecture on non-Euclidean geometry. I admit that they might have been submitted elsewhere with equal propriety; they might indeed have been presented in the introductory lecture for the entire course has for its aim, as you know, the democratizing of scientific knowledge and scientific criticism. Why, then, have I inserted the discussion in the present connection? The explanation, which is a very simple one, is in terms of personal psychology. I had been considering whether I should or should not include in the course a lecture on non-Euclidean geometry, and what, if I discussed the subject at all, it would be best to say. Well, the course of my meditation respecting the first question ran as follows: the birth of non-Euclidean geometry was and is one of the most momentous events in the history of thought; no other has served to reveal in so clear a light the nature of logical Fate and the nature, scope and limitation of intellectual Freedom; no other has so well disclosed the distinction,—which is radical and cannot be obliterated,—between the world of conception and the world of perception,—the world of pure thought and the world of sensuous experience; no other has so clearly defined the great problem of ascertaining how the two worlds are interrelated; the matter in question,—the advent and nature of non-Euclidean geometry,—is one of the few great mathematical matters that professional philosophers have seriously sought to understand and it is one that has at the same time persistently haunted the imagination of the educated portion of the non-mathematical public; many, very many, have been the attempts to explain the subject in the daily press, in spoken lectures, in magazines and in books; nevertheless, outside of mathematical circles, understanding of the matter, it must be owned, is meagre; for making the matter clear to the man-in-the-street adequate means has not yet been found. What, I asked myself, is to be done? Must we in *this* case relinquish the hope of successful popularization? And I answered, no, we must keep on trying; for I vividly recalled Gergonne's noble dream and the world's great hope—Democracy; so my mind was set swarming with the considerations adduced in the prelude; and that is why I have presented it here.

Do I flatter myself with the belief that in this lecture

the nature of non-Euclidean geometry is at length going to be made so plain that he who runs may read and understand? I do not; nothing is farther beyond my hope. Much that others have said I shall omit, and most of what I shall say has been repeatedly said, in one way or another, by them; if I succeed in adding only a little light to that given by the extant literature of the subject, I shall be quite content.

In Lecture VII, as you will remember, I pointed out that the term non-Euclidean has two meanings—one of them specific and usual, the other one generic and less usual; the former meaning always refers to the theory of parallels; the latter does not. In the present discussion, the term will be used in the specific sense only.

It is customary to say that non-Euclidean geometry is a strictly modern discovery, due to the daring genius of a young Hungarian, John Bolyai, and independently to that of a Russian, Lobachevski, both of whom flourished in the first half of last century. The discovery, as I have intimated, was preceded by an immense period of preparation in which geometricians wrestled with a very old puzzle—the so-called problem of parallels. If you will consult Dr. T. L. Heath's superb edition of Euclid's *Elements,* you will find that controversial discussion of that problem began in pre-Euclidean days, was but aggravated instead of terminated by Euclid's handling of the matter, and, though culminating in the birth of the new geometry, has continued (among the geometrically ill informed) down to our own day, a hundred years after the puzzle was virtually solved by the two pioneers I have named. There is no tale more romantic, nor, in the proper sense of the term, more *human,* in the whole history of Thought. A human tale, I have said, distinctively human. I am unable to understand how anyone can ponder its character and its significance intelligently and candidly without seeing clearly that the old zoological conception according to which human beings are a species of *animal* is not only false, as Count Korzybski has pointed out, but is stupid as well; in view of its baneful effects upon the world's ethics, the monstrous misconception deserves indeed to be branded as the Great Stupidity.

I shall not here recount the tale; the story has been often told in all the lands and all the tongues of science. If you desire to learn the story, you will find in the mentioned work of Dr. Heath an ample clue to the literature. For an account that is at once very clear, very succinct and finely critical, I have special pleasure in referring to Dr. George B. Halsted's article ("Geometry, non-Euclidean") in the *Encyclopedia Americana.* I own to the feeling of a little pride,—pardonable, I hope,—in that citation for it was my privilege, as then mathematical editor of *Americana,* and my good fortune, to obtain the article in question. I requested Dr. Halsted to write it because he was specially qualified to do it; no other American scholar knew more than he of non-Euclidean origins and no other has done so much as he has done, by voice and pen, to signalize the importance of non-Euclidean geometry as making and marking a momentous epoch in human Thought.

Of non-Euclidean geometry there are two principal varieties; these are associated respectively with the names of their inventors—Lobachevski (1793-1856) and Bernhard Riemann (1826-66). I am going to tell, as clearly as I can without too great prolixity, how the varieties arise and what they are. And I will begin with that of the Russian.

The point of departure is Euclid's famous postulate,—his postulate *V*,—which I stated in a previous Lecture (VII); this postulate,—or assumption, for that is what it is,—is pretty long; it is known, however, to be equivalent to the briefer assumption:

> *Through any point there is one and but one line parallel to a given line.*

For many centuries geometricians, great and small, tried to *deduce* this assumption,—which may be called the *one*-parallel assumption,—as a *theorem* from Euclid's other assumptions (conscious and unconscious). They failed, and today we know why—the assumption is not implied in the other ones and so is not deducible from them, not even by demons or archangels or gods. Now, what the adventurous spirit of Lobachevski led him to do is simply this: retaining all of the Euclidean assumptions save the one respecting parallels, he replaced the latter by an assumption contradicting it, and then proceeded to deduce the consequences of the set of assumptions he had thus adopted as postulates. What is the assumption with which he replaced Euclid's postulate of the *single* parallel? It may be stated as follows by help of Fig. 30: *If line PF rotate (in the plane of the figure) about P, say counterclockwise, it will come to a position, call it PK, where it first fails to cut line L, and then, without cutting L, it will rotate through a finite angle KPH into a position PH such that, if it rotate further, it will cut L to the left of F, the angles FPK and FPK' being equal.*

According to this assumption the lines of the Pencil

*P* fall into two sets—the set of those that cut *L* and the set of those that do not; the former set consists of all the lines *within* the angle *KPK'*; and the latter set consists of *PK, PK'* and the lines *within* the angle *KPH* (or, what is tantamount, the angle *K'PH'*). The limiting or boundary non-intersectors, *PK* and *PK'*, are the Lobachevski *parallels*—parallel, that is, to *L*. We may, accordingly, call Lobachevski's asumption the *two*-parallel assumption.

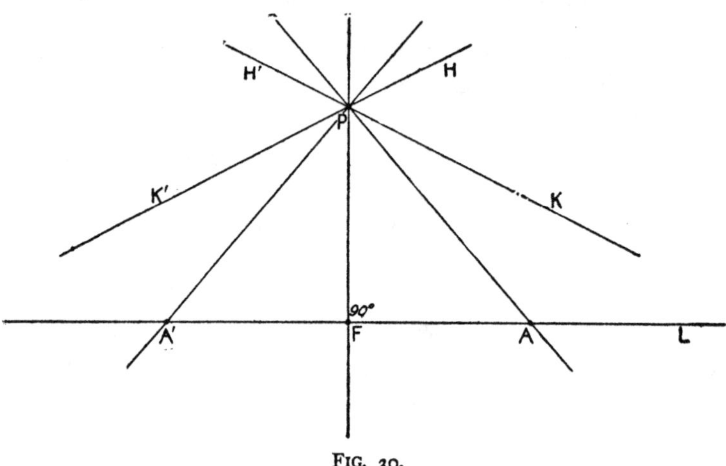


FIG. 30.

There remains a third possibility, pointed out in 1854 by Riemann in his famous inaugural dissertation, with which professional philosophers seem to have but little or no acquaintance, though it involves the complete demolition of the Kantian conception of space and though its thought is sufficient to have immortalized several men had they severally originated its several parts. I refer to the lecture, *On the Hypotheses which lie at the Basis of Geometry*. Translated into English by W. K. Clifford, it is found in the latter's *Collected Works*. In the address Riemann indicated the possibility of constructing a geometry upon a basis of postulates containing the assumption that

> *No two lines of a plane are parallel;* or, what is equivalent, *every two lines of a plane intersect.*

This *no*-parallel assumption contradicts not only the *one*-parallel assumption of Euclid but another assumption of his,—a *tacit* assumption,—namely, that a line has *infinite* extent (see propositions 16 and 28, Book I, in Heath's edition of the *Elements*). The remaining assumptions of Euclid are retained by Riemann.

We have now before us three postulate systems,—one of them Greek,—one of them Russian,—one of them German. Upon them have been erected and now stand three geometries, one of them Euclidean, the other two non-Euclidean; of these two, the former is often called Lobachevskian, and the latter Riemannian. For a good reason, which I will not pause here to explain, Professor Felix Klein has called the three geometries respectively Parabolic, Hyperbolic, and Elliptic, descriptions that are now in common use.

With the elements of the parabolic geometry you are familiar; with those of the other two varieties you are presumably not acquainted. The hyperbolic and elliptic geometries have been built up by various methods, elementary and more advanced, pure and analytic. I credit you with having curiosity to see how the building can be done by the familiar elementary processes of ordinary geometry. To apply them here would detain us too long; but if you have the curiosity, you can gratify it by reading the previously mentioned essay of Professor F. S. Woods on "Non-Euclidean Geometry" (found in *Monographs on Modern Mathematics,* edited by J. W. A. Young).

We have seen in what respects the *bases,* the postulate systems, of the three geometries are alike and in what respects they are unlike. We naturally pass to a comparison of the *superstructures*—to a comparison, that is, of the theorematic contents of the geometries. Do the geometries intersect? Have they, that is, any theorems in common? The answer is obvious: they have in common such and only such theorems as are deducible from the assumptions that are common to the three systems thereof. One such theorem is this: The summit angles of a birectangular quadrilateral are equal; in other words, if $ABCD$ be a quadrilateral having right angles at $A$ and $B$ and having the side $AC$ equal to the side $BD$, then the angle at $C$ is equal to that at $D$. You may wish to make a list of such common theorems as an exercise. More striking are the theorems in which the geometries differ; such differences are of course due to the differences in the postulate systems. Let us notice some of them. The one most commonly mentioned relates to the sum of the angles of a triangle. In the parabolic geometry that sum is *constant* (the same for all triangles) and is *exactly* two right angles, as you know; in the hyperbolic and elliptic geometries the sum is *variable* (depending upon the triangle's size); in the former geometry it is always *less* than two right angles and *decreases* as the triangle's area increases; in the latter, the sum is always *greater* than two right angles and *increases* with the area.

Again: If two lines of a plane are perpendicular to a third, then, in parabolic geometry, the two are *parallel;* in the hyperbolic, they are *not* parallel nor do they *intersect;* in the elliptic, they *meet* at a point whose distance from the third line is *finite,* and all perpendiculars to this line meet at that point.

One more: In parabolic geometry each summit angle of an isosceles birectangular quadrilateral is a *right* angle; in hyperbolic geometry, it is *acute;* and in elliptic geometry it is *obtuse.* So it is seen that neither of the non-Euclidean geometries contains rectangles among its figures.

It would be easy to prolong the list of such differences; the theorems already stated, which are not difficult to prove, are sufficient, however, for illustration and they ought, I think, to challenge the curiosity of any intellectual student.

There are certain questions which you are doubtless bursting to ask, for in a discussion of this subject thoughtful beginners always ask them.

One of the questions is this: Can we be quite certain that neither of the non-Euclidean geometries involves an inner contradiction? In other words, can we be certain with respect to each of them that the propositions constituting it are compatible with one another? The answer is, *yes.* The propositions constituting a geometry consist, as you know, of its postulates and of the propositions logically deducible therefrom; and so, if the postulates are mutually compatible, the whole geometry is self-consistent. The question thus reduces to this: Can we be certain with respect to each of the non-Euclidean geometries that its postulates are mutually compatible? Now, in respect of the Euclidean postulates, we saw in Lecture VI, you will remember, that we can be as certain of their compatibility as it is possible to be of any reasoned proposition, and that is what I mean by "quite certain" and it is what you mean. Well, it can be shown that, *if* Euclidean geometry is self-consistent, then each of the non-Euclidean geometries is self-consistent. And this has been done by various mathematicians in various ways. It has been done very simply by Henri Poincaré in his widely known *Science and Hypothesis;* and it has been done still better, in the sense of greater detail, in the previously cited *Elementare Geometrie* of Weber and Wellstein. Suffice it here to say,—for I am going to leave it to you to examine the proofs,—that the principle or the trick involved in them is that of *showing the postulate systems of the non-Euclidean geometries to be each of them satisfied by suitably selected classes of geometric entities found in Euclidean geometry.* So, you see, if the non-Euclidean geometries have any unsoundness in them, there is a corresponding unsoundness in Euclidean geometry. In respect of soundness—inner consistency—self-compatibility—logical concordance among the parts of each—the three geometries are on exactly the same level, and the level is the highest that man has attained. The three doctrines are equally legitimate children of one spirit,—the geometrizing spirit, which Plato thought divine,—and they are immortal. Work inspired and approved by the muse of intellectual harmony can not perish—it is everlasting.

Another question is: Are these geometries *true?* They are true in the sense in which truth resides in a body of propositions of which some are mutually compatible premises and the rest are inevitable consequences thereof, enchained thereto by the binding threads of logical fate, which is changeless and timeless. Such truth, however, though it is ineffably precious, is only a quality or an aspect of that inner consistency which gives an autonomous body of propositions its peculiar beauty, pure, perfect, and eternal. But it is not this aspect of logical consistency that the "true" of your question is designed to mean.

Perhaps the question you intend to ask is this: Are the geometries *true* in the sense of giving an *exact* account of Space? Or, better, since they contradict one another in cardinal matters, is *one* of them true in the indicated sense? I have already pointed out that the term "space" does not occur in Euclid's *Elements,* and I may add that there is no *necessity* for its occurrence in either of the non-Euclidean geometries. Since it is nevertheless customary to use the term in philosophic discussions of geometry, we, too, must do so here. If we are to do so profitably, we must make and keep steadily in mind a fundamental distinction, which is indeed a pretty obvious one but is commonly neglected; and the neglect of it is always attended with utter confusion. We must, I mean, not fail to distinguish sharply between *perceptual* space and spaces of *conception;* that is, between space in which points, lines, planes, circles, spheres, and so on, are material or physical dots, rods, or ropes, slabs or rough irregular "surfaces" thereof, hoops or rings, globes or balls (of wood or gold or marble), and so on, and a space in which the terms point, line, . . . denote pure *concepts* of which no instance is found, for no instance exists, in perceptual space. Unless we make and keep that radical distinction, we might better abandon the discussion; but if we make the distinction clearly and do not lose it, the question you have put can be answered clearly and rightly.

And the answer? It is found in the following considerations.

The space of a geometry is always a *conceptual* space.

What *is* the space? It is the class of conceptual entities (no matter what we *call* them, points, lines, and so on, or jabberwockies) which satisfy or verify the geometry's postulates and about which the geometry is therefore a reasoned discourse. The geometry *is,* therefore, true in the sense of giving an exact account of space, where by "space" is meant the space of *that* geometry. I say "of that geometry" for, you see, two geometries which contradict one another in one or more respects have different spaces. The answer to your question is, then, this: Euclidean geometry, Lobachevskian geometry, and Riemannian geometry are each of them *true* in the sense that each of them gives an *exact* account of *its own* space, which is a conceptual space.

But what of perceptual space? What, I mean, of that all-enveloping region or room or spread which is revealed to us by touch and sight and hearing and the sense of muscular movement? Is one of the three geometries *true* in the sense of giving an *exact* account of this space? The question is a fallacy of interrogation; it implies, that is, that our perceptual space is a thing of which an exact geometric account is possible; but it is not such a thing—perceptual space is not, rightly speaking, geometrizable. Wherein it fails to be so, is easy to make clear. Consider, for example, three of its "lines," say pencils or rods, $l_1$, $l_2$, $l_3$. Compare their "lengths" perceptually, which is the only way in which *such* "lengths" can be compared. Compare $l_1$ with $l_2$, then $l_2$ with $l_3$, and then $l_1$ with $l_3$. You know what may happen, for it is a common phenomenon of such comparison of perceptual things. It may happen that the "length" of $l_1$ is equal to (indistinguishable from) the "length" of $l_2$, that the "length" of $l_2$ is equal to the "length" of $l_3$, while that of $l_1$ is *un*equal to (distinguishable from) that of $l_3$, so that, in symbols, the record of relations will stand thus:

$$(1)\quad l_1 = l_2,\ l_2 = l_3,\ l_1 \neq l_3.$$

Now reflect that our perceptual space *is such that* the the concurrence of relations like (1) is a familiar and unavoidable phenomenon of "lengths" not only but also of "areas" and "volumes"; then reflect that such situations as (1),—though inherent in the nature of perceptual space,—are utterly and glaringly *illogical;* finally, reflect that all the relations occurring in a geometry must be, unlike those of (1), *logically* consistent; and you will be thus led to see clearly that our perceptual space cannot be geometrized in Euclidean fashion or otherwise, if the term geometry is to retain its essential meaning.

But is there not, among the various current meanings of the term "truth," one meaning which enables us to say that Euclidean geometry, regarded as a doctrine about our perceptual space, is true? The answer is: yes, there is such a meaning. It is the "instrumental" meaning insisted upon by Professor John Dewey—the "pragmatic" meaning first signalized by C. S. Peirce, subsequently interpreted, elaborated and advocated by William James and others. It is the meaning in accordance with which an idea or a proposition or a doctrine is true if it "works," in so far as it "works," so long as it "works." The meaning is not without merits that commend it to all men and women for all human beings have, below their distinctively human qualities, certain animal propensities and animal impulses, and, in the animal world, the end always justifies the means—all ways that "work" equally well, all means that are equally "effective," are equally good. It is the pragmatic meaning of truth that makes treason a crime, if it fail, and a virtue, if it succeed. It is a meaning that is especially congenial to practicians and "politicians," whose "philosophy" never rises above the question: How can I "get there"? How can I put this thing "through," "over" or "across"? What is the means that will "work"? It is a meaning, too, that is especially congenial to an industrial age,—an experimental age,—an age of laboratories,—an adventurous age when men act more than they think. In the headlong rush and hurly-burly of such an age, men and women are not aware of the fact that the world of human affairs would quickly dash upon utter destruction but for the guiding and saving influence of a nobler truth-conception which they do not consciously own,—the conception, I mean, of truth as having its highest meaning in the unchanging relations and eternal laws of Logical Thought.

I have said that in the pragmatic sense of "true," Euclidean geometry may be said to be true of our perceptual space. It may be said because this geometry, when applied (as we say) to this space, "works"—which means that temples, aqueducts, tunnels, bridges, railways, ships and other architectural and engineering structures whose designing is guided by Euclidean formulas are successful—they wear out but they do not perish from any essential defect of structural design.

Does not the fact just stated show that Euclidean geometry is superior to the other two varieties? It does not; for, if the designing and the building of the structures alluded to were guided by Lobachevskian or by Riemannian formulas, the structures, when completed, would not differ *perceptibly* from the former ones, and the railways, ships, and so on, would be equally durable and serviceable; the reason is that, though the formulas of any one of the geometries differ radically from their correspondents in either of the other two, yet they differ in such a way that the difference could not crop out in any physical structure unless the latter were vastly larger than any our small planet admits of. We may say, then, that all three of the geometries are pragmatically, or instrumentally, true of our perceptual space—the space of sensuous experience, though, as we have seen, no one of them and no other geometry is or can be, rightly speaking, the geometry or a geometry of perceptual space.

Since all three of the geometries in question are pragmatically true of our perceptual space, why is it that in *practical* work, like bridge building and the like, the Euclidean variety is employed exclusively? It is because the Euclidean formulas are *simpler,* easier to use than the others. What, if any, is the epistemological significance of this fact? The question seems important. I do not know the answer. Maybe one of you will discover it.

Is non-Euclidean geometry always 3-dimensional? No; like the Euclidean variety, it may have any number of dimensions from *one* up.

What of Einstein geometry? The answer is implicit in what has been said. "Einstein geometry" is not geometry,—not yet, at all events,—it is a figure of speech, convenient for experts, misleading for laymen. That is not a comment upon the doctrine of Relativity regarded as being,—what it is,—a physical theory.

The advent of non-Euclidean geometry is, I have said, one of the gravest events in the history of thought. It has been tragic as well. The two facts are connected. Thirty years ago, I visited a locally eminent professor of mathematics in an excellent middle-West college of the sectarian variety. I was astonished to find him in a sad mental state, worried, distracted, agitated, tremulous, unable to sleep or rest, thinking always about the same thing, and no longer able to do so coherently. What was the trouble? For many years he had been teaching geometry,—Euclidean geometry,—and his teaching had been done in the spirit and faith of a venerable philosophy. Like almost all the educated men of his time and like millions of others in the preceding centuries, he had been bred in the belief that the geometry he was teaching was far more than a body of logical compatibilities; it was not only true internally,—logically sound, that is, —but it was true externally—an exact account of space, the space of the sky and the stars; its axioms were not mere assumptions,—not mere *ifs*,—they were truths, "self-evident" truths, and, like the propositions implied by them, they were not only valid but were known to be valid, and valid eternally; in a word, the geometry of Euclid was a body of *absolute knowledge* of the nature of space,—the space of the outer world,—other space there was none. That was a comforting belief, a congenial philosophy, held as a precious support of religion and life; for, though there are many things unknown and some perhaps unknowable, yet *something,* you see, was known; there was thus a limit to rational skepticism; our human longing for certitude had at least one great gratification—the validity of Euclidean geometry as a description of Space was indubitable. Such was the philosophy in which my dear old friend had been bred, and, with unquestioning confidence, he had devoted long years to the breeding of others in it. At length, he heard of non-Euclidean geometries, in which his cherished certitudes were denied—denied, he knew, by *great* mathematicians, by men of creative genius of the highest order; he could not accept, he could not reject, he could not reconcile; the foundations of rational life seemed utterly destroyed; he pondered and pondered but the great new meaning he was too old to grasp, and his mind perished in the attempt,—killed by the advancement of science,—slain by a revolution of thought.

Pitiless indeed are the processes of Time and Creative Thought and Logic; they respect the convenience of none nor the love of things held sacred; agony attends their course. Yet their work is the increasing glory of a world,—the production of psychic light,—the growth of knowledge,—the advancement of understanding,—the enlargement of human life,—the emancipation of Man.

## LECTURE XVIII

### The Mathematics of Psychology

BACKWARDNESS OF THIS AND OF THE PSYCHOLOGY OF MATHEMATICS—THE LAW OF WEBER AND FECHNER REËXAMINED—SOME OF THE LAW'S UNNOTICED IMPLICATIONS—THOUGHT AS INFINITELY REFINED SENSIBILITY.

Two of the subjects that this course should include are the mathematics of psychology, and the psychology of mathematics. The two things, though closely related, are distinct—they interpenetrate but neither includes the other; it is one thing to mathematicize psychology, and a very different thing to psychologize mathematics; the aim of the former is to express psychological relations in mathematical terms; that of the latter is to study those aspects of mathematics which are psychological as distinguished from logical; both enterprises are immense and important; neither one is far advanced, and the reasons are evident—the tasks are difficult, requiring special preparation, while psychologists have not been mathematicians and the latter have not been psychologists. But psychologists have done as much to mathematicize their subject as mathematicians have done to understand the psychology of theirs, though the former task is perhaps the harder of the two.

I will begin with the mathematics of psychology and will in the main confine my remarks to its most famous achievement, which is also, I believe, its most important one—the so-called Psychophysical Law of Weber and Fechner. When the law was duly announced in its mathematical form, it aroused much interest among psychologists, evoked much admiration, and stimulated research and discussion; at the same time it produced something like fright or consternation for it began to seem that one could not be a *scientific* psychologist without being a mathematician, and that was a fearful thought. But the "interest," the "admiration" and the "fright" were destined to pass or, at least, to suffer much mitigation. A generation ago the law had been often presented and elaborately discussed. At length it was handled by William James in *The Principles of Psychology*. After presenting it with characteristic honesty and with quite as much accuracy as could be expected from one who not only was not a mathematician but knew and owned that he was not, James proceeds to examine the claims that had been made in behalf of the law and then, with absolute candor and great confidence, to estimate its significance. And what is the estimate? It is this: "Fechner's book, *Psychophysik,* was the starting point of a new department of literature, which it would be perhaps impossible to match for the qualities of thoroughness and subtlety, but of which, in the humble opinion of the present writer, the proper psychological outcome is just *nothing*" (Vol. I, p. 534). Again: "The Fechnerian *Maasformel* and the conception of it as an ultimate 'psychophysic law' will remain an 'idol of the den,' if there ever was one" (p. 549). Of that judgment, right or wrong, pronounced by so great an authority, it may not be said that *its* psychological "outcome" or effect—upon psychologists—was "just nothing." On the contrary the effect was great and it was soothing; it dampened interest in Fechner's work; it moderated admiration of the man; and it greatly relieved many a psychologist who had been frightened by what had seemed a serious mathematical invasion of his subject. James's discussion of the matter is very interesting and enlightening—more so than any other I have seen; his presentation of the law in question is marred, however, by some inaccuracies; moreover, respecting the significance of the law, it has, if I be not mistaken, certain implications and important bearings not noticed by James nor, I believe, by others. It has, therefore, seemed to me that a discussion of the matter might be properly included in this course, even though part of my remarks can at most remind you of things you are already familiar with.

*The Rise of Psychology as an Experimental Science.* —For convenience of reference let me place before you, in chronological order, five important names.

Immanuel Kant (1724-1804)
Johann Friedrich Herbart (1776-1841)
Ernst Heinrich Weber (1796-1878)
Gustav Theodor Fechner (1801-1887)
Bernhard Riemann (1826-1866)

I well remember that, when I was a boy, it was customary for people who liked to talk about science to speak of two kinds thereof—the natural, or physical, sciences and the so-called mental sciences. The classification was then an old one but it is not yet without some vogue. The distinction may not have been profound but it was obvious: the former kind of science was *quantita-*

*tive* for it dealt with phenomena that were *measurable;* the latter kind was *qualitative* for it dealt with phenomena that were supposed to be *non-measurable.* Now, owing to its lack of precision, a qualitative science was looked upon as being *scientifically* inferior to one that was quantitative. According to Kant, for example, psychology was not a true science and never could be. What he meant by a "true science" is sufficiently revealed by his saying that a natural science is a science only in so far as it is mathematical. It is noteworthy that this saying of a great philosopher accords perfectly with the saying of one who was at once a great physicist and a great mathematician—Bernhard Riemann: "Natural Science is the attempt to understand nature by means of *exact* concepts." It is "noteworthy" but is not surprising in view of Kant's strong predilection for mathematics and physical science, and of Riemann's early interest in psychology and metaphysics (as shown in the *Anhang* of his *Gesammelte Werke* or in Keyser's translation of the *Anhang* in the *Monist,* January, 1900).

Among the first to reject the Kantian dogma respecting psychology was Herbart, the so-called "exact philosopher." Herbart believed it was possible to build up a statics and a mechanics of elementary ideas,—a mechanics of mind, let us say,—modeled after the classical mechanics of matter, and he boldly essayed the task. But *apriori* reasoning cannot do a work that calls for patient experimentation, and Herbart's mathematical formulæ, though rather impressive to the physical eye, have but little interest except as representing a reaction and a prophecy.

It was not a *professional* philosopher who took the first important steps toward making psychology a laboratory, or quantitative, science; they were taken by men trained in the ways of natural science. These men were Weber, professor of anatomy and physiology in the university of Leipzig, and Fechner, professor of physics in the same institution. What was the new problem they set for themselves and how did they attack it? Well, there is in our world what we call matter and there is what we call mind. Let us not tarry to debate the great present-day question whether the two things are essentially one nor whether they are derived from a "neutral" something—something, that is, that is neither matter nor mind.[1] For our pioneers mind and matter were obviously two, the two were related, and the problem was to ascertain *how*. And their method was that of experiment and observation. Where did they begin? And why *there?*

*Sense Departments and Their Fundamental Problems.*—Imagination, conception, reason, will, all these have to do with matter, but experimental research did not begin with them. Why not? Because it was best to begin at the beginning, and the beginning is sensation—it is in what we call sensation that our "minds" first get into some sort of knowing connection with "matter." It was soon found necessary to distinguish many more "sense departments,"—as they are called,—than the traditional five departments of hearing, sight, and so on; for example, our capacity to feel pressure gives rise to a distinctive class of sensations, and so the pressure-sense is spoken of as a sense department; in like manner, we speak of sense departments corresponding respectively to the capacities for feeling warmths, brightnesses, sizes, sounds, and so on, it being evident that some of the de-

[1] In relation to the question, see Russell's *Analysis of Mind* and Keyser's review of it in *The Literary Review* (N. Y. *Evening Post*).

partments are sub-departments of others; that of length, for example, is a division, branch or species of the generic department of size.

Each department presents three problems—fundamental problems. It is easy to see why this is so, and what the problems are.

Everyone knows that some lights are too dim to be seen, that some tastes are too delicate to be discerned, that some pressures are too slight to be felt, that some lengths are too short to be sensed, and so on for the other sense departments. The form of my statement involves a contradiction in terms but the meaning is clear, and that is enough. Out of the kind of facts stated arises one of the experimenter's problems. The problem is: Given a sense department, to determine the smallest amount of (the appropriate) stimulus that will yield a sensation.

The problem just stated has a complement. Everyone knows that a stimulus may be too great, as well as too small, to produce a sensation. Thus in the sense department of tones there is no sensation answering to 100,000 vibrations per second; a pressure may be so great that we cannot feel it; a light may be so intense as to blind us; and so on. Whenever the stimulus exceeds a certain amount, the nerves are put out of commission, and the scale of sensation reaches an end. Hence the problem: Given a sense department, to determine the maximal stimulus that will produce a sensation. And so we see that the world of sensory experience possible to us humans is a confined world,—walled in by the surrounding presence of limits, a lower and an upper limit in every department of sense. The two problems stated are those of determining the location of all the parts of the wall.

The third one of the experimenter's fundamental problems has to do with what occurs *within* the wall. It is this: Given a sense department and in it a sensation corresponding to the least (or greatest) stimulus that will produce it, to determine how much the stimulus must be increased (or decreased) to beget a new, or different, sensation.

*Some Technical Terms and Symbols.*—In the literature we encounter, as you probably know, the equivalent terms—Threshold, *Limen, Schwelle*—introduced by Herbart about a hundred years ago, and the symbols—*R, L, D, T, RL, TL, DL*—whose meanings are easy to grasp. *R* comes from the German *Reiz,* signifying stimulus; *L* stands for *Limen,* or threshold; and *T* for terminal; *RL* denotes initial threshold,—the least stimulus that will yield a sensation; *TL* denotes terminal threshold,—the greatest stimulus that will yield a sensation; and *DL* denotes difference threshold,—the difference between the least (or greatest) amounts of stimulus that correspond to two just discernibly different sensations. The three fundamental problems may accordingly be restated thus: To determine in *each* sense department its *RL, TL,* and *DL.*

The pioneers, Weber and Fechner, were contemporaries but the work of Weber came first. He dealt mainly with hearing and touch. A rough statement of Weber's Law,—so named by Fechner,—is this: The increase of stimulus necessary to produce a change of sensation is not a constant difference, but is a constant *ratio* of the preceding stimulus. It is, you notice, concerned with the *DL,* the difference threshold. It is often referred to as Fechner's law or the Weber-Fechner law or the psychophysical law. It is to Fechner,—whose great work,

*Elemente der Psychophysik,* appeared in 1860,—that we owe the *first* formulation of the methods which, with many modifications and improvements, are now employed in psychological laboratories throughout the world. That is why Fechner is called the father of modern psychology.

*Symbolic Statement of the Psychophysical Law.*—We are not here concerned to deduce the law nor to verify it, but to state it in symbols and to examine its meaning.

Consider one of the sense departments, say that of pressure. Denote by $S_1$ the sensation produced by a certain stimulus $R_1$. Next suppose the stimulus increased till there is felt a sensation $S_2$ just discernibly different from $S_1$. Denote the new stimulus by $R_2$. Suppose the stimulus to be again increased till a new sensation $S_3$ is felt that is just distinguishable from $S_2$, and denote the third stimulus by $R_3$. We now have the table of correspondents:

$$\begin{array}{lcl} S_1 & \ldots\ldots & R_1 \\ S_2 & \ldots\ldots & R_2 \\ S_3 & \ldots\ldots & R_3 \end{array}$$

The question is: how are the $R$'s related? If we are to suppose them related at all, connected, that is, by some invariant order, or law, the simplest guess would be that

$$R_3 - R_2 = R_2 - R_1.$$

But that guess would be wrong. Experiment shows that it is, not the difference, but the ratio that is constant; that is,

$$\frac{R_2}{R_1} = \frac{R_3}{R_2}.$$

And this is very remarkable, for no one could know in advance that so simple a relation holds. A relation so complex as not to admit of statement in a finite number of words may exist in the world but, for scientific purposes, such a relation is practically equivalent to chaos, to no relation at all. Denote the constant ratio by $K$. Then $R_2 = KR_1$, $R_3 = KR_2$. As $K$ exceeds 1, we may write $K = 1 + r$, where $r$ is positive. Then

$$R_2 = R_1(1+r)$$
$$R_3 = R_2(1+r) = R_1(1+r)^2.$$

Denote by $A$ the amount of stimulus such that the amount $A(1+r)$ is the smallest stimulus that will yield a sensation in the sense department under consideration; in other words, $A(1+r)$ is the $RL$ of that department. Let the sensation corresponding to the stimulus $A(1+r)$ be denoted by the number 1. The absence of sensation corresponding to $A$ may be indicated by zero (0). The experimental results may be shown, as follows, in tabulated form.

| Sensations | Stimuli |
|---|---|
| 0 . . . . . . . . . . . | $A$ |
| 1 . . . . . . . . . . . | $A(1+r)$ |
| 2 . . . . . . . . . . . | $A(1+r)^2$ |
| . | . |
| . | . |
| . | . |
| $n$ . . . . . . . . . . . | $A(1+r)^n$ |
| . | . |
| . | . |
| . | . |

Of course the final continuation marks do not mean *ad infinitum*, for, as we have seen, the sensation scale has an upper limit—an end as well as a beginning. Observe that

$$n = \log_{1+r} \frac{\text{stimulus}}{A}$$

$$= \log_e \frac{\text{stimulus}}{A} \cdot \log_{1+r} e.$$

Let us now write $S = n$, $R = \text{stimulus} \div A$ and $C = \log_{1+r} e$. Then we have

$$S = C \log R$$

Such, then, is Fechner's formulation of what he called the Psychophysical Law. We must not fail, if we are to understand it, to note very carefully, the meaning of the symbols. Observe that $S$, the initial letter of the word sensation, denotes nothing but a cardinal number. Note that the constant $C$ depends on $r$ and that $r$, which is the same for the sensations of a given sense department, differs for different departments. Note also that $R$ is not the stimulus that produces $S$, as it is commonly said to be (by Professor James, for example), but that it is the quotient of that stimulus divided by $A$, and that $A$ depends upon the sense department under investigation.

The sense departments that have been most investigated are those of light, muscular sensation, pressure, warmth and sound. For these the values of $r$, as reported by Professor Wundt, are as follows:

For light, $r$ = about $\frac{1}{100}$
For muscular sense, $r$ = about $\frac{1}{17}$
For pressure, $r$ = about $\frac{1}{3}$
For warmth, $r$ = about $\frac{1}{3}$
For sound, $r$ = about $\frac{1}{3}$

In other departments, where investigation is more difficult, there is wide divergence in the results that have been reported.

*The Literature of the Law.*—"Those," says James, "who desire this dreadful literature can find it." The best of it is cited under the caption, "Weber's Law," in the eleventh edition of *The Encyclopedia Britannica.* For an excellent introduction to the methods of quantitative psychology, I have pleasure in referring to Titchener's *Experimental Psychology.*

*As to the Validity of the Law.*—In the nature of things, no law can be shown to be absolutely valid by means of experiment. Respecting Weber's Law, we may safely make the following statements. Experiment has shown it to be approximately valid in several of the chief departments of sense. As these are the departments most accessible to experiment, it may be that the law will yet be found to be approximately valid in other departments. Such validity has not been disproved for any department. The law is found to hold best in the mid-region of a sense scale; that is, it is least certain near the initial and terminal thresholds. This fact, however, is consistent with the assumption that the law is equally valid throughout the scale, for it is plain that, near the beginning and the end of a scale, where sensation is dim because of defect or excess of stimulus, the distinctions of different sensations are more difficult to detect and record.

*What is Measured.*—Fechner calls the law of psychophysics a *Maasformel.* But what is it that is measured? What is the magnitude? According to Fechner it is sensation. He says:

> Our measure of sensation amounts to this: that we divide every sensation into equal parts, that is, into

> equal increments out of which it is built up from the zero of its existence, and that we regard the number of these equal parts as determined by the number of the corresponding variable increments of stimulus that are able to arouse the equal increments of sensation, just as if the increments of stimulus were the inches upon a yard-stick.

It is evident that, in Fechner's view, two just discernibly different sensations, belonging to a same sense department, differ by a *sensation-unit* (of a sort characteristic of such department). According to him, a sensation denoted by $n$ in the foregoing table is the *sum* if $n$ equal sensation-units, that is, $n$ times the sensation denoted by the number 1 in the table. Thus, according to Fechner's interpretation of the experimental facts, sensation *increases* as the terms in the arithmetic progression, 1, 2, 3, . . . , while the series of corresponding stimuli is an increasing geometric progression. And this logarithmic correspondence between the two progressions Fechner regarded as the *law of correlation between mind and matter*, between the *psychical* world and the *physical* world. He thus judged that he had made a very great discovery and naturally spoke of it with a feeling of triumph. Fechner was at once physicist, mathematician, poet, dreamer and mystic—a magnanimous man who believed that all the animals, the plants, the earth and the stars have souls. For the considerations that led him to this noble belief, see his works entitled *Nanna* and *Zendavesta*. In his view, the souls of human beings are, in the scale of being, intermediate to the souls of plants and the souls of stars, the latter being likened by him to angels. God, he taught, is the soul of the universe, and the uniformities that we call natural laws are simply the ways of God. This view of things he called the "daylight" view of the world in contrast with the "night view" of materialism. It is not to small men, but to great ones, that this sublime conception of our universe appeals. Even so hard-headed a scientific man as Bernhard Riemann, whose philosophic *Fragments* I have already alluded to, there speaks of the conception with deep interest and grave respect. And, as you may be interested to know, Professor William James has dealt with the same subject in one of his latest writings. I refer to an article, entitled "The Earth Soul," which appeared in the *Hibbert Journal*, January, 1909.

Fechner's view that a given sensation is a *sum* of sensation units has not ultimately found favor with psychologists. Why not? Mainly because a sensation is not felt or sensed as a sum of sensations or of sensation units, and the question relates to sensations, not as they may appear *mediately* in our *reflection* upon them, but as they appear *immediately* in *feeling*. A sensation of brightness, for example, is not felt to be composed of so-and-so many units of brightness. A feeling of pink, says James, is surely not a portion of our feeling of scarlet.

And so the question recurs: What is it that the experiments have measured? Or approximately measured, as we ought to say, for it is evident that nothing ever is or can be measured with absolute precision experimentally. Perhaps we may say that what is measured is a *human sensorium's discriminative sensibility to stimulus*. Our sensibility does not detect every difference of stimulus. It does detect some differences. By experiment these have been approximately ascertained for several departments of sense, and the psychophysical law is an approximately accurate statement of the way in which they are related.

*Sensation as a Function of Stimulus.*—In the mathematical meaning of the term "function," sensation is a function of stimulus within the interval between the initial and the terminal thresholds of any department. For to any amount of stimulus in such an interval, there corresponds a sensation. That is to say, in no such interval have there been found any blind spots or gaps or regions or points where the sensorium fails to respond. The question arises: Is sensation, or the intensity of sensation, a *continuous* function of stimulus? Most psychologists answer *yes*. Among these may be mentioned, for example, Titchener, Ward and Stout. James has answered *no*. The men named leave one in doubt whether they know precisely what is meant by a continuous function. Let us recall to mind the idea of functional continuity. Let us remember that if $f(x)$ is to be a continuous function of a real variable $x$ in an interval having $a$ for its beginning and $b$ for its end, the following conditions must be satisfied:

(1) If $x'$ be any value of $x$ in the interval, then $f(x') =$ some definite value.

(2) Limit $f(x' \pm \Delta x) = f(x')$ as $\Delta x$ approaches zero. Condition (1) is indeed included in (2) but it is helpful to state it explicitly. Now we know that the ordinary function, $y = c \log x$, is continuous throughout any interval not containing the value, $x = 0$. But is the Fechner function, $S = C \log R$, a continuous function of $R$? No; for consider a stimulus greater than $A(1+r)^n$ and less than $A(1+r)^{n+1}$; compare the corresponding sensation $S$ with that denoted by $n$ (in the above-given table) and then compare it with that denoted by $n+1$; in the first comparison $S$ appears to be $n$; in the second, it appears to be $n+1$. Condition (1) is, you see, not satisfied. Then, of course, condition (2) is not satisfied. Moreover, when we speak of $f(x)$ as a continuous function, we imply that the

variable $x$ can vary continuously. In the Fechner function, however, the variable $R$, which means stimulus divided by $A$, can not thus vary. For weights, pressures, the number of air or of ether vibrations per second, etc., when they vary, vary discontinuously; the changes may indeed be small, but they are finite, and they occur as a leap or bound. It thus appears that to debate whether or not sensation is a continuous function of stimulus is to engage in the rather meaningless exercise of discussing whether a function of a discontinuous variable is a continuous function of it.

*The Fallacy of the Tangent Galvanometer Experiment.*— Imagine a coil or ring of copper wire placed in the plane of the magnetic meridian. At its center is suspended a small magnetic needle so it may turn in a horizontal plane. In its initial position it points towards the magnetic pole. A current of electricity passed through the wire will cause the needle to turn by an amount depending on the strength of the current. Now notice how handsomely the behavior of the needle resembles that of sensation. The electric current plays the rôle of stimulus. Owing to the presence of friction in the turning of the needle, a certain small strength of current is necessary to start the needle. This amount may be likened to the "initial threshold." And there is a kind of "terminal threshold," too, for no finite strength of current can bring the needle to an angle of 90° with the meridian plane. When the needle starts, it leaps to a certain position and there remains, if the current be steady, till the current has been increased by a certain amount, when the needle leaps again, and again remains in its new position if the current be steady, and so on and on. And so we see there is here involved a kind of "difference threshold." The like phenomena are, of course, observed in the behavior of a pair of scales for weighing. It is evident that, owing to friction, the tangent of the needle's angle with the meridian plane, though it is a function of the current's strength, is *not* a continuous function of it. It is common, however, to idealize the situation, by disregarding the frictional effect, and to say that the function *is* continuous, though as a matter of fact it is not. Misled by such considerations certain psychologists have argued fallaciously as follows: The tangent of the needle's angle really is, they say, a continuous function of the current's strength. They admit that it does not appear to be so, but that, they say, is because the fact or the law is " masked " by the presence of friction. And then they contend that the situation is essentially the same in the case of sensation. Sensation, they say, is indeed a continuous function of stimulus, though it does not appear to be such. For, they say, the continuity is " masked " by what they call the " frictional effect " or the opposition of some chemical or mechanical or physical resistance offered by the sensorium to the action of stimulus. The fallacy is obvious. It consists in ignoring some of the facts. The question is not whether the functions under consideration would be continuous if there were no friction or " frictional effects," but whether they are in fact continuous in a world where friction and " frictional effects " persist as part of reality.

*The Number of Possible Sensations Finite.*—How many different sense departments are there? I know of no way to *prove* that the number is finite, but the assumption that the number is finite seems to be *very* probably correct. Let us make the assumption. We know that the number of discernibly different sensations that any department admits is finite. It follows that the total number of different sensations of which a human being is capable is a finite number. I am not aware that this important consideration has been recognized in the literature of psychology. It would not be strange if it has not, for the distinction of finite and infinite, though it is very important scientifically and in some of its connections is awe-inspiring, has not yet gained much intelligent recognition outside the circle of mathematicians. What I wish to signalize here is this: the fact that the ensemble of possible *concepts* is infinite and the fact that the ensemble of possible *sensations* or *percepts* is finite together confront with a difficult and important problem those psychologists who hold that our mental life is based on sensation in the sense that all ideas arise out of sensation. For others the problem does not exist. Plato, for example, held ideas to be eternal, existing before, during and after sensation or perception, and that the world of sensation or perception is only an imperfect and transitory imitation of the eternal world of ideas or concepts.

*Sense Continua.*—We have seen that the results of experiment do not warrant us in saying that sensation is a continuous function of stimulus in the mathematical sense of the term. Nevertheless, it will be convenient to speak of physical or experiential or sensible or sense "continua" and we shall do so but we shall thereby mean merely that to any amount of stimulus between the initial and terminal thresholds of a sense department there corresponds a sensation in the department. We must at no time confound this meaning of continuous with the mathematical meaning of the term. With this understanding we will speak of the "continuum" of pressures or of weights or of sizes or of sounds and so on for the other departments of sense.

*A Remarkable Property of Sense " Continua."*—Owing to the presence of difference thresholds in sense departments, the sense continuum of any department possesses a remarkable property. It is this: If $S_1$, $S_2$, $S_3$ are sensations corresponding to three different amounts, $R_1$, $R_2$, $R_3$, of stimulus, it may happen that no two of the $S$'s are distinguishable from each other and it may happen that $S_1$ is indistinguishable from $S_2$, and that $S_2$ is indistinguishable from $S_3$ but that nevertheless $S_1$ and $S_3$ are distinguishable from each other. So that in a given department we may have three $S$'s (three in the sense of their being produced by three different amounts of stimulus) such that

$$(1)\quad S_1 \equiv S_2, \quad S_2 \equiv S_3, \quad S_1 \equiv S_3;$$

and we may have three such that

$$(2)\quad S_1 \equiv S_2, \quad S_2 \equiv S_3, \quad S_1 \not\equiv S_3;$$

of course there are other possibilities. Thus if the sense " continuum " be that of pressures, then, if the $S$'s correspond respectively to 10, $10\frac{1}{2}$ and 11 grams, they satisfy relations (1); but if they correspond to 10, 11 and 12 grams, they satisfy relations (2). The matter may, of course, be exemplified in other sense " continua." Thus in the sense " continuum " of lengths the sensations produced by (say) three pencils of three different lengths may satisfy relations (1) or (2).

*Another Aspect of the Matter.*—This matter has another striking and important aspect. If *sensation* be allowed to judge, then we should say that a stimulus $Q$ and a stimulus $Q'$ (of the same kind as $Q$) are *equal* if the corresponding sensations, $S$ and $S'$, are *indistinguishable*, and that the $Q$'s are *unequal* when the $S$'s are *distinguishable*. Accordingly, if sensation be the judge, then three quantities, $Q_1$, $Q_2$, $Q_3$, belonging to a given sense " continuum," may be such as to satisfy the relations

$$(3)\ Q_1 = Q_2, \quad Q_2 = Q_3, \quad Q_1 = Q_3,$$

or such as to satisfy the relations

$$(4)\ Q_1 = Q_2, \quad Q_2 = Q_3, \quad Q_1 \neq Q_3.$$

Now, as we noted in the preceding lectures, the series (2) of relations or the series (4) violates the logical law of Contradiction (or of Non-contradiction, as it is sometimes called). What of it? In response we have only to reflect upon the rôle of that law in the realm of rational life. The law is indispensable to logical thinking, to science, to the *very life* of intellect. No doubt a pig or other animal may treat some object $O_1$ as if it were the same as some different object $O_2$, then react to $O_2$ as if it were the same as some third object $O_3$, and then react to $O_1$ as if it were different from $O_3$; and this the animal may do without feeling any sense of shock or surprise. This sense of shock is a *human* experience,—an idiosyncrasy of a rational being,—a mark of man. Might we not employ it as the definition of man? Instead of saying with Plato that man is a featherless biped, would it not be better to define man as the creature that is capable of feeling the shock of contradiction (2) or (4)? Such a definition would have the merit of excluding from the *genus homo* some featherless bipeds even if it did not include any quadrupeds. However this may be, we are literally driven, on pain of intellectual or logical extinction, to say, in the case of the relations (4), that $Q_1 \neq Q_2$ or that

$Q_2 \neq Q_3$ or that both of these inequalities subsist despite the verdict of sensation to the contrary. But what does that *mean?* It means that we are driven to *assume* that there are quantities of magnitude so small as to be insensible, too small to be sensed or felt or perceived. And what does *that* mean? It means—and the answer is fundamentally important—that we are driven to posit or postulate the existence of purely *conceptual* quantities or amounts of magnitude. And we do it.

*Properties of the Conceptual Magnitudes.*—We must not fail to note some of the properties with which our human minds have been *constrained* to *endow* such conceptual magnitudes. If we were to suppose our discriminative sensibility to be by some means so increased or refined as to bring the assumed insensible quantity $Q_1 \backsim Q_2$ (the difference between the sensible quantities $Q_1$ and $Q_2$) well within the domain of the increased sensibility, we have no reason to doubt that the old phenomenon would recur therein; we should, that is, expect to find quantities $q_1$, $q_2$, $q_3$ within the interval $Q_1 \backsim Q_2$ such that

$$q_1 = q_2, \quad q_2 = q_3, \quad q_1 \neq q_3;$$

that is to say, fatal violence to the mentioned law of logic would remain. And so it would did we suppose our sensibility to be again increased so as to render sensible the newly postulated quantity, $q_1 \backsim q_2$; and so on and on. Observe attentively that the indicated process of intercollating ever smaller and smaller insensible quantities is precisely like that by which symbols called rational fractions are interlarded between the integers and then other rational fractions are inserted between the former ones, and so on and on. Observe also that, in order by this means to rescue the law of non-contradiction from the violence of the relations (4), it would be necessary to suppose our discriminative sensibility to be increased till the difference between quantities (of magnitude) corresponding to any two rational fractions, however slight their difference, would be sensibly discernible, that is, sensation would have to be a continuous function of stimulus where continuity signified continuity defined in the domain of rational numbers. The definition would be the same as that above recalled except that the function and its argument would be restricted to *rational* values.

No such endless refining of sensibility has occurred nor is it possible, but we know that in the long course of time the various kinds of sensible magnitude,—the magnitudes revealed by or in sensation,—have gradually been submitted to a *conceptualizing* process that is, in a very important respect, virtually equivalent to the refining process in question. The various kinds of sensible magnitude,—the various sense " continua " presented respectively in various sense departments of sound, color, weight, taste, warmth, duration, hardness, spatial extent, velocity, acceleration, etc.,—have all of them, in the course of the centuries, got replaced in our thinking with corresponding insensible or conceptual magnitudes having the structure of the system of rational numbers. The process has been partly conscious and partly unconscious. This system, you know, is what has been sometimes called the mathematical continuum of first order, as by the late Professor Henri Poincaré, for example, in *Science and Hypothesis*. So we may say that one of the great achievements of the human intellect in its dealing with the magnitudes revealed in sensation has been the substitution, for such sense " continua," of conceptual magnitudes or conceptual "continua" of the type of the mathematical "continuum" of the first order. As already pointed out, the great replacement or substitution has been logically *coerced* by the mentioned contradiction that inheres in the "continua" of sense. The aim, which has been conscious or unconscious, and the issue have been emancipation, intellectual harmony, increase of freedom from logical discord: we know that, if $q_1$, $q_2$, $q_3$ be quantities of a conceptual magnitude or "continuum" of the type in question, and if $q_1 = q_2$, and $q_2 = q_3$, then $q_1 = q_3$ *without exception.*

*The Need of Further Emancipation.*—But the replacement mentioned is not enough. For starting with the kind of conceptual magnitudes in question, we are destined to encounter contradictions or discords of another sort. This fact may be shown as follows: Consider a sensible line. It is for sensibility a thing like a rope or a cable or a chalk mark. In it inheres the old contradiction (4). Now suppose it replaced by a corresponding conceptual magnitude of the type of the first order "continuum." The new thing can be made as thin and as narrow as the difference between any two rational numbers. But this difference can be made to approach as near as we please to zero. Taking the limit, we have a line having only length, the thickness and the width being zero. Such a line is, we say, geometric; it is not sensible, but is purely conceptual. Let us now take two such sensible lines having a (sensible) bulk or piece in common, as indicated in Fig. 31, and let us suppose them submitted to the conceptualizing process above indicated. The result, we say, is two geometric lines and a geometric point common to them, the point being, we say, the limit of the bulk that the lines have in common as sensible lines and also as conceptual lines before passing to the limit. We are thus led to believe that any two conceptual lines that cross one another have a point in common. (In this connection see the book of Poincaré above referred to.) Such, however, is not always the case if the lines are "continua" of only first order, that is, if they have points that correspond only to rational numbers or, as we say, to rational coordinates. It is sufficient to consider the crossing of a circle and a straight line through its center, if either or both of them be supposed to have *no* points *except* such as have *rational* coordinates. For

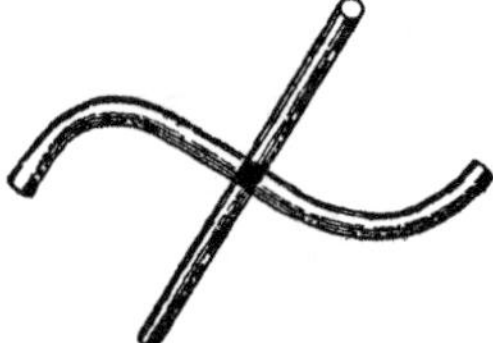

FIG. 31.

let $x^2+y^2=1$ and $y=x$ be respectively such a circle and line. If we undertake to solve for their common points, we immediately find that they have none; for we obtain $x=1:\sqrt{2}$ and $y=1:\sqrt{2}$; but these numbers are not rational and hence do not represent a point common to the crossing line and circle. It might be supposed that such a circle could have only a finite number of points and that their ensemble, if plotted, would not look like a circle. It is easy, however, to prove that the number of the points is infinite and that they constitute a dense set, a set, that is, such that between any two of them there is another one. Obviously, one of the points is the point (0, 1). Suppose $x\neq 0$. Then since $x$ and $y$ are to be rational, $y$ must be of the form, $y=1-rx$, where $r$ is rational. Substitute in $x^2+y^2=1$. We thus get $x^2+(1-rx)^2=1$, whence

$$x=\frac{2r}{1+r^2},$$

and

$$y=\frac{1-r^2}{1+r^2}.$$

Thus it is seen that our circle, which by hypothesis contains no points except such as have rational coordinates, yet contains an infinite number of points—one for each rational value of $r$. It is evident also that, given any one of them, there is another one as near to it as we please.

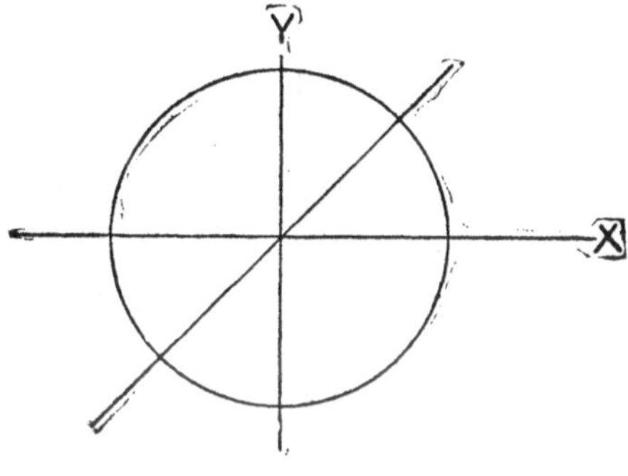


FIG. 32.

Another illustration: if, operating in the conceptual "continuum" of rational lengths, we affirm the existence of a square, we find the "continuum" contains no quantity or magnitude to connect the diagonally opposite corners. For, if the length of the square's side be $s$, the length of the diagonal $d$, if there were a diagonal, would be $d=s\sqrt{2}$; but as $s$ is rational, $s\sqrt{2}$ is not rational and so it is not a length in the "continuum" of operation.

The discovery of the incommensurability of the side and the diagonal of a square, which was a very great discovery, is commonly attributed to Pythagoras. It is said that the discovery made him unhappy. Why? Because he was a union of scientific man and religious mystic, who taught that Number is the essense of all things. But here was something,—the ratio of the side and diagonal of a square,—that no number recognized by him and his sect could express. This fact, should his religious followers find it out, might disturb or destroy their faith. Accordingly, so it is said, Pythagoras decided to keep the fact a secret among a few of his leading disciples, and " one of their number, Hippasos of Metapontion, is even said to have been shipwrecked at sea for impiously disclosing the terrible discovery to their enemies." You should not fail to read what Bertrand Russell has said on this subject in his recently published book entitled *Our Knowledge of the External World as a Field for Scientific Method in Philosophy*. Of philosophic method in science we hear much—the phrase is familiar if the thing itself is not. But what, pray, is meant by " scientific method in philosophy "? That is what Mr. Russell aims to tell us, in outline, in this book. It is in the main a rough, preliminary, semi-popular sketch or adumbration of a method that is, I am told, being employed by Mr. Whitehead in minute detail in his preparation of the fourth volume of the *Principia Mathematica*. I commend the book to your serious attention. Whoever it was that discovered incommensurables, it is certain that their existence was known to the Greeks, as we learn, for example, in the *Metaphysics* of Aristotle. It is a pleasure to be able to say that Aristotle has at length learned to speak English. I refer to W. D. Ross's translation of the

*Metaphysics* and to the English translations of Aristotle's other works published or being published by The Clarendon Press. On page 983a of Ross's translation we find the following:

> For all men begin, as we said, by wondering that the matter is so (as those who have not yet perceived the explanation marvel at automatic marionettes)—whether the object of their wonder be the solstices or the incommensurability of the diagonal of a square with the side; for it seems wonderful to all men that there is a thing which cannot be measured even by the smallest unit. But we must end in the contrary and, according to the proverb, the better state, as is the case in these instances when men learn the cause; for there is nothing which would surprise a geometrician so much as if the diagonal turned out to be commensurable.

That rendering is hard to beat. It is true that to make the thought of the Stagirite quite agree with the modern mathematical conception of the matter it would be necessary to replace the phrase, " by the smallest unit " by some such expression as, by any unit however small, for there is no such thing as the *smallest* unit, just as, zero being excluded, there is no smallest rational number (nor indeed a smallest irrational number). The translation, however, is excellent. Do but compare it with the following translation taken from the *Metaphysics* of Bohn's Classical Library.

> For, indeed—as we have remarked—all men commence their inquiries from wonder whether a thing be so, as in the case of the spontaneous movements of jugglers' figures, to those who have not as yet speculated into their cause; or respecting the solstices, or the incommensurability of the diameter; for it seems to be

> a thing astonishing to all, if any quantity of those that are the smallest is not capable of being measured. But it is necessary to draw our inquiry to a close in a direction the contrary to this, and towards what is better, according to the proverb. As also happens in the case of these, when they succeed in learning those points; for nothing would a geometrician so wonder at, as if the diameter of a square should be commensurable with its side.

No one who had grasped the author's thought even fairly well could have written *that.* And these two specimens are typically representative: they serve to exemplify the comparative merits of the two translations as wholes.

*The Grand Continuum.*—On first encountering the sort of contradiction or discord dealt with a moment ago, our minds are surprised and shocked because we cannot but believe that there must be, or ought to be, a kind of magnitude such that a definite part or amount of it just reaches from a corner to the diagonally opposite corner of any given square and because we cannot but feel that geometric lines or curves ought to be so conceived that, if they cross, they intersect, or have a point (or points) in common. What has the human mind done about it? What has it done to secure release from the kind of discord in question and so to enlarge the sphere of intellectual harmony? What it has done is this: it has assumed or created the kind of magnitude required. This new sort of magnitude is, of course, not sensible; it is conceptual, but it is not of the type of the mathematical continuum of first order, for it is in this type that the difficulties to be overcome have their roots; the structure of the new variety of magnitude is patterned on the structure of the mathematical continuum of *second* order: this latter continuum is the mathematical continuum proper—the

"Grand Continuum" as it was called by Professor Sylvester. It contains such symbols as $\sqrt{2}$, $e$, $\pi$ and countless hosts of others sandwiched between the symbols constituting the first-order continuum, much as the rational fractions of this are sandwiched between the cardinal numbers.

The Grand Continuum has been a subject of profound investigation by mathematicians, especially during the last half-century, and has long been everywhere an important theme of university instruction in what is called the theory of the real variable. This instruction, which has found its way into numerous text-books on function theory, is mainly based, directly or indirectly, upon three classical expositions of the matter. I refer to Dedekind's exposition, which has been translated by Beman and Smith and with another of the author's works has been published under the title *Essays on Number;* to that by Georg Cantor in his creative memoirs on *Mannigfaltigkeitslehre* (or *Mengenlehre*); and to the exposition found in the works of Weierstrass. For our present purpose it will be sufficient to remind ourselves briefly of one way in which the concept of the Grand Continuum may be formed and of its two essential or definitive marks. Consider the following two sequences of rational numbers: first, the sequence of all rational numbers such that each of them is less than 2; second, the sequence of all rational numbers such that the *square* of each of them is less than 2. Each of the sequences approaches, as we say, a *definite somewhat* as a limit. The limit of the first is 2, which is rational; the limit of the second is not rational; we call it *irrational*, denote it by the symbol $\sqrt{2}$, and say that this irrational is given or *defined* by the sequence (or any other sequence) having it for limit. There are infinitely many such sequences,—sequences (of rationals), that is—that approach perfectly definite somewhats as limits which (limits), however, are not rational numbers. The total ensemble of definites thus defined or definable is the system of irrational numbers. And these, taken together with the rational numbers from which they are thus derived or derivable, are commonly said to constitute the system of real numbers. This use of the terms " rational," " irrational " and " real," though dictionally somewhat unfortunate, is historically justified. I need not say that as employed in mathematics, these terms have completely lost whatever metaphysical connotation they may once have had: rational does not signify *reasonable;* nor irrational, *unreasonable;* nor is a " real " number any more real metaphysically than is any other sort of number. Well,—to return from this cautionary digression,—it is the real numbers that constitute the Grand Continuum. Perhaps it were better to say that the system of real numbers is the *basal* instance of the Grand Continuum for other continua essentially like it are derived from it as the model. You are aware that it is common to give the name continuum to any segment of the Grand Continuum, where, by segment, I mean any two real numbers together with all the numbers that lie between them in value. Thus the numbers *zero* and 1, with the real numbers between them, constitute a continuum of second order. The most convenient and vivid example or representation of such a continuum is the ensemble of points constituting a straight line-segment as ordinarily conceived—as conceived, that is, in such a way that by taking an arbitrary point of the line for origin and employing an arbitrary unit of length or distance, a one-to-one correspondence can be set up between the points of the line and the real numbers. Such a continuum of points or of real numbers is a *linear* or *one*-dimensional continuum of second order. The ensemble of *pairs* of real numbers or the ensemble of points of a plane (as ordinarily conceived, in analytic geometry, for example) is a *two*-dimensional continuum of second order; for a three-dimensional one, it suffices to refer to the ensemble of triplets or triads of real numbers or to the ensemble of the points of our familiar geometric space as ordinarily conceived. And it is evident that second-order mathematical continua may have any given dimensionality whatever. For a logically much more refined account of the system of real numbers, you should examine Russell's *Principles* and especially the *Principia*. I am giving here but a sketch of the *usual* account.

*The Definitive Marks of a Grand Continuum.*—What are the characteristic or definitive marks or properties of a mathematical continuum of second order? The answer is: an ensemble of numbers or points or other elements is such a continuum when and only when the ensemble is *compendent* and *perfect*. These are technical terms. What do they mean? Let us answer in terms of points. An ensemble of points is compendent (or *zusammenhängend* as the Germans say or connected as it is common to say in English) if it be such that, given any two points of it, it is possible, by stepping only on points of the ensemble, to pass from one of the given points to the other by a finite number of steps, where each step is equal to or less than some previously assigned distance, however small. An ensemble of points is perfect, provided it be identical with the ensemble of its *limit*- points, where, by a limit-point of an ensemble, is meant a point such that there are points of the ensemble distant from the given point by an amount less than any prescribed distance, however small. It is easy to see that the properties, compendence and perfectness, are independent properties: neither of them implies the other. The ensemble, for example, of the *rational* points of a straight line is compedent, but it is not perfect for many of its limit-points,—that one, for example, whose distance from the origin is $\sqrt{2}$,—are not members of the ensemble. On the other hand, the ensemble composed of the points, 0, 1, 2, 4 and all the *real* points between 0 and 1 and all between 2 and 4, is perfect, but it is plainly not compendent for the passage from, say, point 1 to point 2 cannot be made in the required way. It is clear, however, that the system or ensemble of the real numbers is at once perfect and compendent. The same is true of the segment composed of zero, 1 and the intervening numbers; it is true of the ensemble of points of a straight line or of any segment of it; it is true of the ensemble of the points of a plane or of the ensemble composed of the points inside and of those on the circumference of a circle; and so on and on.

These considerations may, I trust, suffice to give you a general notion of that great mathematical instrument known in modern analysis as the Grand Continuum or as the mathematical continuum of *second* order or simply as *the* mathematical continuum. And let me say that you will miss a main point if you overlook the intimate connection of the matter with the Weber-Fechner law.

I have said that in the long course of time and in the interest of intellectual harmony or freedom the various *sense* " continua " of weights, lengths, sounds, pressures, velocities and so on, have been gradually replaced, in our thinking, *first* by corresponding *conceptual* continua of the type of the mathematical continuum of first order and *second* by conceptual continua of the type of the mathematical continuum of *second* order. Of course, I do not mean to imply that the first replacement was completed before the second began. I mean merely that the order indicated is, roughly speaking, correct. Neither do I intend to imply that the process of substitution has been always a conscious one nor that it has been always accompanied by a realizing sense of its actuating motive. Much of our intellectual life is not attended by consciousness either *that* it is going on or *why* it proceeds in this direction rather than that. That the replacements have been actually made, however, is sufficiently evident in the fact that students of natural science,—physicists, for example, or astronomers or chemists,—habitually and freely employ the real numbers, whether rational or irrational, algebraic or transcendental, to express quantities or amounts of the various kinds of physical magnitude. Are these students aware that they are thus dealing with purely conceptual continua and not with such as are revealed in sense? It must be said that, for the most part, they are not. For the most part these students are not indeed aware that there *are* such continua even in mathematics; much less are they informed regarding the inner structure or constitution of them; they employ them naïvely, as children may handle tools which they have not yet analyzed. This is not said in any spirit of reproach or derogation, for it is only in very recent times that even mathematicians themselves have made the constitution of continua a subject of deliberate investigation, though the matter figured itself vaguely in the background of their thought for more than two thousand years. Even Aristotle in his *Physics* made a stab at the question.

*Some Questions.*—Hereupon certain questions naturally supervene, One of them is this: How can the symbols or elements or terms that constitute the mathematical continuum be effectively employed in studying such a magnitude as pressure, for example, or gravity or velocity? The answer seems to reside in two considerations: one of them, which I have hitherto mentioned in these lectures, is the fact that the various kinds of the magnitude in question are each of them conceived to be composed of, or decomposable into, parts or elements matching in a one-to-one way the elements or terms that constitute the mathematical continuum, and *related* among themselves as the terms of the continuum are related among themselves; the second consideration is the fact that natural science is concerned, not with the *constituents* of a magnitude, but only with the *relations* among them. This second consideration, which so easily escapes attention, is one of those fundamental matters which Professor Poincaré never wearied of insisting upon. See his *Science and Hypothesis*, for example.

Another natural question is this: Does the replacement of the sensible or of the rational continua by continua patterned on the model of the Grand Continuum guarantee us against all difficulties resembling the kind represented by the possibility of two lines crossing without intersecting? The answer is, *no.* This particular kind of difficulty has indeed been overcome by means of the Grand Continuum. But there remain to surprise us other difficulties of a somewhat similar kind. Let us glance at one of them. Consider a sensible curve and a sensible straight line. We can always dispose them so that they will have a *common* part without crossing. Let us now replace them, in thought, by corresponding conceptual magnitudes of the type of the second-order mathematical continuum. Next let the breadth and thickness diminish more and more, keeping always a common part without crossing. It appears that, at the limit, the common part will be a point at which, however, the line and the curve do not cross; that is to say, it appears that the line becomes a *tangent* to the curve at the point. In some such way, it came to be believed that a curve, if continuous at any point, admits a tangent at the point. Nothing could be more natural and the belief was persistent and long-lived. Yet we know to-day that in such matters

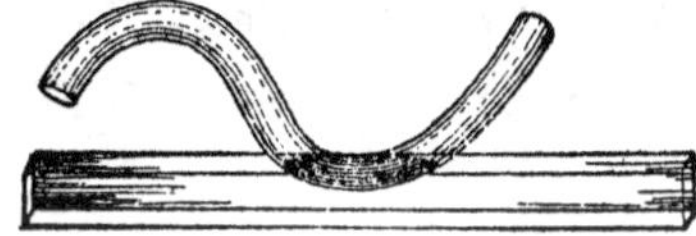

FIG. 33.

our intuition, precious as it is, cannot be implicitly trusted for we know today that the belief in question is false. This fact may be shown by the following classical example. Consider the locus of the equation

$$y = x \sin \frac{1}{x}$$

It is continuous at every point except the point whose abscissa is $x = 0$. At this point $y$ is not defined, since division by zero is meaningless. Hence we may define it as we please. Let us agree that $y$ shall be zero when $x = 0$. This being done, the curve is now continuous at the point, $x = 0$, as well as at all other points. Differentiating, we get

$$\frac{dy}{dx} = \sin \frac{1}{x} - \frac{1}{x} \cos \frac{1}{x}$$

for the slope of the tangent at any point whose abscissa $x$ yields a definite value for $\frac{dy}{dx}$. But at the point, $x=0$ $\frac{dy}{dx}$ has no value; it is indeed meaningless as involving division by zero. Hence the curve, though it is continuous at the point in question, does not admit a tangent at the point.

As you may know, such curves are infinitely numerous. For other examples one may consult a memoir on discontinuous functions by Darboux in the *Annales de l'École Normale superieure*, Vol. IV, 2d series. Some examples of the kind are beautifully discussed in the appendix of W. B. Smith's *Infinitesimal Analysis*. See also W. K. Clifford's remarks on " Crinkly Curves " in Vol. I of his *Elements of Dynamic*. The matter is, of course, dealt with in all up-to-date books of advanced calculus or of the theory of functions of the real variable. By a famous example adduced by Weierstrass, it was shown that a curve may be continuous at every point and yet have no tangent at any point.

Phenomena of the kind above indicated naturally raise another question, which I may state without attempting to discuss it here. The question is: Is it possible to construct continua of higher than second order, continua, that is, whose elements are, so to speak, compacted more closely together than in the case of the Grand Continuum; and what would be the bearing of such a higher continuum upon the sort of phenomena above indicated? The first part of the question is discussed by Paul Du Bois-Reymond. The works of this author are not easy to read. The student may be referred to G. H. Hardy's *Orders of Infinity: the Infinitär-calcül of Paul*

*Du Bois-Reymond*, where the chief ideas of the latter are presented and his works cited.

*Conquest or Transcendence of Sensibility Thresholds.*— By way of emphasis,—for the matter is very important, especially for psychology,—let us state explicitly that the conceptual continua which we have been discussing at so great length are not infected, as are sense "continua," with the presence of thresholds. They are free from initial thresholds, for any such conceptual continuum is composed of parts that continuously decrease in size down to zero. They are free from terminal thresholds because they each of them present a sequence of parts increasing beyond any assigned finite amount, however great. They are free from difference thresholds, for any difference, however small, between portions of a conceptual continuum is conceptually discernible. In a sense "continuum" there are, properly speaking, neither infinitesimals nor infinites, but in a conceptual continuum there are both. In a word we may say that *conception*, or *thought*, is a kind of *infinitely refined sensibility*, for there is no quantity too small for thought to detect and to discriminate from any other. Here this lecture must close. Our topic has been the mathematics of psychology; the discussion, you see, has inevitably led us pretty far into our next topic—the psychology of mathematics—and even into the psychology of science in general.

## LECTURE XIX

### The Psychology of Mathematics

RETARDATION OF MATHEMATICS AND SCIENCE BY BACKWARD PSYCHOLOGY—PSYCHOLOGY OF MATHEMATICS ESSENTIAL TO BEST MATHEMATICAL TEACHING—QUESTIONS FOR PSYCHOLOGISTS—SYMMETRY OF THOUGHT AND ASYMMETRY OF IMAGINATION.

IN the preceding lecture we saw that a little study in the "mathematics of psychology" led us quickly and *naturally* into the "psychology of mathematics" and even into that of science in general. Yet we are now going to discuss the "psychology of mathematics" as if *its* field were well defined and did not run into all other subjects, which, in their turn, penetrate it. Nature does not greatly respect our little academic custom of carving her up and calling the pieces departments of study. Chemistry, physics, mechanics, geometry, metaphysics, psychology, logic, ethics, esthetics, and the rest all penetrate and overflow the walls we surround them with, and mingle their waters in one vast sea. In the great world of Nature,—the subject of all thought,—there are emphases indeed but no fixed divisions corresponding to our pretty *ologies, ographies,* and *ics,* and even the emphases perpetually shift their incidence. And yet there is a sense in which such divisions and walls are not artificial but are natural for they are made by man, and man is a part of nature—the part that studies the whole. In that sense, man-made divisions of nature are natural divisions, made by Nature herself to facilitate the process of self-understanding; but they are not aboriginal and they are not permanent,—they are experimental devices, mere conveniences for the service of a people or an age, and destined to change.

There was a time when what we now call *logic* and what we now call *psychology* were not distinguished—not held apart; today they are; and so, given any scientific or philosophic subject, we habitually speak of its logic *and* its psychology; the two things, though the subject is one, represent different types of interest in it, different emphases or aspects of it. Accordingly, mathematical science presents two fields of interest: the logic of mathematics and the psychology of mathematics. In the former, research has achieved great results; in the latter, but little of solid worth. Why the great disparity? Because the logic of mathematics has interested philosophic-minded logicians who were at the same time mathematicians; while those who have dealt with the psychology of mathematics have not known mathematics well enough to know what it was they were attempting to psychologize or even to know that they did not know—witness, for example, the literature of the "psychology of number." In the introductory lecture I pointed out that modern research in the logic of mathematics has culminated in a really marvelous thesis: rightly understood, mathematics and logic are identical,—the two are one science,—logic (in its tradition sense) being the earlier part of that science, and mathematics (in its traditional sense) being its later part. We could, therefore, say with perfect justice that what we are now to discuss is the psychology of *logic*. For our present purpose it is better, however, to say psychology of *mathematics;* the other title is too unfamiliar.

It is my aim to signalize the importance of the subject, to suggest a few of its problems, and thus possibly to incite some mathematician to acquire sufficient psychological competence, or some psychologist to acquire sufficient mathematical competence, to deal with it effectively. The subject no doubt possesses great interest in itself. "The genesis of mathematical discovery," says Poincaré in *Science and Method,* "is a problem which must inspire the psychologist with the keenest interest." But that is not the main point. The main point is that the general neglect of the subject by competent men throughout the centuries has greatly *retarded* the progress of science; it has, in the first place, retarded the progress of mathematics and it has thus retarded the progress of all the sciences whose prosperity depends upon that of mathematics. That such is the case, a few considerations will make sufficiently evident.

Consider, for example, the birth and development of the concept of hyperspaces, which, as we saw in a previous lecture, is less than a hundred years old. We have noted its great importance for mathematics, for physical science and for philosophy. Why was the advent of this great concept so long delayed? The answer is: because the *psychology* of mathematics was so little understood. For many centuries the concept in question had been knocking at the door but it was not admitted because psychologically ignorant mathematicians and psychologically ignorant philosophers believed that mathematical concepts, if they be not indeed concepts of *sensible* or *perceptible* things, must at all events be concepts of things that we can *imagine*. The evidence supporting my answer is unmistakable and conclusive. Let us consider some of it. I am here greatly indebted to Professor Manning's admirable "Introduction" to his *Geometry of Four Dimensions*. In the following quotations from various authors cited by him, I shall take the liberty, for it will be helpful, to italicise some of their words.

In the *De Caelo* of Aristotle we are told that "The line has magnitude in one way, the plane in two ways, and the solid in three ways, and beyond these there is no other magnitude because the three are *all*." It is plain that the "all" is an "all" for imagination, not for conception. And we are further told that "There is no transfer into another kind, *like* the transfer from length to area and from area to solid." The statement is true for perception and for imagination; but for thought or conception, it is false.

For another instance in point consider the following statement (of the sixth century, A. D.) found in the *Commentaries* of Simplicius: "The admirable Ptolemy in his book *On Distance* well proved that there *are* no more than three dimensions, because of the necessity that distances should be defined, and that the distances defined should be taken along perpendicular lines, and because it is possible to take *only* three lines that are mutually perpendicular, two by which the plane is defined and a third measuring *depth;* so that if there were any other distance after the third it would be entirely *without measure* and *without definition.* Thus Aristotle seemed to conclude from induction that there is no transfer into another magnitude, but Ptolemy proved it." Here it is again evident that the psychology of mathematics had not yet learned to discriminate the conceivable from the imaginable. We now know that in a space of *n* dimensions *n* lines can be mutually perpendicular and we know that such spaces are geometrically just as good as any other. And not only was the progress of geometry thus retarded but that of algebra, too, in as great or greater measure; for not only were lines, surfaces and solids things in imagination's realm but so, too, were *numbers;* it is well known that for the Greek mathematicians and a long series of their successors, numbers were geometric things —one number was a *line* (segment), a product of two numbers was a *rectangle* or a *square,* and that of three a parallelopiped or a *cube.* False psychology thus balked the advancement of *equation* theory, for $x^3$ was indeed a *real* thing,—a geometric cube,—but what, pray, was $x^4$, for example, or $x^5$, and so on? The answer was evident—they were *unreal.* So, in the sixteenth century, we are told by Stifel, reviser of Rudolph's Algebra, that "going *beyond* the cube just *as if* there were *more* than three dimensions" is a thing "against nature." And in the following century John Wallis regards the giving of "ungeometrical" names to the fourth and higher powers of numbers as quite intolerable. They are, he says, a "Monster in Nature, less possible than a Chimaera or Centaure." Why? Because "Length, Breadth and Thickness take up the whole of Space. Nor can Fansie *imagine* how there should be a Fourth Local Dimension beyond these Three." Here note again what the barrier is,—a childishly naïve psychology,—Nature does not transcend the imaginable,—what is merely conceivable is monstrous,—a psychology so rude and so crude that, were it strictly applied, the very possibility of mathematics, rightly understood, would be thereby excluded.

It would not be difficult to produce much more evidence of similar kind. But I will content myself with one further citation—one that is less than a century old and comes from a mathematician of great power. I refer to Möbius, author of *Der barycentrische Calcul* (1827). He saw indeed that, if there were a space of four dimensions, it would be possible to rotate in it a solid figure of ordinary space just as a plane figure can be rotated in the latter space; and he saw that, if such a rotation of solids were possible, we could make two symmetric solids coincide just as we can make two symmetric plane figures coincide by rotation in space immersing the plane. This is perfectly good mathematics, which, however, he rejects because of a false psychology. His statement is this: *"Da aber ein solcher Raum nicht gedacht werden kann, so ist auch die Coincidenz in diesem Falle unmöglich."* He meant that such a space cannot be *imagined*—he could not have meant that it cannot be *conceived,* for he had already conceived it; his blunder was not one in logic; it was a blunder in psychology—the psychology of mathematics; though an able mathematician, he did not know that a conceivable space and a conceivable rotation are perfectly good mathematically, even though they transcend the domain of imagination.

The foregoing facts show clearly that a backward psychology of mathematics not only operated to hamper the progress of algebra, but actually delayed, for more than two thousand years, the advent of the concept of hyperspace and *n*-dimensional geometry.

If you turn to the genesis of non-Euclidean geometry, you find an essentially similar tale. The birth was baffled for over twenty centuries. Baffled by what? By a psychology which recognized no space except our *sensuous* space, which believed that our sensous space is geometrizable, that Euclid's axioms are "self-evident" truths regarding it, and that his *Elements* embodies an exact description of it; by a psychology which, therefore, could not contemplate even the possibility of *non*-Euclidean geometry and which, when such a geometry was at length devised in spite of it, insisted upon trying its claims in the courts of *sensibility* and *perception* and *imagination,* not knowing that, in questions regarding the logical validity of geometric science, those courts are entirely without jurisdiction.

Let me allude briefly to another branch of modern mathematics—*projective* geometry. It was invented, we have seen, in the seventeenth century, lost, forgotten, and re-invented in the nineteenth. Why not before—centuries before? What was in the way? Logic? Not primarily; it was psychology—a false psychology of mathematics. The invention, as you know, required the conception of *infinitely* distant points and the conception of lines and planes such that, if parallel, they *meet* in those points. But how could that happen? How could parallels meet? They could not, it was said,—it was psychologically impossible,—the possibility was denied by *sense,* denied by *perception,* and, most conclusive of all, denied by *imagination.*

I have just now mentioned "infinitely" distant points. We are thus reminded of the modern concept of infinity, —of infinite classes, ensembles, sets, or manifolds,—the subject of Lecture XV. We have seen that this great concept, though it is classic today, was born but yesterday. Why not a thousand or two thousand years ago? Well, it *was* born or well-nigh born, as we have seen, to the genius of Epicurus and had the fortune to inspire the great poem of Lucretius. With this exception, it had no career in science and none in philosophy—it was sterile. Why? The same old trouble—a shallow psychology. For, as you know, an infinite class must have a part containing as many things as the whole class contains. But who ever *saw* such a class or ever *imagined* one? "Nonsense!" exclaimed psychology, and the great conception,—so important for science, for philosophy and for rational theology,—slumbered for twenty centuries.

Passing to another field, we find that the development or generalization of the *number* concept was greatly hampered by the same cause. The descriptive terms,—"surd" (which means *absurd*), "irrational," "imaginary," and "impossible,"—which were applied to large classes of numbers that had been literally forced upon the attention of mathematicians by familiar operations, sufficiently tell the tale. Mathematically those numbers, as we now know, were quite as genuine, quite as legitimate, as the ordinary integers and fractions. Why, then, were they called "surd," "irrational," "imaginary," and "impossible"? Because they encountered a psychology that did not understand the nature,—the mental nature,—of mathematical generalization: a psychology which held that the new "numbers," in order to be legitimate, must conform to the familiar laws of the old ones and must, moreover, like the old ones, admit of interpretation or application in the so-called "actual" world of sense-perception.

The history of many another mathematical development bears similar witness. But we need not pursue the matter further. The evidence now before us is sufficient. It is perfectly clear that in course of the centuries the progress of mathematics has been much retarded, sometimes arrested for long periods or diverted from its natural course, by a psychology which, in things mathematical, often did not know a knee from an elbow.

Is mathematics retarded by that cause today? I believe that mathematical *research* is not much thus retarded—at all events not directly. No doubt the number of mathematicians who are also expert psychologists is very small. But research mathematicians usually, though not always, understand the psychology of their own science well enough to recognize a mathematical idea as being such, wherever and whenever it occurs. If it be a new one and be found to be interpretable in the world of perception or in the world of imagination, they are thereby rejoiced, naturally so; but if it be not thus interpretable, as it may not be, they are not so psychologically unenlightened as to refuse it hospitality on that account. The history of their subject has taught them better.

But mathematical research and the dissemination of mathematical knowledge are very different things. In respect of the latter, I have no doubt that, if teachers of mathematics were better trained in psychology and especially in the psychology of mathematics, their teaching would be far more effective; for questions of logic would then be seen and set in clearer light, less frequently confused with psychological considerations, while the latter, presented as such, would often contribute to the instruction a light of their own. Will you allow me a word of personal experience? I count it a great good personal fortune that as a young man I received mathematical instruction from one in whose teaching the logic, the philosophy, the psychology, and the poetry of the subject mingled together and fortified each other like the parts of an orchestra. I refer to Professor William Benjamin Smith, now of world-wide fame as a Biblical scholar and critic. I am not going to enlarge here upon this important matter of making the psychology of mathematics effective in mathematical instruction, but will merely refer you, for some relevant suggestions, to the earlier lectures where distinctions of logical and psychological were repeatedly indicated and where, especially in Lecture VII, in connection with the psychological discrimination of logically identical doctrines, was introduced the important notion of "excessive meaning."

We have been talking about the neglect and backwardness of the psychology of mathematics. Thus far we have referred mainly to the neglect of it by mathematicians. What are we to say of its neglect by professional psychologists? I have no desire to be fault-finding, querulous or unjust. I am well aware that psychologists have many things to occupy their attention—that their field is vast, diversified and complicate. I know that, like other scientific folk, they are obliged to select. I know that for an outsider to attempt to dictate or prescribe their choice would be presumptuous. At the risk, however, of seeming impertinent,—which usually means a little too pertinent,—I venture, as an interested layman, to suggest that, in neglecting the psychology of mathematics, professional psychologists not only neglect an obligation to mathematics and natural science but also neglect an exceedingly interesting subdivision of their own proper field. For their field is Mind,—psychology, we are told, is the study of mind, the study of mental phenomena,—and I believe we may assume that, where there is mathematics, there is some manifestation of mind, —that mathematics, regarded as an enterprise, is an enterprise of mind,—that, regarded as a body of achievements, it is a body of mental achievements,—that, regarded as a mode of life, it is a mode of mental life, —that, in a word, mathematical phenomena *represent* mental phenomena and are unsurpassed as means in the study of mind. I do not mean that *all* kinds of mental phenomena are thus represented. Lust, for example, is not, nor fear, nor anger, nor hate, nor malice, nor envy, nor many another such amiable propensity of unregenerate souls—of course I am speaking here of mathematics and not of mathematicians, who have many interesting qualities that their science has not. But perception — discrimination — imagination — fantasie — conception — judgment — analysis — synthesis — reasoning — generalization — the energy of will — the restraint of passion—the sensibility and daring of genius—the sense for order, for symmetry, for harmony, for intellectual beauty, for cogency and clarity of thought,—where outside of mathematics do *such* mental phenomena, which it is the psychologist's profession to examine, show themselves in so clear a light? It is indeed obvious that the whole literature of mathematics may be read and interpreted as a commentary upon the nature of the human mind. Select, for example, a well-wrought demonstration and examine it. What can you say of it? You can say this: A normal human mind is *such that,* if it begin with such-and-such principles or premises and with such-and-such ideas and if it combine them in such-and-such ways, moving from step to step in such-and-such an order, it will find that it has thus passed from darkness to light,—from doubt to conviction. Obviously such a proposition is not mathematical; it is psychological—it states a fact respecting the nature of a normal human mind. Such interpretations of mathematical literature are psychologically very illuminating; the possibility of making them is so evident, once it is pointed out, that I should have refrained from mentioning it except for the fact of its being commonly overlooked and neglected.

For another example, consider the phenomenon of generalization,—the process by which the human mind from time to time enlarges the empire of its rational activity. What *is* generalization as a process of mind, as a mental event? What are the mental phenomena involved? How? In what relations? I am not going to attempt to answer here. I wish merely to propose the problem to students of psychology. Generalization occurs in *all* fields of thought but in mathematics it may be seen in its nakedness. There, then, is the best place to study it as a phenomenon of mind. Take, for example, the striking succession of generalizations by which the domain of the number concept, which once contained nothing but our familiar integers, has been gradually extended to embrace positives and negatives, rationals and irrationals, reals and imaginaries, cardinals and ordinals, including the transfinite numbers of Georg Cantor; or take the no less striking series of generalizations by which the conception of geometry has been enlarged. As specimens of generalization, those alluded to are probably the best to be found in the history of thought. I venture to commend them as such to students of mind. Some of you are psychologists. If you will study the great process of generalization by help of the specimens mentioned, then the rest of us will go to you confidently,—as laymen to experts,—for enlightenment. For there are questions to be asked. Generalization seems to be sometimes very simple and sometimes very complicate. We should like to know what mental phenomena,—what sorts of mental activity,—are involved in it. What, if any, is the rôle of imagination in it, and that of conception and that of reasoning? Does generalization transcend the realm of imagination? What is the office of logic therein? Is generalization the *end* of a series of operations or is it the *beginning* of a new series? Is it a *conclusion forced* by *reason* or does it involve a *creative* act of *will* stimulated by motives but *not coerced* by them? What *are* the actuating motives of the process? Are all generalizations essentially alike? If not, what are the kinds, and how do they differ? How do the phenomena of scientific generalization compare with those of idealization in other fields? Such questions are neither primarily mathematical nor primarily metaphysical; they are psychological questions, which it is your proper function as students of mind to investigate for your own enlightenment and for that of others. Let me cite again the statement of that great man, Henri Poincaré: "The genesis of mathematical discovery is a problem which must inspire the psychologist with the keenest interest."

The things I have been saying are submitted as suggestions only; being a layman's suggestions, they are probably very inferior to the best that could be made. I am tempted, nevertheless, to add yet another one. It is that a good way,—perhaps the best way,—for psychologists to advance their own subject would be to coöperate with philosophic mathematicians and philosophic physicists in their efforts to solve the great problem mentioned near the close of Lecture X,—the problem, I mean, of discovering the relations between the data of *sense* and the *conceptual* objects of science,—the problem, in other words, of ascertaining whether, how, and to what extent such conceptual and hypothetical objects (points, instants, space, time, atoms, electrons, ether, etc.) can be replaced by objects actually *constructed* out of sense-given data and having the properties demanded by science. If such constructions can be made, science will be able to dispense with many hypotheses or many "as ifs." That problem, it is evident, is a truly great one.

In this lecture (as also in preceding ones) I have repeatedly emphasized the importance of a certain psychological distinction which I have called the distinction between imagination and conception. It is today well recognized by all mathematicians. They are accustomed to designating it,—not quite happily, I believe,—as the distinction between "intuition" and "analysis." It is the distinction between the power of the mind to *picture* and its power to *think*. We have seen that failure to make it has often retarded the progress of mathematics directly and that of kindred sciences indirectly. It is absolutely essential to the philosophy of science; without it the history of thought cannot be understood. I am here reminding you of the matter because the considerations with which I intend to close will incidentally shed new light upon it.

I wish to call your attention to certain contrasting psychological phenomena that seem not to have found recognition in the literature of psychology. I shall present them without attempting to explain them. What I wish to point out is that, in relation to space, conception or thought is symmetric in its representations and that imagination is not. My theme is: *The symmetry of thought and the asymmetry of imagination.*[1]

Consider the simple algebraic expression

$$u_1x_1+u_2x_2+\ldots+u_nx_n+1$$

The $u$'s precede the $x$'s, but that is here of no importance, for, owing to the commutative law of ordinary multiplication, $ux$ is equivalent to $xu$; if we replace the $u$'s by the corresponding $x$'s and the latter by the former, the expression remains algebraically unaltered. On that account we say that the expression is *symmetric* with respect to the $u$'s and the $x$'s. Such interchange of the $u$'s and the $x$'s may be likened to the interchange of two opposite halves of a perfectly symmetric tree—the figure of the tree as a whole remains unchanged. It will be convenient to denote the expression by the symbol $E(u, x)$—the symbol $E(x, u)$ would, of course, do just as well but let us use the former.

Now consider the equation

$$(1)\quad E(u, x)=0$$

It is, like the expression, symmetric in the sense defined. We may interpret the equation geometrically. To do so, let us view the $x$'s as coordinates of a point in a point-space, $S_n$, of $n$ dimensions. If we suppose the $u$'s to have definite values the equation (1) imposes one condition on the mobility of the point $(x_1, x_2, \ldots, x_n)$; and so the equation represents,—has for its locus, as we say,—a space $S_{n-1}$ of points. If we give the $u$'s another set of values, thus obtaining a new equation of form (1),

[1] A paper on this subject which I presented at the Princeton meeting of the American Philosophical Association (1910) was published in the *Journal of Philosophy, Psychology and Scientific Method,* June 8, 1911.

the new equation will represent another space of $n-1$ dimensions. It is thus plain that the $u$'s serve for coordinates for a variable $S_{n-1}$ in $S_n$ just as the $x$'s serve for coordinates of a point in $S_n$. Since the $u$'s may take as many different systems of values as the $x$'s may take, you see that the space $S_n$, in which we are operating, contains as many $S_{n-1}$'s as it contains points.

We have just now seen that, if the $u$'s be held fixed in value and the $x$'s be allowed to vary subject to condition (1), this equation represents some definite $S_{n-1}$ as the ensemble or *locus* of the points contained in it. Now note very carefully the reciprocal or dual, as it is called, of the fact just stated. The dual is that, if the $x$'s be held fixed in value (thus giving us a fixed point, say, $P$) and the $u$'s be allowed to vary subject to condition (1), the equation represents $P$ as the ensemble or *envelope* (as it is called) of all the $S_{n-1}$'s containing it.

Naturally the two interpretations—one for the $u$'s fixed and the $x$'s variable, the other for the $x$'s fixed and the $u$'s variable—of one and the same equation

$$E(u, x) = 0 = E(x, u)$$

may be significantly described as *symmetric* interpretations. Indeed, as you readily see, if in the conceptual space $S_n$ (of operation) we interchange the notion of point (as an *envelope* of $S_{n-1}$'s) and the notion of $S_{n-1}$ (as a *locus* of points), $S_n$ will as a whole remain, like the initial expression, like our equation, like our symmetric tree, absolutely unchanged. Under the mentioned operation, $S_n$ is an invariant. In the same way, *systems* of equations like the foregoing one admit of symmetric interpretations. But I shall not deal with such systems where $n$ is general. It will be easier for you, and for my purpose it will be sufficient, to begin with the simple case where $n = 2$ and to observe what happens when $n$ is taken larger and larger.

It is essential to note the fact that the above-given symmetric interpretations are conceptual—interpretations by, in and for pure thought. It is equally essential to note that our "spatial" imagination or intuition or picturing power attempts to imitate them,—attempts to make in *its* way parallel interpretations,—interpretations, that is, which correspond to or match in detailed one-to-one fashion the thought interpretations. In other words, imagination endeavors to find in its own domain images, pictures or objects to match the conceptual objects—points, $S_{n-1}$'s, lines, $S_{n-2}$'s, and so on—which figure in the interpretations by thought. We are going to see that this enterprise of imagination succeeds fairly well if $n$ be small, that its prosperity decreases as $n$ increases, and that its failure is well-nigh complete when $n$ is taken very large. At the same time, we shall see that, in the case of interpretations by thought, symmetry never fails in even the least degree, no matter how high the dimensionality of the space in which we are operating.

Let us for convenience denote any two reciprocal thought-interpretations by the symbols $T(u)$ and $T(x)$, the former when the $u$'s are fixed and the $x$'s are variable, and the latter when the $x$'s are fixed and the $u$'s are variable; and let $I(u)$ and $I(x)$ denote the corresponding pair of interpretations essayed by imagination.

Consider first the simple case where $n = 2$; $S_2$, the space of operation, is a plane; equation (1) now is: (1) $u_1x_1 + u_2x_2 + 1 = 0$. What are $T(u)$ and $T(x)$? The former is a conceptual *range* of points; the latter, a conceptual *pencil* of lines. What are $I(u)$ and $I(x)$?

The former is the conceptual range's so-called image or mental picture, commonly represented to the physical eye by a row or series of dots; the latter is the conceptual pencil's so-called image or mental picture, commonly represented to the physical eye by a set of physical lines (or dot rows) having a physical point (or dot) in common. You are familiar with the sensuous figures.

Still keeping $n=2$, let us take a *pair* of equations like (1), writing them, for short, (2) $E'(u, x)=0$, (3) $E''(u, x)=0$; and consider (4) $E'+\lambda E''=0$ where $\lambda$ is a parameter. Denote the ranges represented by (2) and (3) by $R'$ and $R''$, and denote the pencils represented by them by $P'$ and $P''$. What are $T(u)$ and $T(x)$ of (4)? The former is a conceptual *variable range* of the pencil (of ranges) determined by $R'$ and $R''$; the latter is a conceptual *variable point* (or pencil) determined by $P'$ and $P''$; it is plain that $I(u)$ and $I(x)$ are respectively the so-called images of the variable range and variable point (or pencil) just mentioned.

Advancing to the case where $n=3$, we have for field of operation the space $S_3$ of ordinary geometry. Consider (5) $u_1x_1+u_2x_2+u_3x_3+1=0$. $T(u)$ of (5) is obviously a conceptual *plane* of points, while $T(x)$ is a conceptual *bundle* of planes (a *point*, that is, enveloped by the planes containing it); and, of course, $I(u)$ and $I(x)$ are respectively the " images " of the conceptual plane and bundle (or point as the bundle's vertex or carrier).

Let us now take a pair of equations like (5), namely, (6) $E'=0$, (7) $E''=0$; and consider (8) $E'+\lambda E''=0$. Let $\pi'$ and $\pi''$ be the planes, and $B'$ and $B''$ the bundles (or points), represented by (6) and (7). $T(u)$ of (8) is a conceptual variable plane of the axial pencil (of planes) determined by $\pi'$ and $\pi''$; $T(x)$ is a conceptual variable point (bundle vertex) of the line (axis of plane-pencil) determined by $B'$ and $B''$; while $I(u)$ and $I(x)$ are imagination's correspondents of the foregoing concepts.

Finally, let us join with (6) and (7) a third equation (9) $E'''=0$, independent of them and representing a plane $\pi'''$ or a point (plane bundle) $B'''$. Consider the equation (10) $E'+\lambda E''+\mu E'''=0$. What are its $T(u)$ and $T(x)$? The former is a conceptual variable plane of the point (or bundle) determined by $\pi'$, $\pi''$ and $\pi'''$; the latter is a conceptual variable point (bundle vertex) of the plane determined by $B'$, $B''$ and $B'''$; while $I(u)$ and $I(x)$ are the imitating " images " of the same.

Let us now pass to $n=4$; our field of operation is $S_4$, a four-dimensional space of points. Consider (11) $u_1x_1+u_2x_2+u_3x_3+u_4x_4+1=0$. Its $T(u)$ is a conceptual lineoid (an $S_3$) of points, and its $T(x)$ is a conceptual hypersheaf of lineoids (a point enveloped by the $\infty^3$ lineoids containing it). Now scrutinize carefully the results, $I(u)$ and $I(x)$, of imagination's effort to imitate or represent pictorially the concepts $T(u)$ and $T(x)$. You observe at once the following facts: (a) both $I(u)$ and $I(x)$ are inferior to their analogues for $n=3$ or $2$; (b) the defect of $I(u)$ differs in kind from that of $I(x)$. Indeed the two kinds of defect are, in a sense, reciprocal; for $I(u)$, in trying to match $T(u)$, though it succeeds in imaging points and point configurations *interior* to the lineoid or locus, presents no image of the lineoid itself or the locus as a whole; while, on the other hand, $I(x)$, in trying to match $T(x)$, presents an image corresponding to the point or envelope but no image to match the enveloping lineoids. The contrast may be vividly seen as follows: Note that, in the one case, the lineoid is the bond or *lien* of the elements,—points,—of which it is the locus, and that, in the other case, the point is the bond or *lien* of the elements,—lineoids,—of which it is the envelope. And now the fact to be noticed is this: $I(u)$ images elements, but not their bond; $I(x)$ images the bond, but not the elements.

It is plain, too, that $I(u)$ is more satisfactory than $I(x)$. This fact becomes obtrusively evident if we geometrize $T(u)$ and $T(x)$ themselves. The two geometries,—which we must remember are conceptual,—match each other in fact-to-fact fashion perfectly; with respect to each other they are *perfectly symmetric*. In the two geometries a point of $T(u)$ corresponds to a lineoid of $T(x)$, and a line-segment joining two points of $T(u)$ corresponds to the angle of two lineoids of $T(x)$. Now it is evident that the image of a segment is very superior to any image we can form for the angle between two intersecting lineoids. I need not give further examples, which are endless in number and tell the same tale. If you desire to do so, you can pursue the matter in $S_4$ and in higher and higher spaces.

The conclusion is that, in relation to space, conception or thought is perfectly symmetric and that imagination or intuition is asymmetric. As $n$ increases, thought continues to look about in spaces of ever-ascending dimensionality like a binocular being with no impairment of its twofold vision; its light is spread abroad equally everywhere; whilst imagination's eyes not only fail more and more as $n$ mounts higher, but they fail in unequal measure. To change the figure, thought enters and moves about freely in the hyperspaces like an eagle with both wings equally outspread and always adequate for any zone however vast or high, but the movement of imagination there is like the flight of a bird of feeble and failing wings, unable to rise and soar.

## LECTURE XX

### Korzybski's Concept of Man [1]

WHAT TIME-BINDING MEANS—DIMENSIONALITY AND THE MATHEMATICAL THEORY OF LOGICAL TYPES—THE NATURAL LAW OF CIVILIZATION AS AN INCREASING EXPONENTIAL FUNCTION OF TIME—HUMAN ETHICS AS TIME-BINDING ETHICS, NOT THE SPACE-BINDING ETHICS OF ANIMALS.

A FEW years ago our lives were lapt round with a civilization so rich and comfortable in manifold ways, so omnipresent, so interwoven with our whole environment, that we did not reflect upon it but habitually took it all for granted as we take for granted the great gifts of Nature,—land and sea, light and sky and the common air. We were hardly aware of the fact that Civilization is literally a product of human labor and *time*; we had not thought deeply upon the principle of its genesis nor seriously sought to discover the laws of its growth; we had not been schooled to reflect that we who were enjoying it had neither produced it nor earned its goods; we had not been educated to perceive that we have it almost solely as a bounty from the time and toil of by-gone generations; we had not been disciplined to feel the mighty obligation which the great inheritance imposes

[1] Part of this lecture is found in my Phi Beta Kappa address on *The Nature of Man* (Science, Sept. 9, 1921) and some of it in an article by me in The Pacific Review, Dec., 1921.

upon us as at once the posterity of the dead and the ancestry of the yet unborn. We had been born in the midst of a great civilization, and, in accord with our breeding, we lived in it and upon it like butterflies in a garden of flowers, not to say as "maggots in a cheese."

Since then a change has come. The World War awoke us. The awakening was rude but it was effectual. Everywhere men and women are now thinking as never before, and they are thinking about realities for they know that there is no other way to cope with the great problems of a troubled world. They have learned, too, that, of all the realities with which we humans have to deal, the supreme reality is Man; and so the questions that men and women are everywhere asking are questions regarding Man, for they are questions of ethics, of social institutions, of education, of economics, of philosophy, of industrial methods, of politics and government. The questions have led to some curious results,—to doctrines that alarm, to proposals that startle,—and we are wont to call them radical, revolutionary, red. Is it true that our thinking has been too radical? How the question would have made Plato smile—Plato who had seen his venerated teacher condemned to death for radical criticism. No, the trouble is that, in the proper sense of that much abused term, our thought has not been radical enough. Our questionings have been eager and wide-ranging but our thought has been shallow. It has been passionate and it has been daring but it has not been deep. For, if it had been deep, we could not have failed, as we have failed, to ask ourselves the fundamental question: What is that in virtue of which human beings are human? What is the distinctive place of our human kind in the hierarchy of the world's life? What *is* Man?

I have called the question "fundamental"—it *is* fundamental—the importance of a right answer is sovereign—for it is obvious, once the fact is pointed out, that the character of human history, the character of human conduct, and the character of all our human institutions depend both upon what man *is* and in equal or greater measure upon what we humans *think* man is.

Why, then, have we not asked the question? The reason doubtless is that we have consciously or unconsciously taken it for granted that we knew the answer. For why enquire when we are sure we know?

But *have* we known? Is our assumption of knowledge in this case just? Have we really known, do we know now, what is in fact the idiosyncrasy of the human class of life? Do we know critically what we, as representatives of man, really are? Here it is essential to distinguish; we are speaking of knowledge; there is a kind of knowledge that is instinctive,—instinctive knowledge,—immediate inner knowledge by instinct,—the kind of knowledge we mean when we say that we know how to move our arms or that a fish knows how to swim or that a bird knows how to fly. I do not doubt that, in this sense of knowing, we do know what human beings are; it is the kind of knowledge that a fish has of what fishes are or that a bird has of what birds are. But there is another kind of knowledge,—scientific knowledge,—knowledge of objects by analyzing them,—objective knowledge by concepts,—conceptual knowledge of objects; it is the kind of knowledge we mean when we say that we know or do not know what a plant is or what a number is. Now, we do not suppose fish to have this sort of knowledge of fish; we do not suppose a bird can have a just conception,—nor, properly speaking, any conception,—of what a bird is. We are speaking of concepts, and our question, you see, is this: Have we humans a just Concept of Man? If we have, it is reasonable to suppose that we inherited it, for so important a thing, had it originated in our time, would have made itself heard of as a grave discovery. So I say that, if we have a just concept of man, it must have come down to us entangled in the mesh of our inherited opinions and must have been taken in, as such opinions are usually taken in, from the common air, by a kind of "cerebral suction."

If we discover that we have never had a just concept of man, the fact should not greatly astonish us, for the difficulty is unique; man, you see, is to be both the knower and the object known; the difficulty is that of a knower having to objectify itself and having then to form a just concept of what the object is.

In saying that in the thought of our time the great question has not been asked, I have now to make one important exception and, so far as I know, only one.[2] I refer to Count Alfred Korzybski, the Polish engineer. In his momentous book (*The Manhood of Humanity: The Science and Art of Human Engineering*[3]), he has both propounded the question and submitted an answer that is worthy of the serious attention of every serious student, whatever his field of study. It is the aim of this lecture to present the answer and to examine it by help of the Theory of Logical Types, the Theory of Classes,

[2] Since writing the foregoing I have observed a learned discussion of the question by Professor Wm. E. Ritter in an article, *Science and Organized Civilization,* in the Scientific Monthly, Aug., 1917. Professor Ritter once more defines man as a *kind* of *animal* but the *distinctive* marks of the kind, as given by him, are so grave as to make one wonder why he did not altogether drop the "animal" element from the definition.

[3] E. P. Dutton & Company.

and the author's closely allied notion of "Dimensions."

Let me say at the outset that one who would read the book understandingly must come to it prepared to grapple with a central *concept,* a concept whose rôle among the other ideas in the work is like that of the sun in the solar system. It happens, therefore, that readers of the book, or of any other book built about a central concept, fall into three mutually exclusive classes:

(I) The class of those who *miss* the central concept —(I have known a learned historian to miss it)—not through any fault of their own,—they are often indeed well meaning and amiable people,—but simply because they are not qualified for conceptual thinking save that of the commonest type.

(II) The class of those who *seem* to grasp the central concept and then straightway show by their manner of talk that they have not really grasped it but have at most got hold of some of its words. Intellectually such readers are like the familiar type of undergraduate who "flunks" his mathematical examinations but may possibly "pull through" in a second attempt and so is permitted, after further study, to try again.

(III) The class of those who firmly seize the central concept and who by meditating upon it see more and more clearly the tremendous reach of its implications. If it were not for this class, there would be no science in the world nor genuine philosophy. But the other two classes are not aware of the fact for they are merely "verbalists." In respect of such folk, the "Behaviorist" school of psychology is right for in the psychology of classes (I) and (II) there is no need for a chapter on "Thought Processes"—it is sufficient to have one on "The Language Habit."

What is that central concept? What is Korzybski's Concept of Man? I wish to present it as clearly as I can. It is a concept defining man in terms of Time. "Humanity," says the author, "is the *time-binding* class of life." What do the words mean? What is meant by time-binding or the binding of time? The meaning, which is indeed momentous, will be clearer to us if we prepare for it by a little preliminary reflection.

Long ages ago there appeared upon this planet—no matter how—the first specimens of our human kind. What was their condition? It requires some meditation and some exercise of imagination to realize keenly what it must have been. Of knowledge, in the sense in which we humans now use the term, they had none—no science, no philosophy, no art, no religion; they did not know what they were nor where they were; they knew nothing of the past, for they had no history, not even tradition; they could not foretell the future, for they had no knowledge of natural law; they had no capital,—no material or spiritual wealth,—no inheritance, that is, from the time and toil of by-gone generations; they were without tools, without precedents, without guiding maxims, without speech, without any light of human experience; their ignorance, as we understand the term, was almost absolute. And yet, compared with the beasts, they were miracles of genius, for they contrived to do the most wonderful of all things that have happened on our globe—they *initiated,* I mean, the creative movement which their remote descendants call Civilization.

Why? What is the secret? Have you ever tried to find it? The secret is that those rude animal-resembling, animal-hunting, animal-hunted ancestors of ours were a *new kind* of creature in the world—a new kind because endowed with a strange new gift—a strange new capacity or power—a strange new *energy,* let us call it. And it is in the world today. What is it? We know it partly by its effects and partly by its stirring within us for as human beings, as representatives of Man, we all of us have it in some measure. It is the energy that invents—that produces instruments, ideas, institutions and doctrines; it is, moreover, the energy that, having invented, criticizes, then invents again and *better,* thus advancing in excellence from creation to creation endlessly. Be good enough to reflect and to reflect again upon the significance of those simple words: invents; having invented, criticizes; invents again and better; thus advancing, by creative activity, from stage to stage of excellence without end. Their sound is familiar; but what of their ultimate sense? We ought indeed to pause here, withdraw to the solitude of some cloister and there in the silence meditate upon their meaning; for they do not describe the life of beasts; they characterize Man.

We are speaking of a peculiar kind of energy—the energy that *civilizes*—that strange familiar energy that makes possible and makes actual the great creative movement which we call human *Progress,* of which we talk much and think but little. Let us scrutinize it more closely; let us, if we can, lay bare its characteristic relation to Time for its relation to Time is the relation of Time to the distinctive life of Man.

Compare some representative of the animal world, a bee, let us say, or a beaver, with a correspondingly representative man. Consider their achievements and the ways thereof. The beaver makes a dam; the man, a bridge or some discovery,—analytical geometry, for example, or the art of printing, or the Keplerian laws of planetary motion, or the atomic constitution of matter. The two achievements,—that of the beaver and that of the man,—are each of them a product of three factors: time, toil, and raw material, where the last signifies, in the case of purely scientific achievement, the data of sense, in which science has its roots. Both achievements *endure,* it may be for a short while only,—as in the case of the dam or the bridge,—or one of them may endure endlessly,—as in that of a scientific discovery. What happens in the next generation? The new beaver begins where its predecessor began and ends where it ended—it makes a dam but the dam is like the old one. Yet the old dam is there for the new beaver to behold, to contemplate, and to improve upon. But the presence of the old dam wakes in the beaver's "mind" no inventive impulse, no creative stirring, and so there is no improvement, no progress. Why not? The answer is obvious: the beaver "mind" is *such that* its power to achieve is *not reinforced* by the presence of past achievement. The new beaver's time is indeed overlapped, in part or wholly, by the time of its predecessor for the latter time is present as an essential factor of the old dam, but that old-time factor, though present, *produces nothing*—it is as dead capital, bearing no interest. Such is the relation of the beaver "mind,"—of the *animal* mind,—to time.

Now, what of the new *man?* What does *he do?* What he does depends, of course, upon his predecessor's achievement; if this was a bridge, he makes a better bridge or invents a ship; if it was the discovery of analytical geometry, he enlarges its scope or invents the calculus; if it was the art of printing, he invents a printing press; if it was the discovery of the laws of planetary motion, he finds the law of gravitation; if it was the discovery of the atomic constitution of matter, he discovers the electronic constitution of atoms. Such is the familiar record—*improvement* of old things, *invention* of new ones—*Progress*. Why? Again the answer is obvious: the mind of man, unlike animal "mind," is *such that* its power to achieve *is reinforced* by past achievement. As in the case of the beaver, so in that of man, the successor's time is overlapped by the predecessor's time for the latter time continues its presence as an essential factor in the old achievement, which endures; but,—and this the point,—in man's case, unlike the beaver's, the old-*time* factor is not merely present, it *works;* it is not as dead capital, bearing no interest, and ultimately perishing—it is living capital bearing interest not only but interest perpetually compounded at an ever-increasing rate. And the interest is growing wealth,—material and spiritual wealth,—not merely physical conveniences but instruments of power, understanding, intelligence, knowledge and skill, beautiful arts, science, philosophy, wisdom, freedom—in a word, Civilization.

That great process,—involving some subtle alchemy that we do not understand,—by which the *time*-factor, embodied in things accomplished, perpetually reinforces more and more the achieving potency of the human mind,—the process by which mysterious Time thus continually and increasingly augments the civilizing energy of the world,—the process by which the evolution of civilization involves the storing up or involution of time,—it is that mighty process which Korzybski happily designates by the term, Time-binding. The term will recur frequently in our discussion, and so I recommend that you dwell upon its meaning as given until you have seized it firmly. It is because time-binding power is not only peculiar to man but is, among man's distinctive marks, beyond all comparison the most significant one—it is because of that two-fold consideration that the author *defines* humanity to be "the time-binding class of life."

Such, then, is Korzybski's answer to the most important of all questions: what is Man? Do not lose sight of the fact that we have here a *concept* and that it defines man in terms of a certain relation, subtle indeed but undoubtedly characteristic, that man has to time. By saying that the relation is "characteristic" of man I mean that, among known classes of life, man and only man has it. Animals have it not or, if they have it, if they have time-binding capacity, they have it in a degree so small that it may be neglected as mathematicians neglect infinitesimals of higher order.

The answer in question is not one to which the world has been or is now accustomed. If you apply for an answer to the thought of the bygone centuries or to the regnant philosophies of our own time, what answer will you get? It will be one or the other of two kinds: it will be a *zoological* answer—man is an animal a kind or species of animal, the *bête humaine;* or it will be a *mythological* answer—man is a mysterious compound or *union* of animal (a natural thing) with something "supernatural." Such are the rival conceptions now current throughout the world. They have come to us as a part of our philosophical inheritance. Some of us hold one of them; some of us, the other; and no doubt many of us hold both of them for, though they are mutually incompatible, the mere incompatibility of two ideas does not necessarily prevent them from finding firm lodgment in a same brain.

That Korzybski's concept of man is just and important,—entirely just and immeasurably important,—I have no reason to doubt after having meditated much upon it. But the author does not content himself with presenting that concept; he goes much further; he denies outright the zoological conception and similarly denies the ages-old rival, the mythological conception, denouncing both of them as being at once false to fact and vicious in effect.

Why false? Wherein?

Let us deal first with the zoological or biological conception. Natural phenomena are to be conceived and defined in accord with facts revealed by observation and analysis. The phenomena the author is concerned with are the great life-classes of the world: plants, animals, and humans. What, he asks, are the significant facts about them, their patent cardinal relations, their distinctive marks, positive and negative? And his answer runs as follows: Of plants the most significant positive mark is their power to "bind" the basic energies of the world—to take in, transform and appropriate the energies of sun, soil, water and air; but they lack *autonomous* power to move about in space, and that lack is a highly significant negative mark of plants. The plants are said to constitute the "chemistry-binding" or basic-energy-binding class of life; the *name* suggests only the positive mark but it is essential to note that the *definition* of the class is effected by the positive and the negative marks conjoined. What of the animals? These, like the plants, take in, transform and appropriate the basic energies of sun, soil, water and air, taking them in large part as already transformed by the plants; but this power of animals to bind basic energies,—the positive one of the two defining marks of plants,—is not a *defining* mark of animals; the *positive* defining mark of animals is their autonomous power to move [1] about in space,—to crawl or run or fly or swim,—enabling them to abandon one place and occupy another and so to harvest the natural fruits of many localities; this positive mark, you observe, is a relation of animals to *space;* but they have, we have seen, a negative mark, a relation to *time*—animals lack capacity for binding time. Because of the positive mark, animals are said to constitute the "space-binding" class of life, but it is to be carefully noted that the definition (as distinguished from the name) of the class is effected by the positive mark conjoined with the negative one. Finally, what of humans? We have already seen the answer and the ground thereof—humanity is the time-binding class of life. For the sake of clarity let us summarize the conceptions, or definitions, as follows: a plant is a living creature having the capacity to bind basic energies and lacking the autonomous ability to move in space; an animal is a living creature having the autonomous ability to move about in space and lacking the capacity for binding time; a man, or a human, is a living creature having time-binding power.

It is to be noted that, as thus conceived, the great life-classes of the world constitute a hierarchy arranged according to a principle which Korzybski calls life-dimensions or dimensionality, as follows:

The plants, or basic-energy-binders, belong to the lowest level or type of life and constitute the life-dimension *I*.

The animals, or space-binders, belong to the next higher level or type of life and constitute the life-dimension *II*.

[1] Do sessile animals really constitute an exception? It can be shown, I think, that such animals are space-binders in Korzybski's sense.

Human beings, or time-binders, belong to a still higher level or type of life and constitute the life-dimension *III*.

Whether there be a yet higher class of life we do not know and that is why in the conception of man no negative mark is present.

Now, it is, of course, perfectly clear that, according to the foregoing conceptions or definitions, the old zoological conception of man as a species of animal is false, as the author contends. But may we not say that he is here merely playing with words? Is it not entirely a matter of arbitrary definition? Has he not, merely to please his fancy, quite willfully defined the term "animal" in such a way as to exclude humans from the class so defined? The answer is undoubtedly, *No*. Of course, it goes without saying that we could, if we *chose*, define the mere word "animal" or any other noun so as to make it stand for the "class" of plants, elephants, humans, jabberwocks and newspapers. But we do not so choose. Why not? Because we desire our definitions to be *expedient*, to be helpful, to serve the purpose of rational thinking. We want them, in other words, to correspond to facts. Let us, then, forget the word for a little while and look at the facts. It is a fact that there is a class of creatures having space-binding capacity but not time-binding capacity; it is a fact that there is another class of creatures having both kinds of capacity; it is a fact that the difference between the two,—namely, the capacity for binding time,—is not only beyond all comparison the most significant of the marks peculiar to man, but is indeed the most significant and precious thing in the world; it is, therefore, a fact that not only the interests of sound ethics, but the interests of science, demand that the two classes, thus distinct by an infinite difference of *kind* of endowment, be not intermixed in thought and discourse; it is a fact that use of the same term "animal" to denote the members of both classes,—men and beasts alike,—constantly, subtly, powerfully tends to produce both intellectual and moral obfuscation; it is, therefore, a fact that the author's condemnation of the zoological conception as false to fact is amply justified on the best of grounds.

It is indeed true that humans have certain animal organs, animal functions, and animal propensities, but to say that, therefore, humans *are* animals is precisely the same kind of logical blunder as we should commit if we said that animals or humans are plants because they have certain organs, functions and properties in common with plants; and the blunder is of a kind that is fundamental—it is the kind which mathematicians call the confusion of types or of classes and which Korzybski calls the "mixing of dimensions." To say that humans are animals because they have certain animal propensities is logically on a par with saying that geometric solids are surfaces because they have certain surface properties or with saying that fractions are whole numbers because they have certain properties that whole numbers have.

Why is it that people are shocked on encountering for the first time a categorical denial of their belief that man is a species of animal? Do they feel that their proper dignity as human beings is thus assailed? Is it because the animal basis of their space-binding ethics is being thus attacked? Is it that a well-reasoned scientific conviction is suddenly contradicted? I do not think the shock is due to any of these things. It is, I believe, due simply to the fact that an old unquestioned, uncriticized creed of that great dullard,—Common Sense,—has been unexpectedly challenged. For it is evident to common sense,—it is obtrusively evident to sense-perception,—that humans have certain animal organs and animal experience—they are begotten and born, they feed and grow, have legs and hair, and die, all just like animals; on the other hand, their time-binding faculty is not thus evident; it is not, I mean, a tangible *organ;* it is an intangible *function,* subtle as spirit; and so common sense, guided according to its wont by the uncriticized evidence of sense, and thoughtlessly taking for major premise the false proposition that whatever has animal organs and propensities is an animal, concludes that our human kind is a kind of animal. But in this matter, as in so many others, the old dullard is wrong. The proper life of animals is life-in-space; the distinctive life of humans is life-in-time.

But why are mere concepts so important? Our lives, we are told, are not controlled by concepts but by impulses, instincts, desires, passions, appetites. The answer is: Because concepts are never "mere" concepts but are, in humans, vitally connected with impulses, instincts, desires, passions, and appetites; concepts are the means by which Reason does its work, leading to prosperity or disaster according as the concepts be true or false.

I have said that the ancient and modern rival of the zoological conception of man is the mythological conception according to which man is a mysterious compound or hybrid of natural (animal) and supernatural. This conception might well be treated today as it was treated yesterday by Plato (in the Timaeus, for example). "We must accept," said he, "the traditions of the men of old time who affirm themselves to be the offspring of the gods—that is what they say—and they must surely have known their own ancestors. How can we doubt the word of the children of the gods? Although they give no probable or certain proofs, still, as they declare that they are speaking of what took place in their own family, we must conform to custom and believe them."[1] But this gentle irony,—the way of the Greek philosopher,—is not the way of the Polish engineer. The latter is not indeed without a blithesome sense of humor but in this matter he is tremendously in earnest, and he bluntly affirms, boldly and confidently, that the mythological conception of man is both false and vicious. As to its validity or invalidity, it involves, he says, the same kind of logical blunder as the zoological conception—it involves, that is, a fatal confusion of types, or mixing of dimensions. To say that man is a being so inscrutably constituted that he must be regarded as partly natural (partly animal) and partly supernatural (partly divine) is *logically* like saying that a geometrical solid is a thing so wonderful that it must certainly be a surface miraculously touched by some mysterious influence from outside the universe of space. Among the life-classes of the world, our humankind is the time-binding class; and Korzybski stresses again and again the importance of recognizing that time-binding energy and all the phenomena thereof are perfectly *natural*—that Newton, for example, or Confucius, was as thoroughly natural as an eagle or an oak.

What does he mean by "natural"? He has not told us,—at all events, not explicitly,—and that omission is doubtless a defect which ought to be remedied in a future edition of the book.

You are aware that the terms "nature" and "natural"

[1] Jowett's translation.

are currently employed in a large variety of senses—most of them so vague as to be fit only for the use of "literary" men, not for the serious use of scientific men. What ought we to mean by the term "natural" in such a discussion as we are now engaged in? The question admits, I believe, of a brief answer that is fairly satisfactory. Everyone knows that the things encountered by a normal human in the course of his experience differ widely in respect of vagueness and certitude; some of them are facts so regular, so well ascertained, so indubitable that they guide in all the affairs of practical life; they are *known* facts, we say, and to disregard them would be to perish like unprotected idiots or imbeciles; such facts are of two kinds: facts of sense-perception, or of this and memory, and facts of pure thought; the former are familiar in the moving pageant of the world—birth, growth, death, day, night, land, water, sky, change of seasons, and so on; facts of pure thought are not so obtrusively obvious but there are such facts; one of them is—"If something $S$ has the property $P$ and whatever has $P$ has the property $P'$, then $S$ has $P'$." Now, all such facts are *compatible*—each of them fits in, as we say, with all the others. I take it that what we ought to mean by natural is, therefore, this: *Nature (or the natural) consists of all and only such things as are compatible (consistent) with the best-ascertained facts of sense and of thought.*

If that be what Korzybski means by "natural,"—and I think it very probably is,—then I fully agree with him that humans are thoroughly natural beings, that time-binding energy is a natural kind of energy, and that his strenuous objection to the mythological conception of man is, like his objection to the zoological conception, well taken. If it were a question of biological data, mere mathematicians would, of course, like other sensible folk, defer to the opinion of biologists; it is not, however, a question of biological data, these are not in dispute; it is a question of the logical significance of such data; and respecting a question of logic, even biologists,—for they, too, are sensible folk,—will probably admit that engineers and mere mathematicians are entitled to be heard.

In this connection I desire to say that, for straight and significant thinking, the importance of avoiding what Korzybski calls "mixing dimensions" can not be overstressed. The meaning of the term "dimensions" as he uses it is unmistakable; he has not, however, elaborated an abstract theory of the idea; such an elaboration would, I believe, show that the idea is reducible or nearly reducible to that of the Theory of Logical Types, briefly dealt with in a previous lecture and fully outlined in the *Principia Mathematica* of Whitehead and Russell; it is, moreover, very closely allied to, if it be not essentially identical with, Professor J. S. Haldane's doctrine of "categories" as set forth in his very stimulating and suggestive book *Mechanism, Life, and Personality* (E. P. Dutton and Co.) wherein the eminent physiologist maintains that mechanism, life, and personality belong to different categories constituting a genuine hierarchy such that the higher is not reducible to the lower, that life, for example, cannot be understood fully in terms of mechanism, nor personality in terms of life. It is, you observe, an order of ideas similar to that of Korzybski's thesis that humans can be no more explained in terms of animals than animals in terms of plants or plants in terms of minerals. And it is an order of ideas that recommends itself, to me at all events, because it is fortified by the analogous consideration that geometry cannot be reduced to arithmetic, nor dynamics to geometry, nor physics to dynamics, nor psychology to physics. It will, I believe, be a great advantage to science and to philosophy to recognize that there exists, whether we will or no, a hierarchy of categories and to recognize that, to an understanding of the higher categories, the lower ones, though necessary, are not sufficient.

Is there not, indeed, a highly important sense in which the phenomena of a higher category throw as much light upon those of a lower as the latter throw upon the former? Who can deny that, for example, dynamics illuminates geometry quite as much as geometry illuminates dynamics?

In Korzybski's indictment of the zoological and mythological conceptions of man there are, we have seen, *two* counts: he denies that the conceptions are true; and he denounces them as vicious in their effects, contending that they are mainly responsible for the dismal things of human history and for what is woeful in the present plight of the world. Of the former count I have already spoken; respecting the latter one, my convictions are as follows: (1) if humanity be not a thoroughly natural class of life, the term "natural" having the sense above defined, it is perfectly evident that there never has been and never can be a system of human ethics having the understandability, the authority, and the sanction of natural law, and this means that, under the hypothesis, there never has been and never can be an ethical system "compatible with the best-ascertained facts of sense and of thought"; (2) if, although our human kind be in fact a thoroughly natural class, we continue to *think* that such is *not* the case, the result will be much the same—our ethics will continue to carry the confusion and darkness due to the presence in it of mythological elements; (3) on the other hand, so long as we continue to regard man as a species of animal, the social life of the world in all its aspects will continue to reflect the tragic misconception, and our ethics will remain,—what it always has been in large measure,—an animal ethics, space-binding ethics, an ethics of might, of brutal competition, of violence, combat, and war.

Why so much stress upon ethics? Because ethics is not a thing apart; it is not an interest that is merely co-ordinate with other interests; it penetrates them all. Ethics is a kind of social ether which, whether it be good or bad, sound or unsound, true or false, pervades life, private and public, in all its dimensions and forms; and so, if ethics be vitiated by fundamentally false conceptions of human nature, the virus is not localized but spreads throughout the body politic, affecting the character of all activities and institutions,—education, science, art, philosophy, economics, industrial method, politics, government,—the whole conduct and life of a tribe or a state or a nation or a world. I hardly need remind you that only yesterday the most precious institutions of civilization were in great danger of destruction by a powerful state impelled, guided and controlled by animalistic ethics, the space-binding ethics of beasts. This is indeed an unforgettable illustration of the mighty fact, before pointed out, that the character of human history, human conduct and human institutions depends, not merely upon what man distinctively is, but also in large measure, even decisively, upon what we humans *think* man is. If a man or a state habitually regards humanity as a species of animal, then that man or state may be expected to act betimes like a beast and to seek justification in a zoological philosophy of human nature.

In view of such considerations it is a great pleasure to turn to Korzybski's concept of man, for it is not only a noble conception, as none can fail to perceive, but it is also, as we have seen, undoubtedly just. Nothing can be more important. What are its implications? And what are its bearings? You cannot take them in at a glance—meditation is essential; but, if you will meditate upon the concept, you will find that the body of its implications looms larger and larger and that the range of its bearings grows ever clearer and wider. Indeed we may say of it what Carlyle said of *Wilhelm Meister:* "It significantly tends towards infinity in all directions." Let us reflect upon it a little. We shall see that human history, the philosophy thereof, the present status of the world, the future welfare of mankind, are all of them involved.

The central concept or thesis is that our human kind is the time-binding class of life; it is, in other words, that there is in our world a peculiar kind of energy, time-binding energy, and that man is its organ—its sole instrument or agency. What are its implicates and bearings?

One of them we have already noted. It is that, though we humans are not a species of animal, we are *natural* beings: it is as natural for humans to bind time as it is natural for fishes to swim, for birds to fly, for plants to live after the manner of plants. It is as natural for man to make things achieved the means to greater achievements as it is natural for animals *not* to do so.

That fact is fundamental. Another one, also fundamental, is this: time-binding faculty,—the characteristic of humanity,—is not an effect of civilization but is its cause; it is not civilized energy, it is the energy that *civilizes;* it is not a product of wealth, whether material or spiritual wealth, but is the creator of wealth, both material and spiritual.

I come now to a most grave consideration. Inasmuch as time-binding capacity is the characterizing mark,—the idiosyncrasy,—of our human kind, it follows that to study and understand man is to study and understand the nature of man's time-binding energies; the laws of human nature are the laws,—natural laws,—of these energies; to study time-binding phenomena,—the phenomena of civilization,—and to discover their laws and teach them to the world, is the supreme obligation of scientific men, for it is evident that upon the natural laws of time-binding must be based the future science and art of human life and human welfare.

One of the laws we know now,—not indeed precisely,—but fairly well,—we know roughly, I mean, its general type,—and it merits our best attention. It is the natural law of progress in time-binding—in civilization-building. We have observed that each generation of (say) beavers or bees begins where the preceding one began and ends where it ended; that is a law for animals, for mere space-binders—there is no advancement, no time-binding—a beaver dam is a beaver dam—a honey comb a honey comb. We know that, in sharp contrast therewith, man invents, discovers, creates; we know that inventions lead to new inventions, discoveries to new discoveries, creations to new creations; we know that, by such progressive breeding, the children of knowledge and art and wisdom not only produce their kind in larger and larger families but engender new and higher kinds endlessly; we know that this time-binding process, by which *past time* embodied as cofactor of toil in enduring achievements thus survives the dead and works as living capital for augmentation and transmission to posterity, is the secret and process of progressive civilization-building. The question is: What is the Law thereof—the natural law? What its general type is you apprehend at once; it is like that of a rapidly increasing *geometric* progression—if $P$ be the progress made in a given generation, conveniently called the "first," and if $R$ denote the ratio of improvement, then the progress made in the second generation is $PR$, that in the third is $PR^2$, and that made in the single $T$th generation will be $PR^{T-1}$. Observe that $R$ is a large number,—how large we do not know,—and that the time $T$ enters as an exponent—and so the expression $PR^{T-1}$ is called an *exponential function of Time,* and it makes evident, even to the physical eye, the involution of time in the life of man. This is an amazing function, as every student of the Calculus knows; as $T$ increases, which it is always doing, the function not only increases but it does so at a rate which itself increases according to a similar law, and the rate of increase of the rate of increase again increases in like manner, and so on endlessly, thus sweeping on towards infinity in a way that baffles all imagination and all descriptive speech. Yet such is approximately the law,—the natural law,—for the advancement of Civilization, immortal offspring of the spiritual marriage of Time and human Toil. I have said "approximately," for it does not represent adequately the natural law for the progress of civilization; it does not, however, err by excess, it errs by defect; for, upon a little observation and reflection, it is evident that $R$, the ratio of improvement, is not a constant, as above contemplated, but it is a variable that grows larger and larger as time increases, so that the function $PR^{T-1}$ increases not only because the exponent increases with the flux of time, but because $R$ itself is an increasing function of time. It will be convenient, however, and we shall not be thus erring on the side of excess, to speak of the above-mentioned law, though it is inadequate, as the *natural* law for the progress of time-binding, or of civilization-making.

Hereupon, there supervenes a most important question: Has civilization always advanced in accord with the mentioned law? And, if not, why not? The time-binding energies of mankind have been in operation long—300,000 to 500,000 years, according to the estimates of those most competent to guess—anthropologists and paleontologists. Had progress conformed to the stated law throughout that vast period, our world would doubtless now own a civilization so rich and great that we cannot imagine it today nor conceive it nor even conjecture it in dreams. What has been the trouble? What have been the hindering causes? Here, as you see, Korzybski's concept of man must lead to a new interpretation of history—to a new philosophy of history. A fundamental principle of the new interpretation must be the fact which I have already twice stated,—namely, that what man has done and does has depended and depends both upon what man distinctively is and also, in very great measure, upon what the members of the race have *thought* and *think* man is. We have here two determining factors—what man *is* and what we humans *think* man is. It is their joint product which the sociologist or the philosophic historian must examine and explain. In view of the second factor, which has hardly ever been noticed and has never been given its due weight, Korzybski, in answer to our question, maintains that the *chief* causes which have kept civilization from advancing in accord with its natural law of increase are man's misconceptions of man. All that is precious in present civilization has been achieved, in spite of them, by the first factor—by what man is—the peculiar organ of the civilizing energies of the world. It is the second factor that has given trouble. Throughout the long period of our race's childhood, from which we have not yet emerged, the time-binding energies have been hampered by the false belief that man is a species of animal and hampered by the false belief that man is a miraculous mixture of natural and supernatural. These are cave-man conceptions. The glorious achievements of which they have *deprived* the world we cannot now know and may never know, but the subtle ramifications of their *positive* evils can be traced in a thousand ways. And it is not only the duty of professional historians to trace them, it is your duty and mine. Whoever performs the duty will be appalled, for he will discover that those evils—the evils of "magic and myth," of space-binding "ethics," of zoological "righteousness"—for centuries growing in volume and momentum—did but leap to a culmination in the World War, which is thus to be viewed as only a bloody demonstration of human ignorance of human nature.

We are here engaged in considering some of the major implicates and bearings of the new concept of man. The task demands a large volume dealing with the relations of time-binding to each of the cardinal concerns of individual and social life—ethics, education, economics, medicine, law, political science, government, industry, science, art, philosophy, religion. Perhaps you will write such a work or works. In the closing words of this lecture I can do no more than add to what I have said a few general questions and hints.

Korzybski believes that the great war marks the end of the long period of humanity's childhood and the beginning of humanity's manhood. This second period, he believes, is to be initiated, guided, and characterized by a right understanding of the distinctive nature of Man. Is he over-enthusiastic? I do not know. Time will tell. I hope he is not mistaken. If he is not, there will be many changes and many transfigurations.

I have spoken of ethics and must do so again, for ethics, good or bad, is the most powerful of influences, pervading, fashioning, coloring, controlling all the moods and ways and institutions of our human world. What is to be the ethics of humanity's manhood? It will not be an ethics based upon the *zoological* conception of man; it will not, that is, be animalistic ethics, space-binding ethics, the ethics of beasts fighting for "a place in the sun," the ethics of might, crowding, and combat; it will not be a "capitalistic" ethics lusting to *keep* for self, nor "proletarian" ethics lusting to *get* for self; it will not be an ethics having for its golden rule the law of brutes—survival of the *fittest* in the sense of the *strongest.* Neither will it be an ethics based upon a *mythological* conception of man; it will not, that is, be a lawless ethics cunningly contrived for traffic in magic and myth. It will be a natural ethics because based upon the distinctive nature of mankind as the time-binding,—civilization-producing,—class of life; it will be, that is, a scientific ethics having the understandability, the authority, and the sanction of natural law, for it will be the embodiment, the living expression, of the laws,—natural laws,—of the time-binding energies of man; human freedom will be freedom to live in accord with those laws and righteousness will be the quality of a life that does not contravene them. The ethics of humanity's manhood will thus be natural ethics, an ethics compatible with the best-ascertained facts of sense and of thought—it will be time-binding ethics—and it will grow in solidarity, clarity, and sway in proportion as science *discovers* the laws of time-binding,—the laws, that is, of civilization-growth,—and *teaches* them to the world.

And so I am brought to say a word respecting education. In humanity's manhood, education,—in home, in school, in church,—will have for its supreme obligation, and will keep the obligation, to teach the young the distinctive nature of man and what they, as members and representatives of the race of man, essentially are, so that everywhere throughout the world men and women will habitually understand, because bred to understand, what time-binding is, that their proper dignity as humans is the dignity of time-binding life, and that for humans to practice space-binding ethics is a monstrous thing, involving the loss of their human birthright by descent to the level of beasts.[1] It is often said that ethics is a thing which it is impossible to *teach*. Just the opposite is true —it is impossible *not* to teach ethics, for the teaching of it is subtly carried on in all our teaching, whether consciously or not, being essentially involved in the teacher's "philosophy of human nature." Every home or school in which that philosophy is zoological is, consciously or unconsciously, a nursery of animalistic ethics; every home or school in which there prevails a mythological philosophy of human nature is, consciously or unconsciously, a nursery of a lawless ethics of myth and magic. From

[1] In a recent bulletin of the Cora L. Williams Institute for Creative Education, Miss Williams has said, with fine insight, that "time-binding should be made the basis of all instruction and *The Manhood of Humanity* a textbook in every college throughout the world."

time immemorial, such teaching of ethics, for the most part unconscious, the whole world has had. And we have seen that when such teaching becomes conscious, deliberate, and organized, a whole people can be so imbued with both the space-binding animal ethics of might and the mythical ethics of *Gott mit uns* that their State will leap upon its neighbors like an infuriated beast. Why should we not learn the lesson which the great war has so painfully taught regarding the truly gigantic power of education? If the accumulated civilization of many centuries can be imperiled by ethical teaching based upon a false philosophy of human nature, who can set a limit to the good that may be expected from the conscious, deliberate, organized, unremitting joint effort of home and school and press to teach an ethics based upon the true conception of man as the agent and organ of the time-binding, civilizing energy of the world? I cannot here pursue the matter further; but in closing I should like to ask a few general questions—pretty obvious questions —indicating roughly the course which, I believe, further enquiry should take.

What are the bearings of the new concept of man upon the social so-called sciences of economics, politics, and government?

Can the new concept transform those ages-old pseudo-sciences into genuine sciences qualified to guide and guard human welfare because based upon scientific understanding of human nature?

In view of the radical difference between the distinctive nature of animals and the distinctive nature of man, what are likely to be the principal differences between

Government of Space-binders, by Space-binders, for Space-binders and

Government of Time-binders, by Time-binders, for Time-binders?

Which of the two kinds of government best befits the social régime of autocrats, or plutocrats, and slaves? And which best befits the dream of political equality and democratic freedom?

Which of them most favors the prosperity of "Acquisitive Cunning"? And which the prosperity of Productive Skill?

Which of them is the most friendly to the *makers* of wealth? And which of them to the *takers* thereof?

Which of them most favors "boss" repression of others? And which makes the best provision for intelligent self-expression?

Which of them depends most upon might and war? And which upon right and peace?

Which of them is government by "politics," by politicians? And which of them by science, by honest men who know?

If man's time-binding energy, which has produced all the wealth of the world, both material and spiritual wealth, be *natural* energy, and if, as is the case, the wealth existing at a given moment be almost wholly a product of the time and toil of the by-gone generations, to whom does it of right belong? To *some* of the living? To *all* of the living? Or to all of the living and the yet unborn? Is the world's heritage of wealth, since it is a natural product of a natural energy and of time (which is natural), therefore a "natural resource" like sunshine, for example, or a newfound lake or land? If not, why not? What is the difference in principle?

Are the "right of conquest" and the "right of squatter

sovereignty" time-binding rights? Or are they space-binding "rights" having their sanction in animalistic "ethics," in a zoological philosophy of human nature?

What are the bearings of the new concept of man upon the theory and practice of medicine? Man, though not an animal, has animal organs and animal functions. Are all the diseases of human beings animal diseases or are some of them *human* diseases, disorders, that is, affecting humans in their distinctive character as time-binders? Can Psycho-analysis or Psychiatry throw any light upon the question?

And what of the power that makes for righteousness? Religion, it would seem, has the seat of its authority in that time-binding double relationship in virtue of which the living are at once posterity of the dead and ancestry of the unborn,—in the former capacity inheriting as living capital the wealth of civilization from the time and toil of by-gone generations,—in the latter capacity holding the inheritance in trust for enlargement and transmission to future man.

A final reflection: under the doctrine outlined there lies an assumption—it is that, when men and women are everywhere bred to understand the distinctive nature of our human kind, the time-binding energies of man will be freed from their old bondage and civilization will advance, in accord with its natural law, in a warless world, swiftly and endlessly. If the assumption be not true, great Nature is at fault and the world will continue to flounder. Of its truth, there can be only one test—experimentation, trial. The assumption appears to be the only scientific basis of hope for the world. Must not all right-thinking men and women desire ardently that this noble assumption be tried?

# LECTURE XXI

## Science and Engineering

CHANGE OF EMPHASIS FROM NON-HUMAN TO HUMAN ENERGIES—SCIENCE AS ENGINEERING IN PREPARATION—ENGINEERING AS SCIENCE IN ACTION—MATHEMATICS THE GUIDE OF THE ENGINEER—ENGINEERING THE GUIDE OF HUMANITY—HUMANITY THE CIVILIZING OR TIME-BINDING CLASS OF LIFE—THE FOUR DEFINING MARKS OF THE GREAT ENGINEER OF THE FUTURE—ENGINEERING STATESMANSHIP.

I AM not a professional engineer. What, then, is my apology for daring to speak of engineering? It is not, I fear, a quite convincing one. For it is the apology of a layman who can only plead that for more than twenty-five years he has taught mathematics to engineering students; that during these years he has associated a good deal both with practicing engineers and with professors of engineering science and art; that, like all who think of the matter, he has been deeply impressed in beholding and contemplating engineering achievements, from the great pyramids and aqueducts and roads of what we call antiquity down to the rapidly multiplying marvels wrought on every hand by the engineering prowess of our own day; that he has examined some of the writings of engineers, ancient, mediæval, and modern—the work of Frontinus, the engineering speculations of Leonardo da Vinci, especially the famous books recently produced by "wizard" apostles of "efficiency"; and that he has been thus led to reflect a good deal upon the opportunities, the functions and the obligations of engineering, rightly conceived, in the great affairs of our human world. There is, moreover, the general consideration that a layman, viewing a profession from the outside, seeking thus to ascertain its proper relations to the common weal, may bring to the task a certain freedom, which, were he a member of the profession, he might have lost. "Men trained in a profession," said Professor David Swing, "come by degrees into the profession's channel, and flow only in one direction, and always between the same banks. The master of a learned profession at last becomes its slave. He who follows faithfully any calling wears at last a soul of that calling's shape. You remember the death scene of the poor old schoolmaster. He had assembled the boys and girls in the winter mornings and had dismissed them winter evenings after sundown, and had done this for fifty long years. One winter morning he did not appear. Death had struck his old and feeble pulse; but, dying, his mind followed its beautiful but narrow river-bed, and his last words were: 'It is growing dark—the school is dismissed—let the girls pass out first.' " Finally, it is not my intention to deal with the technique of engineering nor with that of any branch thereof, but rather with its general aspects, with what is essentially common to its branches, with the science viewed as a whole. I shall not be so much concerned with the present status of the science as with its potence and promise. Of individual engineers the ideals may be high or low, worthy or unworthy; but of engineering itself the ideal is great and mighty. It is of that ideal that I intend to speak.

What is engineering? It is evident that the term stands, or ought to stand, for a great conception. What is that conception? Many attempts have been made to define it. Most of them throw more light upon the character and outlook of those who have made them than upon the nature of engineering itself.

To say that an engineer is one who "knows what to do, when to do it and how to do it" may be true,—the formula is very neat,—but it can hardly be said to be quite definitive, seeing that it applies equally well to the wisdom of a wise philanthropist and to the cunning of a cunning thief.

To define engineering in terms of *aim* is no doubt feasible; but to say that the aim is "maximum production with minimum outlay of time, effort and resources" sounds like the "efficiency" cry of brute industrialism, appears to regard *quantity* as the *summum bonum,* seems to ignore the spiritual autonomy of men and women, and to idealize a "system" in which "laborers" are reduced to the level of machines.

To say that the aim of engineering is the "mastering of natural forces and materials for the benefit of mankind" is far better in one respect because it is humane—it represents engineering, I mean, as having for its aim "the benefit of mankind." But what do its sponsors mean by "natural forces"? Do they intend the term to cover the personalities of individual men and women, their perfectly natural civilizing impulses and aspirations? Do they include among "natural forces" the spiritual energies of our human kind—those time-binding powers in virtue of which human beings are human? If they do not, why not? And if they do, what do they mean by "the mastering of natural forces"? The questions are important and sometime the philosophers of engineering must answer them.

The most famous conception of engineering and, in the judgment of many, the best one to be found in the literature is almost a century old. It is due, I believe, to the English engineer, Thomas Tredgold (1788-1829) and is found in the charter of the Institution of Civil Engineers (London, 1828). Engineering is there called an art—"the art of directing the great sources of power in nature for the use and convenience of man, as the means of production and of traffic in states, both for external and internal trade, as applied in the construction of roads, bridges, harbours, moles, breakwaters, and lighthouses, and in the art of navigation by artificial power for the purposes of commerce, and in the drainage of cities and towns." The gist of the matter is in the first eighteen words: *the art of directing the great sources of power in nature for the use and convenience of man.* For our purpose it will be well worth while to reflect upon them a little. Though found in a charter for *civil* as distinguished from *military* engineering, they apply no better to what we today call civil engineering than to any other of the numerous varieties into which, since the words were written, engineering has branched; moreover, they apply no better to a branch of engineering than to Engineering itself, regarded as one great Enterprise, and that is why it will repay us to reflect upon them.

In view of their date (1828) it is not strange that they represent engineering as an "art" instead of a science, as we call it today, or a science *and* art, as, I think, we ought to call it. But that is a trivial matter.

What is not trivial,—what is indeed of the gravest importance,—are the major *emphases* in the conception. These are two: one of them is upon the ultimate aim, purpose, or *end* of engineering; the other is upon *means* thereto.

What is the former? What is engineering for? Is it for "the use and convenience" of engineers? Or for that of a "shop"? Or that of a manufacturing establishment? Or that of an industry? Or that of a group of "capitalists"? Or that of a class of "laborers"? No; it is for no such restricted good; it is infinitely higher and nobler and more embracing—engineering is for "the use and convenience of *man*"; and "man" does not mean this group or that; it means all the people of the world, not only those now living but an unending succession of generations to come. The appeal is thus to an imagination great enough to grasp and represent the *race*. We are wont to say that in things human there can be no perfection. I believe we may say, however, that the ultimate aim, purpose, or end of engineering as presented in that century-old conception of it is a perfect ideal and can never be improved.

But of the other major emphasis in the famous statement the same cannot be said—far from it. For note its incidence. Where does the emphasis fall? It falls upon "directing the sources of power in nature," and the answer is important as indicating the psychology, the attitude and temper, the social philosophy, of an age—an age that still lingers, but is being outgrown and is destined to pass. For what is there meant by "nature"? It is evident that what is meant is physical nature, the external universe, the *non*-human part of the world, and it is evident that the term "power" refers to the blind forces of that "nature"—to wind and wave and tide and gravity and heat and so on. Such things, we say, are non-human, —man has great interest in them but they have no interest in man,—and when they are made, as they can be made, to serve human welfare, what is it that makes them serve? Everyone knows the answer: what makes them serve is human thought,—it is human intelligence and purpose and will,—it is the power that invents,—the power that observes and remembers and imagines and conceives and reasons and creates,—it is, in a word, what we may for convenience call the *spiritual energies* of our human kind. These energies are just as natural as Tredgold's "power." The reflection is no doubt just but it is very obvious. Why, then, insist upon it so? Because, as you must see, it fundamentally alters the traditional point of view. We are seeking a just and worthy conception of the science and art of Engineering, and the reflection in question radically shifts the incidence of the major emphasis. It shifts it from the non-human to the human. For it is clear that what requires "directing"—what requires to be *engineered*—is *primarily,* not the blind forces of external nature, but those other natural forces—the spiritual energies of Man. It is perfectly evident that the ultimate aim and ideal of engineering,—the welfare of our human kind,—not only demands the conquest of physical nature, not only demands subjugation of the non-human forces of the world, but also demands, as even more essential, world-wide enlightenment of human beings, world-wide coördination of human effort, world-wide establishment of Justice; and it is perfectly evident that the *sole* means to these great ends is the understanding and "directing,"—the "engineering," if you please, —of what we have called the spiritual energies of man.

These are the energies with which we dealt in the preceding lecture; they are the energies which, in Korzybski's fine phrase, constitute humanity the "time-binding" class of life; they are the human energies in virtue of which the distinctive life of man is life-in-time; they are the energies that make man the creator of Civilization; man is their sole agency, their sole instrument, their sole organ; characteristic of humankind, they are present in some measure wherever human beings are found. Upon the effectiveness of these energies depends the creation of material and spiritual wealth—the advancement of civilization—the well-being of man. To be effective, however, they must be understood, they must be organized, they must be coordinated, they must be brought into world-wide coöperation—in one word, they require to be engineered. And so I propose to define Engineering to be

> *The science and art of directing the time-binding energies of mankind,—the civilizing energies of the world,—to the advancement of the welfare of man.*[1]

That conception does not represent engineering as it has been practiced in the past nor as it is practiced today. It represents an *Ideal* which engineering will approximate more and more just in proportion as it becomes more and more humanized and enlightened. The ideal is an inspiring one; but it ought not to flatter the vanity of

[1] My friend, Mr. Robert B. Wolf, has pointed out to me that the preamble of the Constitution (1920) of The Federated American Engineering Societies says: "Engineering is the science of controlling the forces and of utilizing the materials of nature for the benefit of man, and the art of organizing and of directing human activities in connection therewith." I hope the reader will compare that conception *critically* with the one which I have submitted. The preamble dedicates the federation "to the service of the community, state, and nation." Why not to the service of the World?

professional engineers; it ought rather to give them a feeling of humility. For consider its spirit and its scope.

Its spirit is not a self-serving spirit nor a class-serving spirit nor any provincial spirit; it is a world-serving spirit—the spirit of devotion to the well-being of all mankind including posterity.

And what is its scope? Is it confined to the kinds of work done today by professional engineers in the name of engineering? It is by no means thus confined; its scope is immeasurably greater; for, over and above such work, which no one could wish to belittle, it embraces whatever may be intelligent, humane, and magnanimous in the promotion of science, in the work of educational leadership, in the conduct of industrial life, in the establishment and administration of justice—in *all* the affairs of a statesmanship big enough to embrace the world.

I am facing the future, and I say "in all the affairs of statesmanship" because I do not doubt that the affairs of state,—which are the affairs of man,—will at length be rescued from the hands of "politicians" and be committed to a statesmanship which will be an engineering statesmanship because it will guide itself and the affairs of state in scientific light by scientific means.

Engineering statesmanship will know enough to know that scientific knowledge cannot be applied to the conduct of human affairs if such knowledge does not exist; it will have sense enough to know also that knowledge which does not exist cannot be suddenly called into existence at the moments when it is needed. Engineering statesmanship will, therefore, be sagacious enough to make ample provision in advance for scientific research; not only for technological research, but,—primarily and especially,—for that kind of research which does not consciously aim at utility or applications. What kind *is* that? It is the kind whose "only purpose," in the clear words of President Nichols,[1] "is the discovery of new knowledge without thought of any material benefit to anybody"; it is the kind which Simon Newcomb[2] had in mind when he said that, "The true man of science has no such expression in his vocabulary as 'useful knowledge' "; it is the kind of which Henri Poincaré said that, if there can be "no science for science's sake," there can be "no science";[3] it is, in a word, the kind of research which springs out of pure scientific curiosity,—out of wonder, as Aristotle said,—and which, just because it is thus disinterested, just because it seeks the True, is the principal source of the Useful also.

The subject of such research will be Nature,—non-human nature and human nature,—the nature of the non-human world and the nature of man,—for we can know nothing else. Engineering statesmanship will have sense enough to know that its work cannot be done without scientific knowledge of both kinds of nature; it will, therefore, provide every means for promoting the advancement of the physical sciences and of those biological sciences that deal with the non-human world; and it will especially provide every means for promoting those researches which have for their aim the understanding of Man. I have said "especially" because engineering statesmanship will have sense enough to know that, of all the things it must deal with, man is the supreme reality, and that, therefore, the understanding of man,—scientific

[1] The Inaugural Address of the President of the Massachusetts Institute of Technology, *Science,* June 10, 1921.

[2] *Congress of Arts and Sciences,* vol. I, p. 137.

[3] Poincaré: *Science and Method* (translation by Francis Maitland), p. 16.

knowledge of human nature,—is absolutely essential to its enterprise.

And here we must say a word respecting the relation of engineering, as it is here conceived, to Education. The science and art of human engineering,—the science and art of engineering statesmanship,—is based upon a most important assumption. The assumption is that when and only when men and women are everywhere bred in the knowledge and the feeling of what man distinctively and naturally is, it will be possible so to organize, to co-ordinate, and to direct the time-binding powers of mankind,—the civilizing energies of the world,—that Civilization will advance in accord with its natural law, which is that of a swiftly increasing exponential function of Time. And so engineering statesmanship will not only provide, as said, for the scientific study of man, but it will provide a system of education whereby the children of man will be taught the results of such study,—an education which will have for its supreme obligation to teach boys and girls and men and women what, as representatives of man, they really and naturally are,—not a higher species of animal, nor a lower species of angel,—but humans, whose proper life is time-binding life, civilizing life, life-in-Time.

In view of such considerations it is easy to see what the defining marks of a great engineer are destined to be. They will not be the marks of mere "efficiency" nor of mere technological knowledge nor of technological skill—they will not be mere engineering technique of any kind, whether "civil" or "mechanical" or "marine" or "architectural" or "sanitary" or "chemical" or "electrical" or "industrial"; these things will be important, as they are now, they will indeed be indispensable, but they will not constitute, and they will not define, the great engineer. The characteristic marks of the great engineer will be four: Magnanimity—Scientific Intelligence—Humanity—Action.

He will be religious and he will be patriotic: "to do good" will be his religion, and his love of country will embrace the world. For he will be the scientific organizer and director of the civilizing energies of the World in the interest of all mankind.[1]

[1] Readers interested in what may be called the *Humanization* of engineering will find it profitable to examine the folowing works:

F. W. Taylor: *The Principles of Scientific Management* (Harper and Brothers, 1916).

H. L. Gantt: *Work, Wages, and Profits* (The Engineering Magazine Company, 1916).

W. N. Polakov and others: *The Life and Work of Henry L. Gantt* (The American Society of Mechanical Engineers, 1920).

Dr. Walter N. Polakov: *Mastering Power Production* (The Engineering Magazine Company).

Robert B. Wolf: "Individuality in Industry" (*Bulletin of the Society to Promote the Science of Management,* 1915); *Non-Financial Incentives* (American Society of Mechanical Engineers, 1918); "Securing the Initiative of the Workman" (American Economic Review Supplement, 1919); and *Modern Industry and the Individual* (A. W. Shaw Company, 1920).

Especially Messrs. Polakov and Wolf deserve the highest commendation and the thanks of all men and women for their insistence upon bringing the theory and practice of engineering under the *control* of humane considerations, upon basing engineering on the time-binding principles characteristic of humans instead of the space-binding principles characteristic of animals, and upon thus making engineering the chief of *civilizing* agencies, devoted to the promotion of Freedom and Justice throughout the World. These men have the vision to see that the time is coming when to call one a "mere space-binder" will be to call him a brute but to call one a *time-binder* will be to call him a *man,* a *human.*

# INDEX